Particle Technology Forum 2020

Held at the 2020 AIChE Annual Meeting

Online
16-20 November 2020

ISBN: 978-1-7138-2298-1

Particle Technology Forum 2020

Held at the 2020 AIChE Annual Meeting

Online
16-20 November 2020

AIChE
The Global Home of Chemical Engineers

Printed from e-media with permission by:

Curran Associates, Inc.
57 Morehouse Lane
Red Hook, NY 12571

CURRAN ASSOCIATES INC.
proceedings
.com

Some format issues inherent in the e-media version may also appear in this print version.

Printed with permission by Curran Associates, Inc. (2021)

For permission requests, please contact AIChE
at the address below.

AIChE
120 Wall Street, FL 23
New York, NY 10005-4020

Phone: (800) 242-4363
Fax: (203) 775-5177

www.aiche.org

Additional copies of this publication are available from:

Curran Associates, Inc.
57 Morehouse Lane
Red Hook, NY 12571 USA
Phone: 845-758-0400
Fax: 845-758-2633
Email: curran@proceedings.com
Web: www.proceedings.com

TABLE OF CONTENTS

603735 Post-Formulation 64-Cu Labeled Nanotracers to Investigate Size and Surface Property Effects on Biodistribution and Tumor Localization in a Mouse Pancreatic Cancer Model

Abstract:

Positron emission tomography (PET) is an extremely sensitive diagnostic tool for use in the early detection of cancers and other diseases. FDG-PET utilizing the PET active glucose analog, 18-fluorodeoxyglucose (18-FDG), has shown promise in clinical cancer imaging but still has several shortcomings. For example, since FDG utilizes glucose metabolism to detect tumors, it is less effective for slow-growing tumors and for differentiating cancers from the surrounding inflammatory tissue. Here we present the formation of nanoparticles using the continuous and scalable flash nanoprecipitation (FNP) process. Phthalocyanine-based dyes are encapsulated inside NPs through a rapid mixing and precipitation process that is stabilized by an amphiphilic block copolymer. After incubation of these NPs with copper ions in solution, the encapsulated dyes exhibit absorbance shifts indicative of the chelation of copper ions into the macrocyclic ring of the phthalocyanine. This absorbance shift was employed to quantify chelation and optimize reaction conditions. Thus, using this same method but instead incubating the NPs with positron-emitting 64-Cu yields PET active nanotracers. Under very mild conditions in a pH 5.5 aqueous buffer, the radiolabeling yield was quantitative after 3hr of incubation at 37C. This technology, when combined with the flexibility of FNP, allows for long timepoint biodistribution tracking of NPs with a variety of different properties such as size and surface charge. In particular, when PS-b-PEG is used as the stabilizing polymer, NPs exhibit long circulation times and enhanced tumor uptake due to the enhanced permeability and retention (EPR) effect. This was demonstrated using our PET nanotracers in a xenograft pancreatic tumor model, which showed high levels of PET activity in circulation at 3hrs and 6hrs post-injection and increasing tumor accumulation over 24hrs (Fig. 1). The development, optimization, and validation of a scalable method to produce nanotracers promises to expand the applications of NP-based PET imaging.

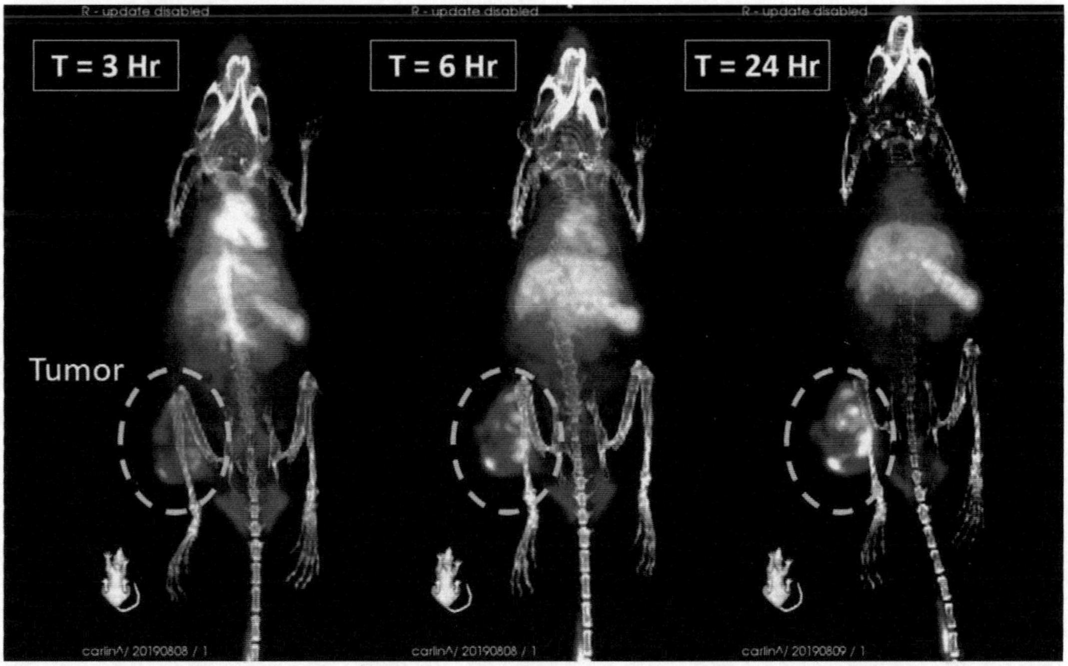

Figure 1 - PET/CT overlay showing biodistribution of PET activity at various timepoints after injection of 64-Cu loaded nanotracers in a pancreatic cancer mice model. Nanoparticles show increased accumulation in the tumor over 24hr.

611215 Optimization of Nanoparticle Delivery Platform for mRNA

Abstract:

The delivery of functional gene into cells to replace mutated genes holds great potential for treatment of many type diseases. In this line of investigation, mRNA therapy has gathered great interest because of its potential role for the transient modification of immune cells for treatment of an array of diseases. However, the main obstacle preventing routine mRNA immune cell transfection is the lack of a high viability, high efficiency, high throughput delivery system. In this study, we propose the use of a physical mixture of sterilized hollow gold nanoshells (HGN), that can be irradiated with picosecond pulsed near infra red light to induce vapor nanobubbles around the HGN which rupture the endosomes, thereby releasing the mRNA to the cell cytoplasm, where the mRNA can produce the proteins of interest. In Summary, we present extensive analysis into HGN sterilization techniques to minimize degradation of mRNA thus allowing the prolonged availability of mRNA for efficient translation as well as increased transfection efficiency in Jurkat and HEK-273 cells. The HGN/mRNA mixture had successful cell transfection with no negative effects on cell viability post transfection. Experiments are underway to determine the stability of genetic material in biological environment and cell extracts using Fluorescence Correlation Spectroscopy. Furthermore, we will investigate the use of Pluoronics to increase structural stability and resealing of cell membrane after irradiation with infrared light.

This Study is funded by the Heltzer Chair at U of M and Regenerative Medicine Minnesota.

610095 Hemoglobin Nanoparticles Made By Desolvation Precipitation May Function As a Therapeutic Artificial Oxygen Carrier

Abstract:

Purpose/Introduction: Blood transfusion is a lifesaving treatment for hypoxia, when organs and tissues are starved of oxygen. Unfortunately, blood cannot be available in all situations, including pre-hospital, battlefield, and remote care settings. Despite substantial research, no red blood cell substitute has ever been approved for clinical use. This work describes a new type of oxygen carrier based on the combination of hemoglobin (Hb), the oxygen carrying component of blood, and nanoparticle technology originally designed for drug delivery.

Experimental Design/Method: Hemoglobin nanoparticles (Hb-NPs) were formed by the desolvation process resulting in a suspension of uniform particles with no blood type limitations. During desolvation, ethanol, in which Hb is poorly soluble, is steadily added to a Hb solution resulting in effective supersaturation. At a critical saturation, precipitation into nanoparticles occurs, and a protein cross-linker is added to stabilize their size and shape. Finally, tangential flow filtration is used to rapidly exchange the final Hb-NPs into a physiological buffer solution. These Hb-NPs were characterized using traditional methods such as DLS, as well as more advanced assays for Hb activity and oxygen carrying potential.

Major Findings/Results: Hb-NPs made by this process had a mean diameter of 245 nm, similar in size to liposomes and other particles approved for intravenous use. Oxygen release kinetics revealed a dissociation constant of similar order of magnitude to that of a red blood cell. The oxygen binding equilibria of Hb-NPs showed high affinity and reversible binding and release of oxygen.

Summary of your interpretation of the data or conclusion: These findings suggest that Hb-NPs can be readily synthesized with a high Hb content and uniform size. They retain the oxygen carrying properties of the Hb protein, mimicking the red blood cell without constraints such as blood type or strict storage requirements. The uniform size and shape of Hb-NPs may overcome some barriers facing traditional oxygen therapeutics such as polymerized Hb products.

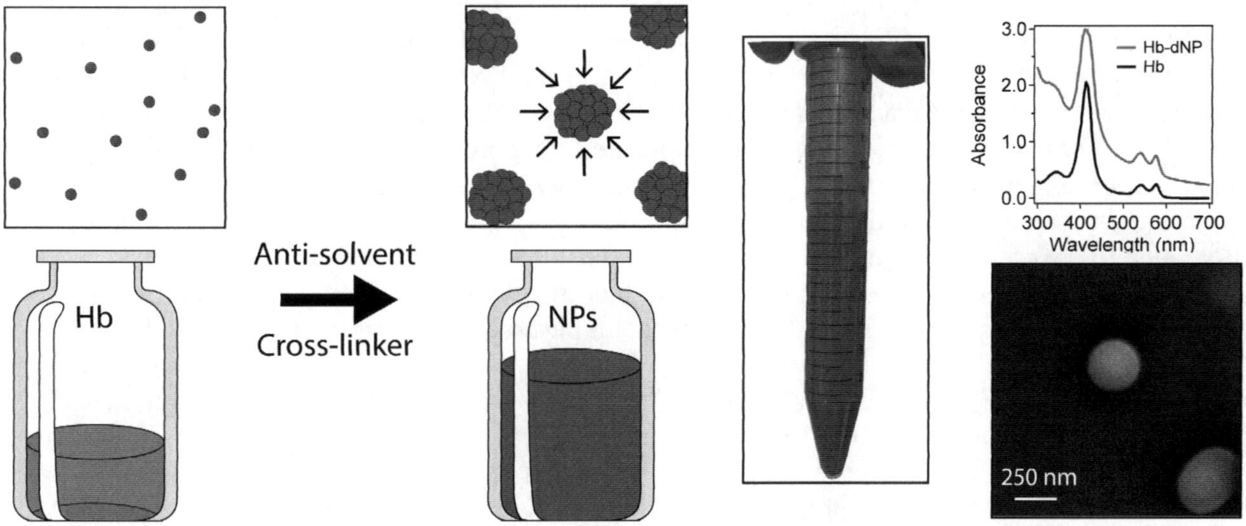

599222 Scale-up of Lumefantrine Nanoparticles Production

Abstract:

Purpose

Lumefantrine is an anti-malaria drug developed in the 1960s in China to address the increasing resistance of falciparum malaria to chloroquine and other previously used malaria drugs. However, due to the hydrophobic nature of Lumefantrine, it suffers from low bioavailability, and therefore a high frequency of dosing is required, increasing the risk of patient non-compliance. Patient non-compliance leads to the development of resistance to anti-malarial drugs. Given the above concerns, and in collaboration with Princeton University, a scaled-up manufacturing process for Lumefantrine nanoparticles has been developed (based on the polymer-directed, rapid precipitation technique, Flash NanoPrecipitation, developed by Princeton University). The impact of formulation on drug product performance and process scale up was assessed, meeting the theme of the ECI conference. The scaled up Lumefantrine nanoparticles have the potential to demonstrate enhanced bioavailability compared with the commercial product and, therefore, serves as a promising new anti-malaria formulation to address the limitations of the current Lumefantrine drug products.

Method

A schematic process flow diagram for the manufacturing process is presented in Figure 1. Four unit operations are included in the manufacturing process, which are flash nanoprecipitation (FNP), flash evaporation (FE), tangential flow filtration (TFF) and spray drying (SD). The FNP step was carried out by mixing four streams of liquid in a multi-inlet vortex mixer (MIVM) designed in-house (Feng, 2019). A total of 300 L of FNP suspension was processed in order to obtain a theoretical quantity of 675 g Lumefantrine nanoparticles (equivalent to 225 g of Lumefantrine).

Results

As can be seen from the stability data of FNP suspension in Table 1, the stability of FNP suspension was not long enough to withstand the total running time of TFF and SD. In addition, it was also anticipated that the 10% THF content in the FNP suspension would cause significant particle size enlargement during TFF (due to Ostwald ripening effect) if the FNP suspension were to be processed in TFF straightaway. For these reasons, a FE process was applied to increase the stability window. A decrease of THF content from 10% in the FNP suspension to approximately 1.2% in the THF lean suspension was obtained from this FE step, reducing the Ostwald ripening effect and prolonging the stability window to 4 days. Given the long running time of the FE step, three sublots were employed in the FNP step to ensure no stability issues of the FNP suspension would be encountered and a limited number of disruptions to the continuous manufacturing process would be made. It can also be noticed in Table 1 that there was no significant change in particle size from the FNP to the FE step. The TFF step was necessary, since otherwise there would be a tremendous volume to be spray dried, making the manufacturing process uneconomical. The slight increase in particle size during the TFF step, as shown in Table 1, was a result of the transmembrane pressure exerted on nanoparticles against the membrane, causing a limited degree of agglomeration. Nanoparticles were also consistently sheared in parallel direction to the membrane by the high velocity inlet flow to reduce such particle deposition. The redispersion of Lumefantrine nanoparticles was also investigated. As can be seen in Table 1, the particle size distribution upon redispersion was not hugely different from that before spray drying, being in the lower half of the nano size range. A low water content and acceptable yield were obtained as well. The solubility and kinetic stability of the final Lumefantrine nanoparticles can be seen in Table 2. The sustained release over 24 h at a high degree of supersaturation (~300 μg/mL compared with ~15 μg/mL for crystalline Lumefantrine) suggested potential enhanced dissolution of the Lumefantrine nanoparticles.

Conclusions

In conclusion, a scale-up continuous manufacturing process (FNP-FE-TFF-SD) to produce Lumefantrine nanoparticles has been built with an overall process yield of around 60%. Lumefantrine nanoparticles manufactured from this process were capable of being reconstituted into a nanosuspension and delivering a sustained high level of supersaturation as compared with crystalline Lumefantrine. A potential route to address growing concerns associated with low bioavailability of Lumefantrine has therefore been demonstrated.

Reference

Feng, J., Zhang, Y., McManus, S.A., Qian, R., Ristroph, K.D., Ramachandruni, H., Gong, K., White, C.E., Rawal, A. and Prud'homme, R.K., 2019. Amorphous nanoparticles by self-assembly: processing for controlled release of hydrophobic molecules. Soft matter, 15(11), pp.2400-2410.

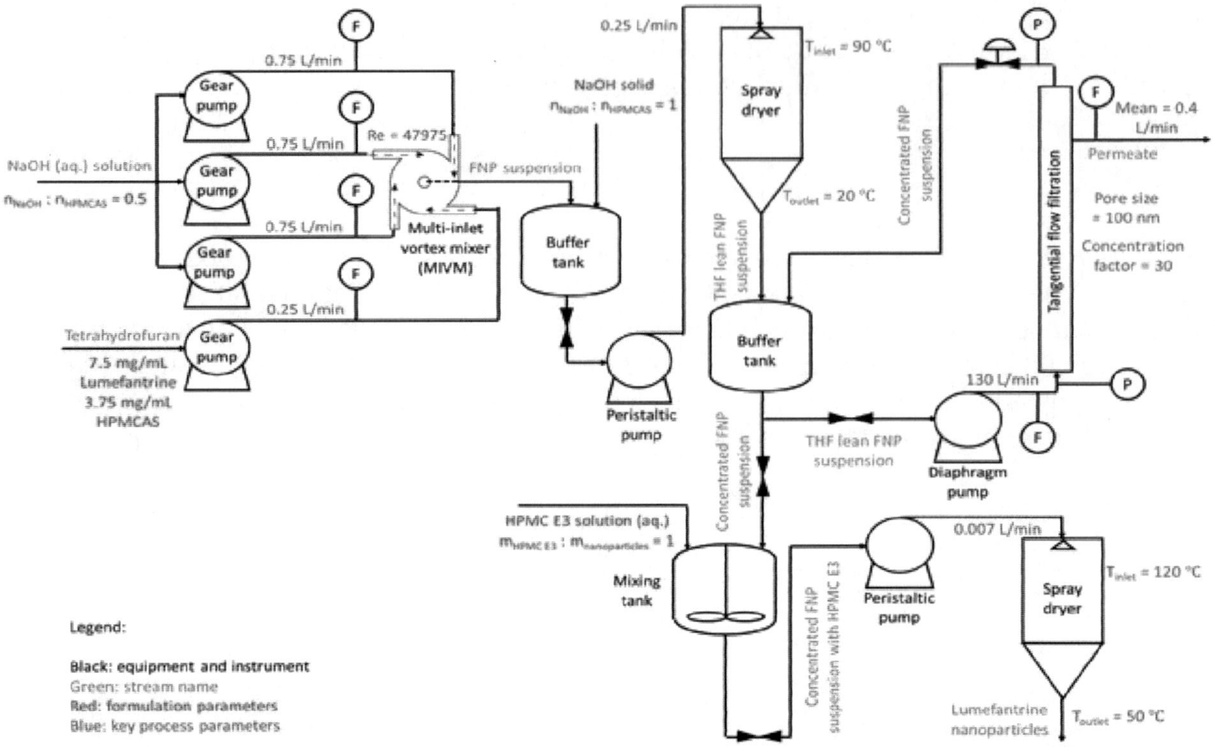

Figure 1. Schematic process flow sheet for the manufacturing process of Lumefantrine nanoparticles

Table 1. Process parameters and characterization of process streams

Unit operation	Process stream	Running time (h)	Characterization				
			Particle size			pH	THF residue (vol%)
			Time point	d_{50} (nm)	PDI		
FNP	FNP suspension	~ 0.6 per sublot	Initial	196.2	0.169	7.49	10
			After 1 day	232.2	0.123	7.40	
			After 30 h	786.1	1.055	7.04	
FE	THF lean FNP suspension	~ 22 continuous	Initial	214.8	0.182	7.73	1.22
			After 1 day	208.6	0.156	7.25	
			After 4 days	238.7	0.178	7.35	
TFF	Concentrated FNP suspension	~ 10 continuous	Initial	333.6	0.367	6.55	N/A
			After 1 day	377.0	0.086	6.54	
			After 3 days	328.8	0.194	6.57	
SD	N/A	~ 27 continuous	Redispersion	452.2	0.242	Yield (%)	Water content (%)
						58	4.77

Table 2. Solubility of Lumefantrine nanoparticles and crystalline Lumefantrine in different dissolution media

Different biorelevant media	FaSSGF		FaSSIF		FeSSIF	
Solubility of crystalline Lumefantrine (μg/mL)	0.51		4.8		14	
Time (hr)	0	1	2	4	6	24
Concentration of Lumefantrine nanoparticles in pH=6.8 phosphate buffer (μg/mL)	0	343	328.8	277.9	421.9	329.6

608728 Surface-Functionalized Calcium Phosphate Nanoparticles for Selective Cancer Cell Targeting

Abstract:

Over the past few decades, rapid growth of nanotechnology has addressed major challenges that traditional technologies cannot sufficiently address. Nanomedicine has shown promising results in disease treatment and nanoparticles have attracted interest in the medical field as effective carriers for drug delivery. Particularly for cancer, which is one of the most lethal diseases throughout the world, the employment of nanoparticles as drug carriers is a rapidly evolving strategy to overcome the difficulties in therapeutic delivery to the tumor site [1]. Due to their high surface-to-volume ratio (large surface area/small size) and tunable surface chemistry, nanoparticulate carriers present novel properties compared to their bulk counterparts and these properties are harnessed to achieve specific advantages, such as high drug loading, controlled drug release, low toxicity, tissue targeting, long circulation time etc.

In this regard, here we present an approach utilizing nanoparticle delivery of chemotherapeutic drugs to tumor cells. We fabricate and characterize a delivery system based on calcium phosphate (CaP) nanoparticles as the carriers, epidermal growth factor (EGFR) antibodies as targeting agents and doxorubicin (DOX) as an anticancer drug. CaP nanoparticles are synthesized by flame spray pyrolysis [2]. Calcium phosphates represent an important family of biomaterials in drug delivery because of their excellent biocompatibility, low toxicity and non-immunogenicity. Upon their synthesis, their surface is modified by polyethylene glycol (PEG) with carboxyl- and amine-terminated functional groups. The ligand density is optimized along with the dispersion protocol before and after functionalization, in order to have suitable hydrodynamic sizes and ensure long circulation times, effective transport across the tumor vessels and deep penetration into the tumor. The functional end-groups of the ligands are then linked with anti-EGFR antibody to enhance tumor specificity. Finally, DOX is incorporated in the nanoconjugates. The cytotoxicity of these nanocarriers on lung epithelial cells is evaluated *in vitro*.

This strategy could be used as a generalized therapeutic treatment for a variety of cancers characterized by overexpression of EGFR.

609420 Understanding and Predicting the Efficacy of Tailor Made Additives on the Crystallisation of p-Aminobenzoic Acid

Abstract:

Unwanted crystallization is a major concern in a number of chemical industries, affecting fuels, agrochemicals, pharmaceuticals and food. The ability to overcome this issue by disrupting the crystallization process, through the use of additives, has been a long studied area of crystallization science and engineering [1]. Since the work of Lahav, Leiserowitz and colleagues at the Weizmann Institute of Science[2,3], a large number of research studies have focussed on tailor-made additives (TMAs), whereby an additive with a very similar molecular structure to a target crystallization compound is used to inhibit aspects of the crystallization process. The majority of work has focussed on crystal growth inhibition, however, more recently the ability of TMAs to inhibit nucleation has been a growing area of research.

This work focuses on the nucleation of alpha-p-aminobenzoic acid (pABA) from ethanol solution, in the presence of TMAs. pABA in ethanol solution is known to nucleate through carboxylic group H-bonding dimer formation, with the crystal structure formed also containing strong amine-carboxyl group H-bonding and π-π stacking interactions [4]. TMAs that could potentially disrupt pABA-pABA dimer formation as well as the formation of the previously described synthonic interactions were chosen to study their effect on pABA nucleation inhibition form ethanol solution.

Seven TMAs were chosen and molecular modelling using an intermolecular grid search was undertaken to screen and understand the dominant molecular interactions between pABA and the TMAs, in terms of synthonic interaction energies. Two TMAs were found to have stronger interactions to pABA than pABA itself, forming competitive carboxylic H-bonding dimers, these were 4-amino-2-methoxybenzoic acid (AMBA2), 4-amino-3-nitrobenzoic acid (ANBA3). These TMAs had the same molecular structure of pABA, but with methoxy or nitro groups attached, respectively. From calculated atomic charge energies, the nitro group of ANBA3 was assessed to be able to disrupt the subsequent binding of pABA molecules into the crystal structure through electrostatic interactions.

A polythermal experimental screen was performed to assess the intermolecular grid search's ability to predict TMA efficacy. The experimental results corroborated findings from the modelling results, with ANBA3 being the most effective nucleation inhibitor followed by AMBA2.

Nucleation kinetics were assessed for pABA vs. pABA-ANBA3 in ethanol solution using the KBHR approach. It was found that pABA nucleates by instantaneous nucleation, whereas the pABA-ANBA3 nucleates by progressive nucleation. The interfacial tension was found to also increase in the presence of ANBA3.

References

1) J. W. Mullin, Crystallization, Butterworth-Heinemann, Oxford, 4th ed., 2001.

2) L. Addadi, Z. Berkovitch Yellin, N. Domb, E. Gati, M. Lahav and L. Leiserowitz, Nature, 1982, 296, 21–26.

3) I. Weissbuch, L. Addadi, M. Lahav and L. Leiserowitz, Science, 1991, 253, 637–645.

4) D. Toroz, I. Rosbottom, T. D. Turner, D. M. C. Corzo, R. B. Hammond, X. Lai and K. J. Roberts, Faraday Discuss., 2015, 179, 79–114.

606384 Multiscale Computer Modeling to Investigate Polymeric Micelle Based Nanocarriers for Anticancer Drug Delivery

Abstract:

Purpose:

Typical micelle-based drug delivery carriers are formed from amphiphilic block copolymers which self-assemble into core-shell nanoarchitectures with a hydrophobic core being able to encapsulate drug molecules and a biocompatible hydrophilic shell. An innovative turbulent co-flow platform of water and ethanol (with drug and polymers) was previously developed to continuously process drug loaded micelle nanocarriers. Polymer aggregation is the result of the intermolecular forces among molecules and along with the jet flow characteristics impact to the formation of micelles. However, the underlying mechanism and the detailed effects from material attributes are only partially understood. In order to investigate the underlying mechanism and quantitative estimation of the effect of material attributes and process parameters on the quality of the polymeric micelle formulation, we implemented a multiscale computational approach to study micelle formation using a coaxial turbulent jet flow.

Methods:

All-atom (AA), coarse-grained molecular dynamics (CG-MD), as well as computational fluid dynamics (CFD) simulations have been conducted to not only reveal the effects to material attributes and processing parameters during the micelle formation, but also as parametric case studies for this process. The initial conditions applied into the simulations were based on polymeric micelle experiments in which PEG-PLA (2kD-1.7kD) was used as model drug carrier. The CFD simulation was implemented using Large Eddy Simulation (LES) model in COMSOL Multiphysics incorporating energy equation and high-resolution mesh in mixing area. The MD simulation trajectories and their analysis were carried out using GROMACS package. We applied both CHARMM and MARTINI force-fields for all-atom as well as coarse-grained simulations, respectively. The initial coordinates of models and their force field parameters were generated by CHARMM software and MARTINI forward mapping approach. The steepest decent algorithm was used in energy minimization with a 20 fs time step followed by 10 ns equilibration step in isothermal−isochoric NVT ensemble at temperature of 300 K, then the production runs were performed beyond 1 μs in the NPT ensemble using the Nosé−Hoover thermostat and the Parrinello−Rahman barostat at pressure of 1 atm with constant temperatures at every 20 fs time steps. The simulations were carried out using periodic boundary conditions. Computations were performed in High Performance Center Supercomputer Cluster at the University of Connecticut.

Results:

The self-assembled process of polymeric micelle was studied in the system with and without drug molecules, and in each case, block copolymer was randomly packed in the simulation box with explicit water and ethanol for all simulations. With optimized CG-MD forward mapping strategy and force field parameters incorporating MARTINI standard beads and recent developed S-beads, we observed that polymeric micelles formed successfully and fell into the experimental size range. The simulation of micelles formation and the size distribution the micelles indicate that CHARMM, MARTINI force fields (FFs) can effectively capture the experiment behavior and results . By adjusting Smagorinsky coefficient (Cs) of LES model companied with energy and mass transfer equations, CFD simulations successfully modeled flow patterns and formation temperatures of co-axial turbulent jet flow from experiments, as well as successful comparison and verification. Our CG-MD and CFD simulations are in good agreement with experimental results.

Conclusions:

In this present work, the model predictions provide an understanding of the impact of drug-polymer interactions, drug-polymer ratio, initial organic phase polymer concentration, as well as maintaining the agreement between the computational predictions and experimental findings highlight the power of the multiscale approach to be able to cover the long and short length scale predictions. Moreover, the details of the formation process was revealed in micro, macro, and meso scales by connecting discrete and continuum computational modeling work which are found to be complementary to one another. The multiscale approach used in this work can be effectively utilized as a powerful tool in the process of discovery, development, and optimization of new drug delivery systems in the co-flow continuous processing.

Acknowledgements: FDA Grant# 1U01FD005773-01.

Disclaimer: This article reflects the views of the authors and should not be construed to represent FDA's views or policies.

609574 Simulation of Drying of Droplets and Its Effect on Particle's Morphology in Spray-Drying

Abstract:

Spray-drying is a manufacturing process utilized by a wide range of industries such as food, agricultural, and pharmaceutical for converting a liquid feed containing insoluble solid particles into particulate form. By removing the moisture, it is intended to achieve a final product with desired chemical composition, morphology (hollow particles, solid cores, agglomerates, etc.), and particle size distribution. There are many variables that affect the drying process, and due to its complexity, traditional trial-and-error methods have been the primary technique in achieving the desired product.

The drying dynamics affect the internal flow patterns within a droplet in spray-drying. The flow patterns are dominated by surface tension and evaporative effects which ultimately disperse the suspended particles in the droplet and finally determine the final product morphology. In the presented work, the drying and internal transport of insoluble particles within a droplet were simulated using a volume of fluid method coupled with a Lagrangian solver for tracking particle movement. The simulations were performed within the open-source framework, OpenFOAM. The solver has been heavily modified to include interfacial mass transfer by employing a coupled Eulerian-Lagrangian framework capable of simulating drying of a droplet and the solid particles suspended inside. Modeling droplet's internal flow created through evaporative effects, and tracking movement of solid particles, drying and the morphological evolution of the product is simulated and analyzed.

606888 Population Balance Model Comparison of Growth, Secondary Nucleation By Attrition and Ripening

Abstract:

Introduction

Secondary nucleation is a phenomenon present everywhere in nature and of fundamental importance for crystallization processes [1,2]. There are two families of mechanisms of secondary nucleation: the first occurs at the interface of the seed crystal and the solution, and it is a combination of an activated process and the effect of shear, i.e. surface- or shear secondary nucleation. The second family results from mechanical collision happening to the seed crystal, i.e. secondary nucleation by attrition. Attrition is the mechanism through which fragments form after an impact of a crystal with a stirrer, other crystals or the wall of the reactor. Those fragments, if small enough, can be considered secondary nuclei. For both families, there is not a clear understanding of the governing physics, and secondary nucleation rates are usually given by an empirical relationship depending on the extent of mixing, on the amount of crystals suspended and on the supersaturation level [1].

Methods and Models

In this contribution, we have extended the well-established mechanistic attrition model developed by Gahn and Mersmann [3] to two different population balance equation (PBE) models, which are used to simulate secondary nucleation processes. The traditional approach describes the formation of attrition fragments due to collisions as a secondary nucleation rate, which is included in the model as a boundary condition (we call it nucleation model). In the alternative approach, the formation of attrition fragments is described as a breakage expression and the growth rate is the result of size-dependent solubility (we call it breakage model) [4,5]. Conversely to the traditional model, this formulation takes into account the size distribution of attrition fragments and their evolution due to growth, see the figure for a schematic representation of the two models. The two approaches have specific challenges and they have different pros and cons.

In both models, the number of parameters to estimate has been limited by using expressions, where the physical and mechanical properties of the system, together with operative conditions, have been employed. In particular, for the breakage frequency, a physical model suitable for PBE has been derived and used in the simulations. The traditional attrition distribution of Mersmann has been adapted to a mechanistic breakage daughter distribution, that we were able to use in a PBE. For the secondary nucleation rate, a new expression has been derived and then compared with standard ones[5].

Results

The physical-based model of Mersmann was successfully integrated into two different population balance models to describe secondary nucleation by attrition for a population of crystals. In order to compare the models, two simulations of a batch, isothermal system have been performed. In both cases, the simulated time was very long, thus resulting in an almost complete depletion of supersaturation. The two models lead to very similar results in the region where supersaturation is high enough to ensure growth and secondary nucleation. However, the breakage model can describe the mechanism of Ostwald ripening, as a result of the embedded size-dependent solubility, when supersaturation is very close to 1, and the simulated time is very long.

The two models are solved in two different numerical ways since the population balance in the case of the breakage model is more complex and numerically intensive than the implementation of the boundary condition, which is the only necessary thing to solve for the nucleation model. This required the implementation of two different numerical schemes: the nucleation model has been solved by following a finite volume scheme with boundary conditions for nucleation. The attrition model has been solved by discretizing the growth part with a finite volume scheme and the breakage term with the fixed pivot technique [6,7].

Conclusions

The two models are very similar in the growth regime, thus where secondary nucleation and growth are the dominant phenomena. At extremely low values of supersaturation, thanks to size-dependent solubility, the first model yields to further development of the crystal population, e.g., Ostwald ripening and aging. The main result is that secondary nucleation by attrition can be described as a birth/death term or as a source term according to the final application of the model. Since the two approaches have very different computational intensities, one can choose the right model based on the objective of the simulation study.

Acknowledgments

This project has received funding from the European Research Council (ERC) under the European Union's Horizon 2020 research and innovation program under grant agreement No 2-73959-18.

References

1. Agrawal, S. G., & Paterson, A. H. J. (2015). *Chemical Engineering Communications*, 202(5), 698–706.
2. Mersmann, A. (Ed.). (2001). *Crystallization technology handbook.* CRC Press.
3. Gahn, C., & Mersmann, A. (1999). *Chemical Engineering Science*, 54(9), 1273–1282.
4. Iggland, M., & Mazzotti, M. (2011). *Crystal Growth and Design*, 11(10), 4611–4622.
5. Bosetti, L., & Mazzotti, M. (2019) *Crystal Growth & Design*1, 307-319.
6. LeVeque, R. J. (2002). *Finite volume methods for hyperbolic problems* (Vol. 31). Cambridge university press.
7. Kumar, S., & Ramkrishna, D. (1996) *Chemical Engineering Science*, 51(8), 1311–1332.

Figure. Results of an isothermal desupersaturation experiment. On the l.h.s., schematic representation of the two models. On the center, the evolution of supersaturation (green), and the increase of the zeroth moment (blue for breakage model, red for nucleation model) corresponds to the formation of attrition fragments: in this region, the two models are almost indistinguishable. On the r.h.s., evolution of the initial population with a zoom on the population of fines, where the difference in the mechanisms of formation of nuclei is visible.

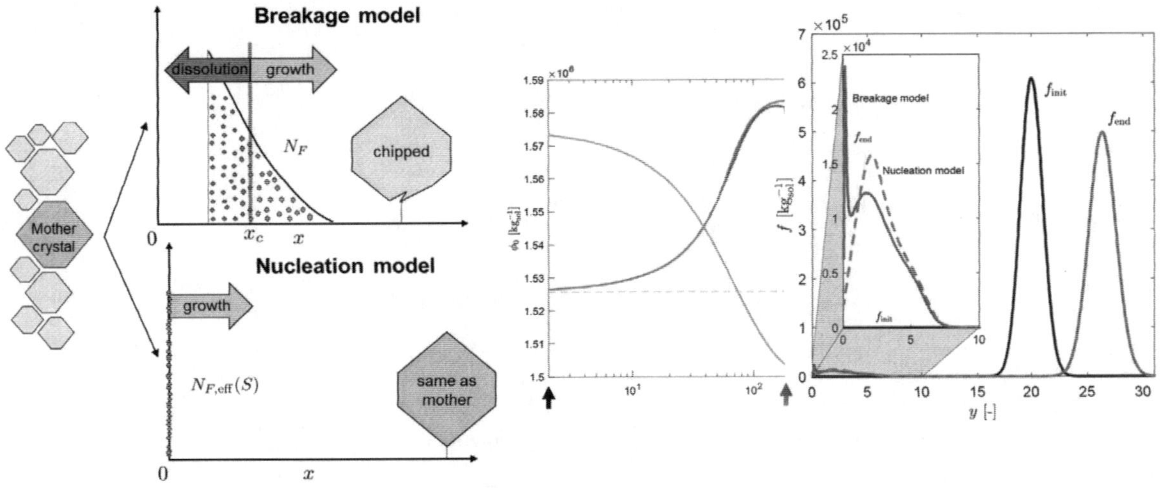

610983 Solubility and Nucleation Kinetics of a Series of Representative Single and Mixed Diesel *N*-Alkanes in Dodecane and Toluene

Abstract:

The formation of *n*-alkane waxes in solution during cold weather is a recurrent problem in the automotive industry. In diesel fuels, crystallisation of long chains *n*-alkanes can result in the formation of large flat plate-like crystals leading to flow impairment, blockage of filters and engine fuel starvation. Clearly the unwanted formation of these waxes and their associated problems can cause significant issues for the manufacturer and end-users.

Current wax mitigation strategies involve the development of chemical additives to control or inhibit the nucleation and growth of crystals so that they no longer form large deposits or block filters. However, the variable nature of fuel composition, specifically the fractionation of the *n*-alkanes, results in a complex challenge for the additive provider to meet operational needs[1]. In order to understand crystallisation of waxes in fuels, it is important to understand the crystallisation of *n*-alkanes in mixtures and singularly.

Here we present the results from an experimental study on the nucleation kinetics and solubility behaviour single and mixtures of 8 *n*-alkanes representative of diesel fuel, ranging from $C_{16}H_{34}$ to $C_{23}H_{48}$ in toluene and dodecane. Analysis of the experimental polythermal data making, use of the analytically derived KBHR approach[2–4] reveals, that in all bar two cases the preferred nucleation mechanism for single n-alkane solutions is instantaneous. i.e. all the nuclei are formed at once at a certain level of supersaturation, to further grow . Furthermore, the influence of chain length on solubility was observed between odd and even alkanes in both toluene and dodecane. By analysing the nucleation and solubility behaviour within the set of n-alkanes tested, suggestions of the factors controlling wax crystallisation in fuel systems are presented and discussed from first thermodynamic principles.

References

[1] A. R. Gerson; K. J. Roberts; J. N. Sherwood, *Powder Technol.*, **65** (1991) *243.*

[2] D. Kashchiev; A. Borissova; R. B. Hammond; K. J. Roberts, *J. Cryst. Growth*, **312** (2010) *698.*

[3] D. Kashchiev; A. Borissova; R. B. Hammond; K. J. Roberts, *J. Phys. Chem.*, **114** (2010) *5441.*

[4] D. M. Camacho Corzo; A. Borissova; R. B. Hammond; D. Kashchiev; K. J. Roberts; K. Lewtas; I. More, *CrystEngComm*, **16** (2014) *974.*

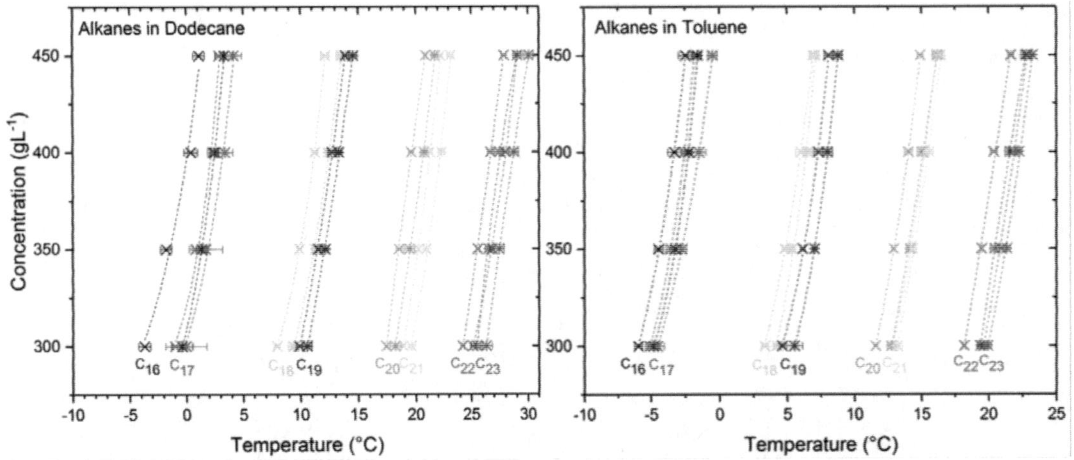

Figure 1 – Solubility curves of 8 *n*-Alkanes $C_{16}-C_{23}$ in toluene and dodecane showing their metastable zones

611896 Multi-Scale Modeling of Soft Matters: A Quality By Design of Drug Delivery Carriers By a Computational Tool

Abstract:

In this work we investigate computationally the relationship between transient motion of multi-compartment microcapsules and their mechanical properties by flowing them in a microfluidic device. Different active ingredients can be encapsulated separately in inner (daughter) capsules that are surrounded by a larger (mother) capsule. The capsule shell (membrane) can be fabricated with materials that are stimuli (e.g. pH, temperature) responsive. Despite significant experimental advance in the fabrication of multi-compartment capsules [1-4], their deformation mechanisms and rheological behavior are poorly understood. Multi-scale modeling is a tool that facilitates the quality by design (QBD) for the fabrication of pharmaceutical capsules (e.g. Chitosan–alginate microcapsules [5]). QBD of microcapsules increases the probability in targeted delivery and the release of encapsulated materials at desired time and location, and mitigates the risk of cross-contamination between the ingredients. Here, we present the effects of different physical and geometrical parameters such as viscosity ratio, number of inner capsules, their sizes and arrangements on the overall dynamical behavior of multi-compartment capsules, by utilizing our membrane spectral boundary element method [6-9]. We further establish the link between the mechanical properties of these capsules and their deformational behavior, which leads to enhanced manufacturing of these class of drug carriers.

[1] Kim, Shin-Hyun, et al. "Multiple polymersomes for programmed release of multiple components." Journal of the American Chemical Society 133.38 (2011): 15165-15171.

[2] Adams, L. L. A., et al. "Single step emulsification for the generation of multi-component double emulsions." Soft Matter 8.41 (2012): 10719-10724.

[3] Dams, Stanley S., and Ian M. Walker. "[5] Multiple emulsions as targetable delivery systems." Methods in enzymology. Vol. 149. Academic Press, 1987. 51-64.

[4] Vasiljevic, Dragana, et al. "An investigation into the characteristics and drug release properties of multiple W/O/W emulsion systems containing low concentration of lipophilic polymeric emulsifier." International journal of pharmaceutics 309.1-2 (2006): 171-177.

[5] Ghaffarian, Rasa, et al. "Chitosan–Alginate Microcapsules Provide Gastric Protection and Intestinal Release of ICAM-1-Targeting Nanocarriers, Enabling GI Targeting In Vivo." Advanced functional materials 26.20 (2016): 3382-3393.

[6] Dodson, W. R., and P. Dimitrakopoulos. "Dynamics of strain-hardening and strain-softening capsules in strong planar extensional flows via an interfacial spectral boundary element algorithm for elastic membranes." Journal of fluid mechanics 641 (2009): 263-296.

[7] Koolivand, A., and P. Dimitrakopoulos. "Deformation of an elastic capsule in a microfluidic T-junction: settling shape and moduli determination." Microfluidics and Nanofluidics 21.5 (2017): 89.

[8] Koolivand, Abdollah, and Panagiotis Dimitrakopoulos. "Dynamics of Elastic Multi-Compartment Capsules in Microfluidic Channels." 2019 AIChE Annual Meeting. AIChE, 2019.

[9] Koolivand, Abdollah, and Panagiotis Dimitrakopoulos. "An Advanced 3D Computational Model for Designing Next Generation Drug Carriers." 2019 AIChE Annual Meeting. AIChE, 2019.

605435 Investigating Mechanisms of Bimetallic Nanocrystal Formation with Liquid Phase Transmission Electron Microscopy

Abstract:

Liquid phase transmission electron microscopy (LP-TEM) has been widely used to investigate the formation mechanisms of metal nanoparticles in liquid, providing guidance to synthesize nanocrystals with desired structures and functional properties.[1, 2] During LP-TEM imaging, the electron beam acts both as the imaging tool and the reducing agent to reduce either metal precursors to form nanocrystals, which enables visualizing nanocrystal formation in real time with nanometer scale spatial resolution. Despite numerous insights gained from these experiments, it is unclear how electron beam induced nanocrystal formation compares to nanocrystal formation during flask-based synthesis. Indeed, a systematic study comparing the chemistry and kinetics of nanocrystal formation during LP-TEM and flask-based colloidal synthesis has not been undertaken.

Here we establish ranges of LP-TEM experimental conditions that are representative of flask-based nanocrystal synthesis. We investigate Au/Cu bimetallic nanoparticle synthesis as a model system due to its promising application as a CO_2 reduction electrocatalyst.[3] Controllable synthesis of alloyed bimetallic nanocrystals is difficult due to differing kinetics and thermodynamics of precursor reduction of different metal species.[4] We utilized poly(ethylene glycol) methyl ether thiol (PEG-thiol, Mw=800 g/mol) to complex Au/Cu metal salts into prenucleation complexes (PNC) prior to nanoparticle synthesis.[5] Our results showed that flask-based synthesis using sodium borohydride to reduce PNCs formed ~2-3 nm Au/Cu alloy nanocrystals. LP-TEM synthesis, which utilizes radiolysis to produce aqueous electrons and hydrogen radicals as reducing agents,[6] formed 3 - 7 nm nanocrystals at low electron dose rates. Imaging at high electron dose rates (high image magnification and beam current) formed irregular aggregated particles. High electron dose rates increase the nucleation rate and create more oxidizing radical species,[7] which oxidizes the sulfur group on the PEG-thiol ligand, leading to formation of uncapped aggregated particles. Our study identified key differences between flask-based and LP-TEM synthesis methods, giving important guidance on how to set experimental conditions during LP-TEM to study nanocrystal formation mechanisms at conditions relevant to flask-based synthesis.

1. Zheng, H.; Smith, R. K.; Jun, Y.-w.; Kisielowski, C.; Dahmen, U.; Alivisatos, A. P., Observation of Single Colloidal Platinum Nanocrystal Growth Trajectories. *Science* **2009,** *324* (5932), 1309-1312.
2. Loh, N. D.; Sen, S.; Bosman, M.; Tan, S. F.; Zhong, J.; Nijhuis, C. A.; Král, P.; Matsudaira, P.; Mirsaidov, U., Multistep nucleation of nanocrystals in aqueous solution. *Nature Chemistry* **2016,** *9*, 77.
3. Kim, D.; Resasco, J.; Yu, Y.; Asiri, A. M.; Yang, P., Synergistic geometric and electronic effects for electrochemical reduction of carbon dioxide using gold–copper bimetallic nanoparticles. *Nature Communications* **2014,** *5*, 4948.
4. Wang, D.; Li, Y., Bimetallic Nanocrystals: Liquid-Phase Synthesis and Catalytic Applications. *Advanced Materials* **2011,** *23* (9), 1044-1060.
5. Marbella, L. E.; Chevrier, D. M.; Tancini, P. D.; Shobayo, O.; Smith, A. M.; Johnston, K. A.; Andolina, C. M.; Zhang, P.; Mpourmpakis, G.; Millstone, J. E., Description and Role of Bimetallic Prenucleation Species in the Formation of Small Nanoparticle Alloys. *Journ American Chemical Society* **2015,** *137* (50), 15852-15858.
6. Schneider, N. M.; Norton, M. M.; Mendel, B. J.; Grogan, J. M.; Ross, F. M.; Bau, H. H., Electron–Water Interactions and Implications for Liquid Cell Electron Microscopy. *The Journal of Physical Chemistry C* **2014,** *118* (38), 22373-22382.
7. Wang, M.; Park, C.; Woehl, T. J., Quantifying the Nucleation and Growth Kinetics of Electron Beam Nanochemistry with Liquid Cell Scanning Transmission Electron Microscopy. *Chemistry of Materials* **2018,** *30* (21), 7727-7736.

606075 Characterization of Starch Products in Continuous Direct Compression Process

Abstract:

Direct compression (DC) is a highly desirable technique to manufacture tablets, and raw material attributes directly influence its effectiveness. Therefore, the selection of specific grades of excipients to achieve desirable powder flow and compression properties is of importance. As industry increasingly shifts toward continuous direct compression, this requires enhanced performance, reliable quality, and consistency directly from the starting ingredients. Powder feeding, as an initial and critical component of the continuous processing, affects the uniformity and quality of the final product and the reliability of the manufacturing system.

In this study, powder property measurements, including Shear Cell Response, Compressibility, Permeability, Dynamic Flow Energy, Bulk and Tapped Density, and Particle Size have been used to fully characterize starch ingredients (i.e., native maize starch, Starch 1500®, Partially Pregelatinized Maize Starch and StarTab®, Directly Compressible Starch (Colorcon, US). In addition, feeder characterization and refill studies were performed to compare the ability of feeders to perform consistent and accurate feeding of the starch ingredients. Principal Component Analysis (PCA) was then applied to examine the performance of the analyzed starch ingredients relative to other common excipients previously analyzed.

Results showed that compared with the other starch materials, StarTab has the largest particle size, most spherical shape, lowest Hausner ratio, lowest cohesivity, lowest compressibility and the greatest permeability. Thus, it is suggested that StarTab provides the best flowability and packing efficiency compared to native maize starch and Starch 1500. In the PCA model, all three starches displayed higher density than other excipients. Furthermore, StarTab was located in a more free-flowing region than the other two starch ingredients and thus consistent with previous results.

Due to sufficient feeder self-control, all three starches have comparable feeding performance. Markedly in feeder refill study, StarTab showed considerable smaller the total deviation of material fed in excess of the set point.

In conclusion, there was an increasing trend in improved flow and feeder refill performance from native starch to Starch 1500 to StarTab products. Based on the results obtained, StarTab is considered as an excellent candidate for use in pharmaceutical continuous direct compression (PCDC).

611967 Impact of Powder Composition on Processing-Relevant Properties of Pharmaceutical Materials: An Experimental Study

Abstract:

In this work, we studied the influence of powder composition on packing density and other processing-relevant properties of binary mixtures, including powder flowability. Binary mixtures of pharmaceutical powders with different particle diameter ratios, α and varying fractions of large and small particles were analyzed systematically. Mixtures of three excipients and one API with different composition (2, 5, 10, 30, 50, 70, 90, 95 and 98 %wt.) were prepared in a Turbula mixer. Powders with different properties and particle size distribution were chosen, in order to obtain three binary mixtures with different size ratios. Then, macroscopic powder properties including bulk (poured) and tapped density (BD and TD) were measured. A powder rheometer was used to measure the flow function coefficient (ffc), cohesion, compressibility and permeability of the binary mixtures. We considered experimentally three classes of binary mixtures, which are characterized by two critical ratios of particle diameter: the critical size ratio of entrance (α_C) and the critical size ratio of replacement (α_r), where $\alpha_C = 0.154$ and $\alpha_r = 0.741$. Below the critical size ratio of entrance (α_C), the particle asymmetry (ratio between large and small particle diameters) is high and small particles can fill the voids between larger ones. Below the critical size ratio of replacement (α_r), the particles asymmetry is low and small particles cannot Above α_r, the particles are more or less symmetric in size and overall packing structure does not change by mixing the particles. Our experiments show that there is a non-linear and non-monotonic dependence of all relevant properties on composition for powder mixtures that have an $\alpha < \alpha_r$. This non-linear behavior is even more significant for strongly asymmetric binary mixtures with $\alpha < \alpha_C$. We argue that this behavior is related to the composition dependence of random packing of particulate systems. Our results have relevance to pharmaceutical particle processing operations where constant powder mixture properties are needed to ensure quality standards are met; such operations include capsule or die filling during tableting, and the continuous feeding of powders via screw feeders. Our results suggest that for pharmaceutical particle processing operations, where constant powder mixture properties are a prerequisite for process robustness, the diameter ratio of API and excipient particles, a should not be smaller than $\alpha_r = 0.741$.

596483 Vibratory Mixing of Pharmaceutical Powders on a Single-Tablet-Scale

Abstract:

While many things change in pharmaceutical industry, tablets remain the most important way of administering Active Pharmaceutical Ingredients (APIs) to a patient. Due to the high production volume, tablets are usually produced in highly optimized processes for fixed doses, dependent on complex supply chains. Security of the supply chain, as well as an observable trend to patient-tailored medicines, call for more flexible production options.

To meet these needs, we aim to develop a small-scale production system for individual tablets. In this context, composition, mixing procedure and compaction step can be adjusted for every single tablet. Such a device will be a powerful tool to cope with the above-mentioned challenges, and furthermore, speed up formulation and process development for new drug products, in the context of translational pharmaceutics.

The system we have designed is based on a single tablet direct compression process, consisting of gravimetric dosing, followed by vibratory mixing and compaction. In our talk we will answer the question of how to set up a vibratory mixing process, to achieve a sufficiently well-mixed powder blend within a reasonable time, i.e., below 10 seconds. An experimental setup is presented which allows to continuously monitor the mixing quality of a 400 mg powder bulk during the mixing process, using highspeed camera image analysis. The vibratory motion is created by an electrodynamic shaker capable of supplying accelerations of up to 100 G for the used setup.

Two microcrystalline cellulose fillers were used as model substances. Virgin Emcocel 90 M having an average particle size of 130 µm, showing slightly cohesive behavior due to its particle shape, and dyed Avicel PH-102 with an average particle size of 100 µm and free-flowing characteristics. The container used as mixing vessel had an inner diameter of 10 mm and a height of 28 mm. Applied frequencies ranged from 100 to 300 Hz and led to different flow patterns within the mixing container, and therefore to different mixing times. Results on the mixing efficiency will be presented.

598582 Scale-up of Dry Granular Material Flow in a Bladed Mixer

Abstract:

Drying of granular materials is a common unit operation for a wide range of industries. In the pharmaceutical industry, drying of Active Pharmaceutical Ingredient (API) crystals is often carried out in an agitated filter dryer. During the drying process, wet particles are heated in an agitated mixer which consists of a jacketed cylindrical vessel with a rotating impeller which promotes convective drying. Incomplete understanding of the process, results in operating problems such as under or over drying, non-uniform drying, polymorphism changes, agglomeration, and particle breakage. Moreover, it is difficult to choose the operating conditions to achieve the desired product due to several phenomena including mass transfer, heat transfer, and changes in material properties happening simultaneously in an agitated dryer. In addition, it is challenging to scale up the process from the laboratory scale to the pilot scale to the manufacturing scale. It is of interest to study the agitated drying process in order to optimize operating conditions for a particular material as well as develop and understanding of factors that affect scale up. In this study, we simplify the problem and we don't examine flow, mixing, heat transfer and mass transfer together but instead we focus on the granular material flow in the agitated mixer. Dry, monodisperse, spherical glass beads were used as model materials in the mixer and the torque on the rotating impeller was measured in order to characterize the material flow. The experiments were designed such that flow occurred in the slow, frictional flow regime or quasi-static regime and we observed that impeller torque was independent of the shear rate (rotation rate of the impeller). We found that the height of the powder bed had a great effect on impeller torque whereas the particle size had less effect on impeller torque. We also found that a portion of the powder bed could be replaced with metal weights (in order to keep the same hydrostatic pressure) without changing the impeller torque. Previous work has suggested that shear stresses can be related to normal stresses for flowing granular materials. By plotting the normal stresses versus the shear stresses in the bed, we were able to compute a bulk friction coefficient and cohesiveness of the material, which are characteristic properties of the material, For scale-up studies, the bulk friction coefficient and cohesiveness of the material were used to predict the impeller torque. We illustrated the prediction of impeller torque of the material by examining a bed of large height and observed good agreement between our predictions and measurements in the lab. We will discuss how this work serves as an efficient way to predict the granular material flow in a larger scale using data from the lab scale. We will also discuss the application of these results to scale up of agitated drying of pharmaceutical materials.

604983 Predictive Approach to Understand and Eliminate Tablet Breakage during Film Coating

Abstract:

Pharmaceutical tablets can be susceptible to damage such as edge chipping or erosion of the core during the tablet coating process. The intersection of certain process parameters, equipment design, and tablet properties may induce more significant tablet damage such as complete tablet fracture. In this work, we develop a hybrid predictive approach using discrete element method (DEM) modeling and lab-based experiments to identify conditions that may lead to tablet breakage events. The approach is extended to examine potential modifications to the coating equipment and process conditions *in silico* to mitigate the likelihood of tablet breakage in future batches. The approach is shown to enhance process understanding, identify optimal process conditions within development constraints, and de-risk the manufacture of future tablet coating batches.

All authors are employees of AbbVie and may own AbbVie stock. AbbVie sponsored and funded the study; contributed to the design; participated in the collection, analysis, and interpretation of data, and in writing, reviewing, and approval of the final publication.

593980 Using Powder Rheology to Assess the Effect of Microcrystalline Cellulose Sphere Size on Performance in a Wurster Fluid-Bed Coater

Abstract:

A fluid-bed coater with a Wurster cylinder is a widely used unit operation for pellet coating. It is a circulating process and doesn't contain a fluid-bed in the traditional sense. Instead, a Wurster cylinder-containing (two-compartment) design is intended to reduce weight gain variability in comparison to a one-compartment design.

A Wurster cylinder contains four distinct zones: the coating zone, the expansion chamber, the downflow bed zone, and the horizontal transport zone. Individual particles are propelled into the coating zone via a high velocity air flow where droplets are sprayed from a nozzle to coat the particles. Once through the Wurster cylinder, the particles settle in the expansion chamber and into the down-flow bed. Between the bottom part of down-flow bed zone and coating zone is the horizontal transport zone. The horizontal transport zone is analogous, to some extent, to pneumatic conveying. The fluidised-bed region is only present within the horizontal transport zone and reduces the friction between particles, helping to convey them to the coating zone. The adjustable position of the Wurster cylinder (the distance between the air distribution disk and the bottom edge of the Wurster cylinder) controls powder flow rate.

The main objective of this study [1] was to investigate the impact of the particle size (d50 ranging from 94 to 424μm) of microcrystalline cellulose (MCC) spheres on the flow behaviour within a Wurster cylinder containing fluid-bed coater. Standard pharmacopeia methods were employed, as well as measuring dynamic and bulk powder flow properties via powder rheology (FT4 Powder Rheometer®, Freeman Technology Ltd, UK). Samples were also measured using alternate powder flow characterisation techniques the results from which are compared to powder rheology.

Specific powder rheology methods are proposed to characterise powder behaviour in the down-flow bed, horizontal transport and coating zones. Thereby providing an understanding of what powder properties could deliver optimum performance.

This study also underpins the requirement for multivariate characterisation when assessing powder flow. Meaning that several characterisation methodologies are required, the results from which can be correlated with process ranking to find the most influential parameters on process behaviour. Rather than relying on single number characterisation to describe behaviour across all processes, a multivariate approach using powder rheology allows for a direct investigation of a powder's response to various process conditions.

[1] Mohylyuk et al., Journal of Drug Delivery Science and Technology, 54, 101320, 2019

603373 In silico Design Space Investigation of a Novel Tablet Coating Process Using Advanced Coupled Simulations

Abstract:

Continuous manufacturing is increasingly used in the pharmaceutical industry, as it promises to deliver better product quality while simultaneously increasing production flexibility. GEA developed the ConsiGma line(s) to be integrated into granular downstream processes. The specially developed tablet coater is unique in its design and process mechanics. It uses high rotation rates compared to traditional drum batch coaters.

Computational methods are increasingly used in the pharmaceutical industry as well and facilitate a deeper understanding of the process mechanics. The discrete element method (DEM) and computational fluid dynamics (CFD) are two methods that allow transition from empirical process design to a mechanistic understanding of the individual process units. This work used CFD-DEM simulations to increase the understanding of the ConsiGma® tablet coater through a multi-variant design study. The results are split into two categories: first, the tablet bed behavior (mechanics) will be investigated and second, the thermodynamically behavior will be studied.

The goal of this work is to present a comprehensive overview of the influence the varying process and material input parameters have on the process output responses. The simulations show how the drum load, rotation rate, and material properties all influence the tablet bed behavior. The same can be said that the drying air flow rate, spray rate, and inlet temperature can influence the coating quality which is tracked through the tablet temperature and wetness. Overall it could be shown that simulations allow for the full coating process to be recreated numerically which, in return, saves material and equipment time typically required for process development.

602865 Expanding the Mechanical Property Assessment of Pharmaceutical Tablet Dosage Forms to De-Risk Manufacturing

Abstract:

Following pharmaceutical tablet compression, physical forces from de-dusting, coating, and packaging can adversely affect tablet integrity, potentially introducing defects such as chipping, erosion, and fracturing that may result in broken tablets. For these reasons, the physical strength of compacted powder is critical to ensuring the quality of pharmaceutical tablets. Thus, its measurements help assess the friability of tablet formulations and the corresponding risk associated with each step of the tablet manufacturing process.

In pharmaceutical manufacturing, the strength of a tablet is most often quantified through its breaking force per USP <1217>. The method outline therein involves producing a tablet using a standard punch and die, placing the tablet between two parallel platens, and compressing the tablet with the platens until it breaks. The force observed at the point of breaking is then taken as the tablet breaking force. In this presentation, we expand upon the generally accepted means for assessing the strength of compacted pharmaceutical powders.

The first part of the presentation explores different methods of producing a tablet and its impact on the tablet strength. Here, we consider two different means of achieving tablet relaxation when it remains in the die following the maximum densification: uni-axial and tri-axial decompression. Taking the ratio of the strength of the tablet made under uni-axial decompression to that made under tri-axial decompression allows assessment of the material sensitivity to the unloading and ejection process during a standard compaction process.

The second part outlines different methods for assessing the quasi-static failure of compacted powder. Herein, we present the effects of anisotropy measured by the ratio between a compact's shear strength and its tensile strength. In the case of inherently anisotropic materials whose shear strength is demonstrably lower than its tensile strength, we pose that the former quantity is a better indicator of strength for risk-assessment purposes. In addition, we also evaluate the tensile strength ratio of the tri-axially decompressed compacts made with and without a small pinhole (i.e. the brittle fracture index) to assess the effects of sensitivity to stress concentration on the tablet's propensity to ejection-related defects (e.g. capping, lamination, etc.).

For both parts, we present the values for commonly used pharmaceutical excipients and show that significant deviations from these values result in tablet defects generally associated with the ejection process. Additionally, from our collection of experimental trends, we propose a regime map constructed using the three different ratios outlined above and demonstrate how this map may be used for assessing the risks of a tableting process.

611898 CFD Simulation of the Regenerator Section of a Carbon Capture Unit Using Amine Based Solid Sorbent

Abstract:

Carbon capture using amine-based solid sorbents from existing flue gas of fossil-fuel-fired power plants has been considered as a promising approach in comparison with the conventional liquid-based systems. National Energy Technology Laboratory (NETL) designed an integrated adsorption and regeneration system of fluidized beds to investigate the capacity of amine-based sorbents. In this study, the focus has been on the computational fluid dynamics (CFD) simulation of the NETL regenerator fluidized bed reactor. After exiting the carbonator, CO_2-laden solid sorbents enter the regenerator at a higher temperature, which causes CO_2release from carbonated amine sorbents.

In this study, the Eulerian-Eulerian two-fluid model is used to conduct two-dimensional CFD simulationof the regeneration section of the NETL circulating fluidized bed system. Kinetic theory was used as the constitutive relation to analyze the particulate phase flow model along with frictional stress and viscosity for the dense solid flow regimes. In addition to continuity and momentum conservation equations, species conservation and energy conservation equations were also solved to calculate the CO_2desorption rate, regeneration efficiency of the sorbents, and flow temperature distribution in the system.

602135 Transient Reacting Flow Simulation of Biomass Combustion in a Pilot-Scale Circulating Fluidized Bed Combustor

Abstract:

Interest in circulating fluidized bed (CFB) combustors as a power generation technology has sky-rocketed in recent years because of several advantages this technology offers over conventional boilers such as increased gas-solid mixing resulting in higher efficiency and the ability to use lower quality fuels. CFB combustors are operated at lower temperatures than conventional boilers, which also mitigates NO_x and SO_2 emissions. A comprehensive computational fluid dynamics (CFD) model of a CFB combustor is developed employing the multiphase particle-in-cell (MP-PIC) module in the open-source MFiX Software Suite to understand and analyze the complex gas-solid hydrodynamics, chemical processes, energy conversion in such a system, and the scale-up considerations required to develop the industrial-scale CFB combustors of the future.

The experimental basis for the simulations is the 50 kW_{th} CFB combustor riser with a diameter of 0.15 m and a height of 5 m designed, built, and operated at CanmetENERGY in Ottawa, Canada. The MFiX-PIC model parameters for the simulation are tuned against cold flow experiments from Canmet using 9 kg of olivine sand as the inert bed material. It is shown that for the relatively coarse fluid meshes and large parcel sizes necessitated by the scale of the simulation, filter size dependent corrections to the drag law must be incorporated to ensure accuracy of the simulation results.

The validated cold-flow model is extended to simulate reacting flow with torrefied hardwood as the feedstock using an established reaction scheme from previous work. The effect of cooling air flow through heat exchanger tubes to maintain the bed temperature in the Canmet riser is represented by prescribing the heat flux across the tube surfaces in the simulation. The pressure and temperature profiles in the riser as well as the outlet gas compositions are compared against Canmet's experiment and show satisfactory agreement. The simulations demonstrate the ability of MFiX-PIC to accurately capture the physics and chemistry of a circulating fluidized bed combustor at pilot plant scales, which can be further extended to industrial-scale systems. The detailed species information in the riser obtained from the simulation is also used to predict NO_x formation based on a sub-grid model.

600260 Measuring Particle Dynamics of Fluidized Bed Using Digital in-Line Holography

Abstract:

The quantification of three-dimensional (3D) particle behavior in a fluidized bed is crucial for updating the basic knowledge and thus implementing better designs for such complicated multiphase system. Current work reports recent development and application of digital in-line holography (DIH) on the *in situ* measurement of (a) solid circulation rates, (b) particle size distributions, (c) 3D location, and (d) velocity distributions in a dense particle fields. A Hough-transform-based hologram processing algorithm have been developed and tested over a wide range of flow conditions, solid fractions, and particle characteristics. The performance of the processing algorithm is firstly benchmarked using synthetic holograms with varying particle size and shape, generated using known particle trajectories from DEM-CFD simulations. Then, using a hopper system, real-time solid flow rates and particle size distributions are measured using DIH and are compared to high-precesion balance and QIPIC results, respectively. Finally, DIH is employed to measure the standing pipe (12.7 mm ID) particular flows in a mini-CFB system. Solid circulation rates evolution are measured and its mean value agrees reasonably well with pressure-based estimations. Moreover, DIH is demonstrated to be capable of resolving fine particle trajectories resulting from attrition (~10 μm, with Stokes number, $St \ll 0$). The motion of these fine particles is useful in determining flow conditions around large particles, and estimate the slip velocities.

595984 The Screened Waveguide for Intrusive Acoustic Emission Detection and Its Application in Circulation Fluidized Bed

Abstract:

Passive acoustic emission (AE) has been widely used in various chemical industries, as it has the characteristics of real time, on-line and environmental-friendly. However, most of the acoustic signals are only collected from the surface of measured device, which are superimposed signals of the acoustic wave from the detection area of the sensor. Besides, the flowing state near the wall is usually quite different from that inside the reactor due to wall effects. It is undeniable that the information near the wall is important, however, detailed information inside the system is definitely more significant for the whole system. Therefore, in order to achieve local information detection and obtain the spatial distribution of particle motions inside systems, a new waveguide with screening structures was proposed based on the theory of sound wave generation and propagation. Subsequently, an intrusive acoustic emission detection based on the screened waveguide was developed to detect the particle flow patterns inside a circulating fluidized bed. Results showed that the noise energy accounted for only 2.1% of the total signal energy received by the waveguide, indicating that under real experimental conditions, the designed waveguide had a good shielding effect. Radial acoustic energy distribution detected via the waveguide was highly consistent with the radial solid flux distribution measured by extraction probe, validating the feasibility of this method. Furthermore, the radial flow pattern of Geldart D particles in a riser of circulating fluidized bed was systematically investigated based on this method, which was also an important supplement to the experimental study of the riser.

607533 Particulate Flow Dynamics in Structured Gas-Solid Fluidized Beds

Abstract:

The performance of bubbling fluidized reactors relies highly on the macroscopic flow of bubbles, which is usually unstable, due to the complex dissipative interparticle collisions and interphase friction occurring at multiple scales. To control the system hydrodynamics, a lot of effort has been spent on investigating how to impose extra flexibilities, including internals, force fields, etc. The improvement associated with these designs is often dependent on the particle type, operation conditions and reactor type, which makes it difficult to obtain a universally valid standard across all scales [1,2]. Nature provides many examples of ordered granular patterns, such as the sandy ripples on dunes, as a consequence of complex interplays between the solid and surrounding airflows. Drawing inspiration from it, we have demonstrated, at specific conditions, that an oscillatory perturbation can suppress instabilities and induce a periodic flow structure at which bubbles self-organize into a regular array of triangular tessellations.

Such a regular array of bubbles is fully scalable and predictable, providing the potential to bypass operational challenges and intensify many solid processes. We have demonstrated the capability to replicate similarly structured bubbles in a cylindrical geometry, extending its appearance from a flat-2D to an annular-3D configuration, while maintaining its degree-of-order and controllability over gas bubbles. Over the last few years, results from our experiments and simulations have demonstrated that operation under such a flow structure would allow us to control bubble properties, such as size, residence time, and separation with unusual precision [3-5]. On the other hand, particles move and bridge bubbles in different arrays, but also impede coalescence of adjacent bubbles, Overall, an ordered macroscopic network of particulate flow paths emerges, which is coupled to the movement of the bubbles.

In this contribution, experimental results are shown to investigate the spatiotemporal coupling between solids and rising bubbles, as seen in Fig.1. Comparisons of particle image velocimetry (PIV) measurements across oscillations at different conditions and bed heights, as well as CFD-DEM simulations, provide insights into the processes of formation, propagation, and collapse of dense solid regions. We also discuss the evolution of short-range and long-range solid motion over different arrays, and their impacts on the stability of macroscopic patterns. These results form a theoretical basis for future developments in process intensification of fluidized beds.

[1] M.-O. Coppens, J.R. van Ommen, Structuring chaotic fluidized beds, Chem. Eng. J., 96 (2003) 117-124

[2] J.R. van Ommen, J. Nijenhuis, & M.-O. Coppens, Reshaping the structure of fluidized beds, CEP, (2009), 49-57.

[3] K. Wu, L. de Martín, M.-O. Coppens, Pattern formation in pulsed gas-solid fluidized beds - the role of granular solid mechanics. Chem. Eng. J., 329 (2017) 4-14.

[4] K. Wu, L. de Martín, L. Mazzei, M.-O. Coppens, Pattern formation in fluidized beds as a tool for model validation: a two-fluid model based study, Powder Technol., 295 (2016) 35-42.

[5] K. Wu, V. Francia, & M.-O. Coppens. Dynamic viscoplastic granular flows: A persistent

challenge in gas-solid fluidization, Powder Technol., 365 (2020) 172-185.

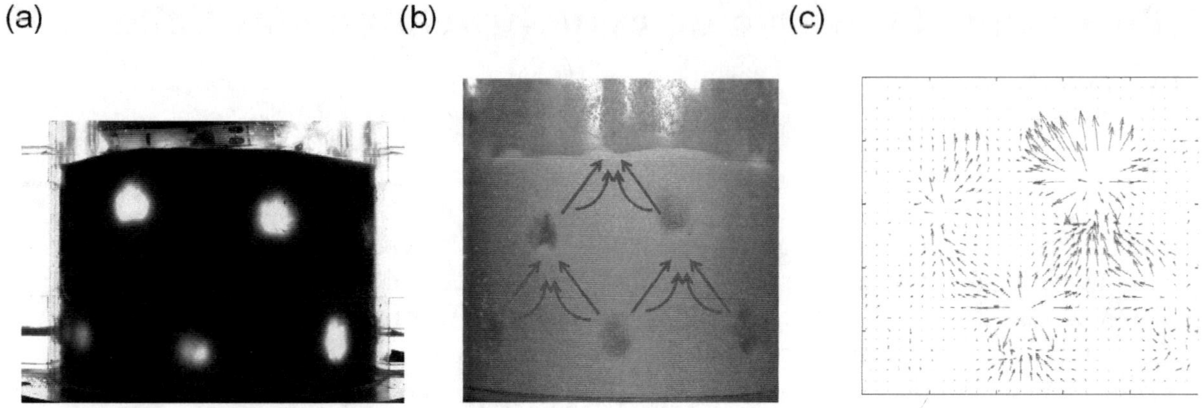

Figure 1 – Representative frame of the structured flow of bubbles in a pulsed annual-3D bed set-up under back (a) and front illumination (b), and its associated solid motion profile (c).

609230 Real-Time Magnetic Resonance Velocimetry and Thermometry of Gas-Solid Systems

Abstract:

Particle systems are challenging to study experimentally. While optical techniques are limited to image the surface of most three-dimensional (3D) systems, intrusive probes only provide measurements at a single spatial location and tend to interfere with the flow field. Hence, a variety of sophisticated tomographic modalities have been developed to provide insights into 3D particle systems. These include x-ray tomography [1], electrical capacitance tomography [2], position emission particle tracking [3] and magnetic resonance imaging (MRI). While the former three modalities can only detect particle positions, MRI additionally allows to measure particle velocities, diffusion, temperature, and chemical reactions [4]. Due to this unparalleled variety of achievable imaging contrasts, MRI is a promising tool for chemical engineering research. However, owing to the sequential nature of data acquisition in MRI, the temporal resolution has been notoriously low [5].

We have recently developed a real-time MRI methodology [6], that increases the temporal resolution of particle velocity measurements by more than four orders of magnitude compared to previous measurements. The methodology relies on a combination of scan acceleration techniques as well as MRI hardware and signal source engineering (Figure 1, left).

In this work, we extend the capabilities of real-time MRI to include temperature measurements of the particle phase (Figure 1, bottom right). Spatially resolved temperature-sensitivity was generated by exploiting the temperature-dependent proton frequency (PRF) shift of water molecules [7,8] combined with MR imaging gradients.

A cylindrical fixed bed reactor model of diameter and height of 80 mm was filled with core-shell thermal tracers. The tracers were manufactured from hollow polypropylene spheres of diameter 10 mm filled with an aqueous solution of paramagnetic rare earth salts. Hot air (60 °C) was injected through a central orifice at the bottom of the bed. 3D thermal maps of the bed were acquired during heating and cooling of the system from and to room temperature. These measurements were compared against optical reference thermometers placed inside the bed. The MR-based temperature measurements exhibited good agreement with these reference measurements.

With an acquisition time of a few seconds for the whole volume, the MR thermometry methodology presented here constitutes a powerful technique to characterize thermal dynamics in a large variety of gas-solid reactors. We are confident that the presented technique will contribute to a better understanding of mass and heat transfer inside gas-solid reactors. It can further be used to test numerical simulations and to aid reactor design.

References

(1) Mudde, R. F. Double X-Ray Tomography of a Bubbling Fluidized Bed. *Industrial & Engineering Chemistry Research* **2010**, *49* (11), 5061–5065. https://doi.org/10.1021/ie901537z.

(2) Chandrasekera, T. C. C.; Li, Y.; Moody, D.; Schnellmann, M. A. A.; Dennis, J. S. S.; Holland, D. J. J. Measurement of Bubble Sizes in Fluidised Beds Using Electrical Capacitance Tomography. *Chemical Engineering Science* **2015**, *126*, 679–687. https://doi.org/10.1016/j.ces.2015.01.011.

(3) Parker, D. J. Positron Emission Particle Tracking and Its Application to Granular Media. *Review of Scientific Instruments* **2017**, *88* (5), 051803. https://doi.org/10.1063/1.4983046.

(4) Gladden, L. F.; Sederman, A. J. Magnetic Resonance Imaging and Velocity Mapping in Chemical Engineering Applications. *Annu. Rev. Chem. Biomol. Eng.* **2017**, *8* (1), 227–247. https://doi.org/10.1146/annurev-chembioeng-061114-123222.

(5) Stannarius, R. Magnetic Resonance Imaging of Granular Materials. *Review of Scientific Instruments* **2017**, *88* (051806). https://doi.org/10.1063/1.4983135.

(6) Penn, A.; Tsuji, T.; Brunner, D. O.; Boyce, C. M.; Pruessmann, K. P.; Müller, C. R. Real-Time Probing of Granular Dynamics with Magnetic Resonance. *Science Advances* **2017**, *3* (9), e1701879. https://doi.org/10.1126/sciadv.1701879.

(7) Hindman, J. C. Proton Resonance Shift of Water in the Gas and Liquid States. *J. Chem. Phys.* **1966**, *44* (12), 4582–4592. https://doi.org/10.1063/1.1726676.

(8) Rieke, V.; Pauly, K. B. MR Thermometry. *Journal of Magnetic Resonance Imaging* **2008**, *27* (2), 376–390. https://doi.org/10.1002/jmri.21265.

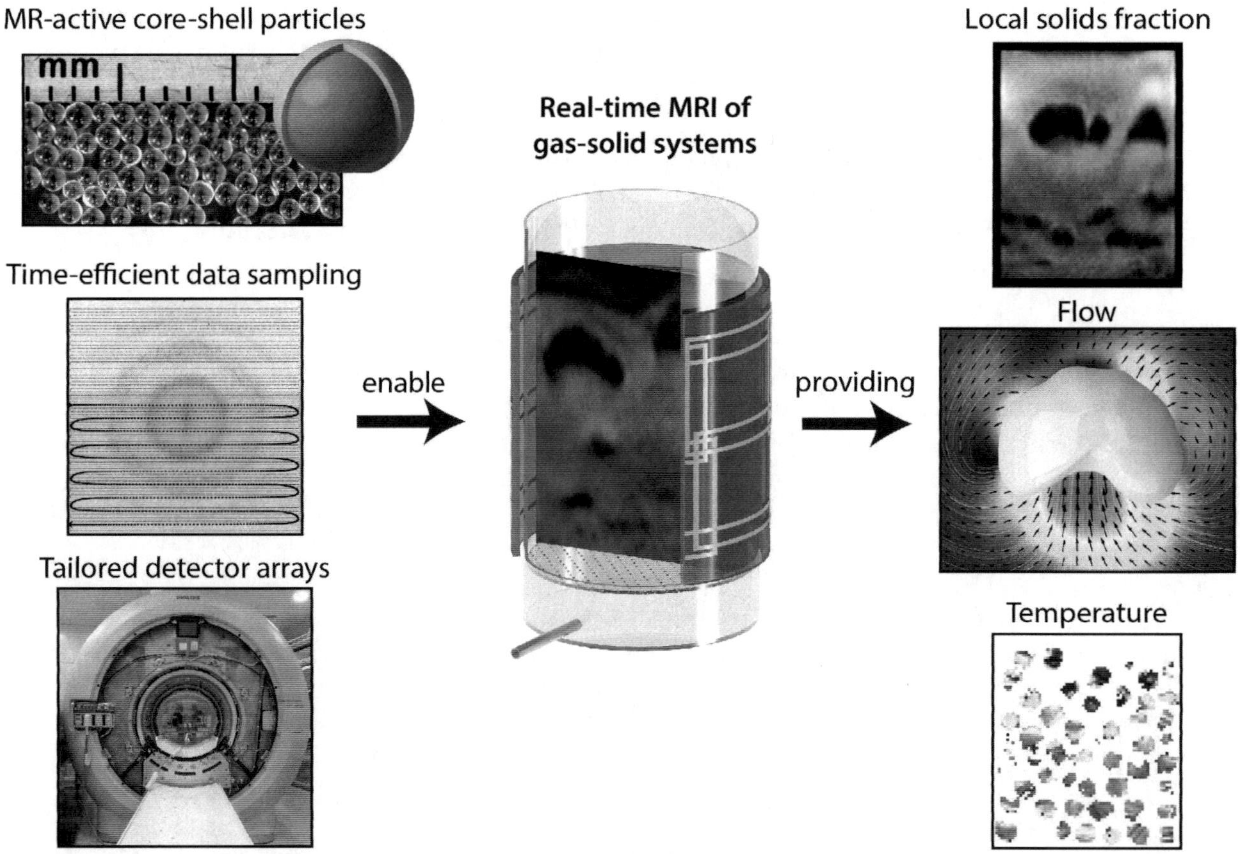

Figure 1. Real-time MRI of gas-solid systems is enabled by a concerted interplay of materials engineering, time-efficient pulse sequences and custom-built hardware. The methodology provides temporally and spatially resolved measurement of the local solids fraction, particle velocity, and temperature.

602561 Experimental Study on Flow Regimes Transition and Bubble Characteristics in a Pressurised Fluidised Bed

Abstract:

Gas-solids fluidised beds have been widely used in process industries due to their excellent mixing and heat/mass transfer performance. Many industrial processes require high operating pressure to improve the overall throughput and reduce the size of the reactors. However, despite considerable studies on atmospheric fluidised beds have been undertaken, understanding the flow characteristics and regime transition at elevated pressure conditions is still insufficient.

In this research, two non-intrusive measurements, i.e. electrical capacitance tomography (ECT) technique and pressure fluctuation measurement, are introduced in a pilot-scale fluidised bed operated at four pressure conditions from 0.2 to 1.3 MPa. A wide range of fluidisation velocities, covering flow regimes from incipient fluidisation to fast fluidisation, is enabled in each operating pressure condition. The flow regime transition, bubble characteristics and effect of pressure, by means of ECT image reconstruction and statistical analyses based on the particle volume fraction (PVF) and pressure fluctuations, are presented and discussed.

The results show that for Geldart B particles, the flow regime transition velocities, namely, the minimum fluidisation velocity (U_{mf}), the onsets of turbulent flow (U_c) and fast fluidisation flow (U_k) show significant decrease as the pressure increases. However, the corresponding fluidisation dimensionless numbers (i.e. U_c/U_{mf}, and U_k/U_{mf}) are almost independent of the operating pressure.

Both the bubble diameter and bubble rise velocity, estimated by image reconstruction and cross-correlation analysis on the dual-plane ECT measurement, increase with the increase of the fluidisation velocity under the same pressure condition. Furthermore, the pixel-by-pixel correlation gives the bubble velocity distribution image across the monitoring cross-section, from which the uneven distribution feature between the centre and near-wall region is clearly observed.

This work demonstrates the high suitability of the ECT technique, applied to a gas-solids pressurised fluidized bed, for the real-time measurement without intruding the flow field. The experimental results, in terms of the pressure effect on gas-solids flow characteristics, bubble properties and regime transition are believed to provide useful information for the proper design and optimisation of pressurised fluidised beds.

608778 Experimental Investigation of Segregation for Non-Spherical Particles in a Fluidized Bed Solids Mixer

Abstract:

A binary mixture of particles varying in size, shape or density tends to segregate in many different flow situations. Removal of segregation is an important goal as it is often detrimental to process efficiency and product quality. The focus of the current work is to see the effect of the shape of particles on segregation in the particular case of a fluidized bed solids mixer. Fluidized bed solids mixer is quite often used in the food and pharmaceutical industry to mix granular particles. The present work describes an experimental method to study the evolution of segregation and its dependence on both the shape and size of the particles. Glass particles of different shapes like spherical, ellipsoids (prolate and oblate), cubical, elongated needle have been used. Results indicate that particle shape plays an important role in determining the extent of segregation. Overall segregation can be reduced by using non-spherical particles.

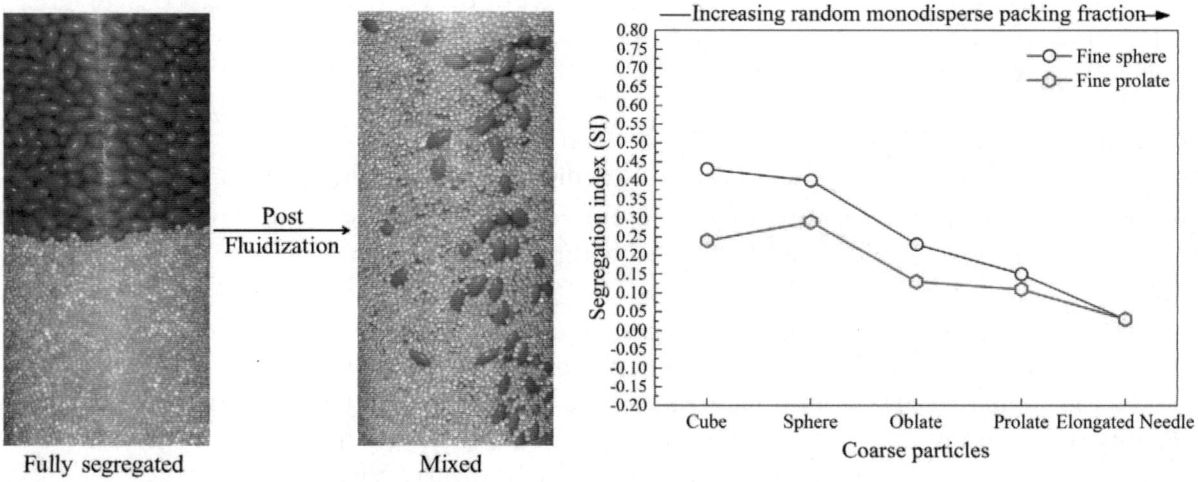

613994 Functional Chiral Nanostructures

Abstract:

The early observation of strong circular dichroism for individual nanoparticles (NPs) and their assemblies have developed into a rapidly expanding field of chiral inorganic nanostructures. They encompass a large family of mirror-asymmetric constructs from metals, semiconductors, ceramics, and nanocarbons with multiple chiral geometries with characteristic scales from *Ångströms* to microns (**Figure 1**). Versatility in dimensions and polarizability of the inorganic materials enables their multiscale engineering to attain a broad range of optical and chemical properties. These capabilities as chiral materials enabled their fast technological translation for biosensing and optoelectronics, which, in turn, opened new venues for scientific inquiry into the unifying role chirality at the interface of materials science, biology, chemistry, and physics.

This talk will address (**1**) the origin of the uniquely high values of optical anisotropy; (**2**) mechanisms of chirality transfer in inorganic materials; and (**3**) differences/similarities with chiral supramolecular, liquid crystal, and biological assemblies. Emerging venues for practical realizations of chiral inorganic nanoassemblies related to circular polarization spectroscopy in terahertz spectral window will be discussed.

References :

Kotov, N.A. Inorganic Nanoparticles as Protein Mimics *Science*, **2010**, 330, 188.

Chen, W. et al. Nanoparticle Superstructures Made by PCR: Collective Interactions of Nanoparticles and a New Principle for Chiral Materials. *Nano Lett*. **2009**, 9(5), 2153.

Srivastava, S.; et al Light-Controlled Self-Assembly of Semiconductor Nanoparticles into Twisted Ribbons. *Science*, **2010**, 327, 1355–1359.

J.Yeom, et al Chiral Templating of Self-Assembling Nanostructures by Circularly Polarized Light, *Nature Mater*. **2015**, 14, 66.

Jiang, et al. Chiral Ceramic Nanoparticles of Tungsten Oxide and Peptide Catalysis, *JACS*, **2017**, 139, 13701.

Yeom et al, Chiromagnetic Nanoparticles and Gels, *Science*. **2018**, 359 (6372), 309-314.

Choi et al, Kirigami Modulators for Terahertz Chiroptical Spectroscopy, *Nature Materials,* **2019**, 18, 820–826.

W.Jiang, et al Emergence of Complexity in Hierarchically Organized Chiral Particles, *Science*, **2020**, 368, 6491, 642-648.

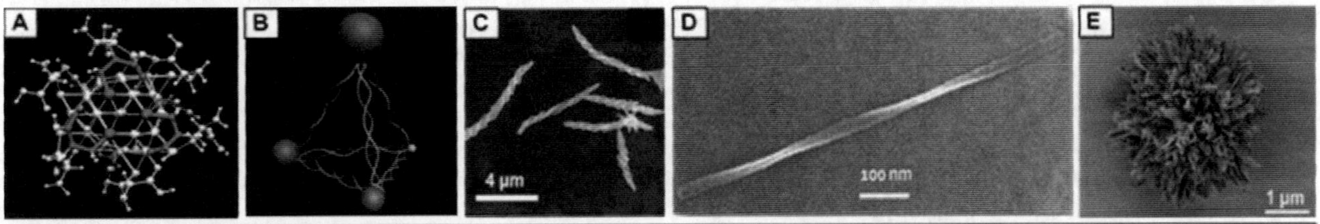

Figure 1. Typical examples of chiral inorganic materials. (A) Co_3O_4 NPs with twisted crystal lattice ; (B) tetrahedral assembly of four different NP; (C) mesoscale helices and (D) twisted ribbons self-assembled from CdTe NPs; (E) chiral hedgehog particles self-assembled from Au-S nanoplatelets.

609360 Continuous Millifluidic Surface Modification of Synthesized Silver Nanowires

Abstract:

Silver nanowires (AgNWs) hold promise as the conductive material in conductive inks suitable for printing processes to create conductive patterns thanks to their three combined characteristics including high conductivity, transparency, and mechanical flexibility. The cost associated with raw material production for this application is among the most important factors in material selection and AgNWs can be synthesized simply in small scale batch process. However, rapid synthesis of high aspect ratio of AgNWs with high concentration is still a challenge that makes a huge gap in their large-scale practical applications. Moreover, generally a mixture of AgNWs and other silver nanostructures in organic solvent are obtained that need to be separated which makes the cost of purification and recovery of synthesized AgNWs a further barrier towards their practical applications. Application of a millifluidic reactor is a method to overcome the challenges in batch AgNWs synthesis through control of a uniform chemical and thermal environment in a small reaction volume to produce high quality corresponding to high aspect ratio, and high yield or purity. Their chemical stability is also critical in practical applications as their oxidation can not only weaken mechanical stability but also increase the sheet resistance of transparent conductive films (TCFs). To eliminate or mitigate the oxidation concern, the surface of AgNWs can be alloyed or coated with graphene, graphene oxide, titanium oxide, zinc oxide, and noble metals such as gold and palladium. The formation of high reduction potential alloys on the surface of AgNWs is a promising technique to improve the performance and stability of AgNW-based TCFs. It has been shown that the chemical stability of AgNWs towards oxidation can be achieved with very small amount of Pd as the noble metal modifier, which is promising for improving the chemical stability of the synthesized AgNWs.

In this study, the AgNWs are synthesized using millifluidic polyol process. The galvanic replacement reaction using sugar substitutes as reducing agent at room temperature is used to modify the surface of synthesized AgNWs continuously. The outlet polytetrafluoroethylene (PTEF) tubing of the millifluidic reactor of polyol process is coiled and replaced in ice bath to quench and cool down the synthesized AgNWs suspension. Another PTEF tubing is connected to the flow of cooled synthesized AgNWs suspension through a T-junction. This PTEF tubing is connected to two millifluidic tubes and syringe pumps that inject the Pd precursor solution ($Pd(NO_3)_2.2H_2O$ aqueous solution) and reducing agent (raw Maui Turbinado brown sugar aqueous solution) to the flow of cooled synthesized AgNWs suspension. After the completion of the galvanic replacement reaction, the Pd treated AgNWs are washed and separated and redispersed in deionized water for further TEM, SEM, and EDX characterizations to illustrate the formation of PdNPs on the surface of AgNWs.

A novel low-cost synthesis of modified AgNWs with high chemical stability against oxidation ready to be coupled with conductive ink formulation and writing machine to create TCFs is illustrated, which is capable to be scaled up for large-scale industry-relevant demand production. This study is supported by National Science Foundation.

603584 Molecular Layer Deposition of Amine-Functional Groups on Fumed Silica Nanoparticles for CO_2 Capture

Abstract:

Traditional liquid CO_2 capture techniques have relied on the chemical reaction between the acidic CO_2 and a basic absorbent, usually liquid amines such as monoethanolamine. However, more recently, solid adsorbent materials have received much attention due to the high regenerative energy requirements and physical limitations of liquid systems. Popular materials have included zeolites, metal organic frameworks (MOFs), microporous organic polymers (MOPs), and supported amines. Polymers and porous silica have served as common support materials for amines due to their high surface area and low production cost. Current processes of loading the amines have focused on impregnation (physical mixture of support material and amine-containing compound) and grafting (bonding amine-containing compounds to the surface, usually through wet reactions). However, sorbents made through impregnation face poor regeneration stability and transport limitations; while grafted sorbents tend to have low adsorption capacities. Here, Particle Molecular Layer Deposition (MLD) is used as a loading method to improve the regeneration stability and adsorption capacity of supported amine sorbents. Two MLD chemistries have been confirmed: (3-aminopropyl)triethoxysilane (mono-amine) and N^1-(3-trimethoxysilylpropyl)diethylene triamine (tri-amine). Both precursors were deposited at 150°C. Adsorption capacity and regeneration stability were investigated using a thermogravimetric analyzer (TGA). Adsorption capacity of the amine functional groups increased to ~2mmol/g as the number of MLD cycles increased, prompting further investigation for optimal loading. Regeneration temperatures are known to be low for supported amines and were demonstrated by both the mono-amine and tri-amine. The mono-amine and tri-amine particles with 10 cycles of MLD underwent 25 regeneration cycles with adsorption at 30°C and desorption at 80°C. Both materials appear stable and robust over these cycling conditions, although the tri-amine may need a longer desorption time due to transport limitations. These materials will be stable for adsorption/desorption cycling temperatures below the deposition temperature of 150°C.

607266 Functionalizing Fe_3O_4 Nanoparticles for Local Luminescence Probing and Mediation of Heat Transfer in Induction Heating Catalysis

Abstract:

Nanoparticle surface chemical transformations have emerged as technological solutions to catalysis, bio-sensing, drug-delivery, and optical enhancement. Among these applications, the utilization of magnetic nanoparticles, such as iron oxide (Fe_3O_4), provide an extra degree of freedom to particle functionality. For example, magnetically recoverable catalysts facilitate catalyst regeneration and recycling, in addition to providing local heating via Radio Frequency (RF) conversion to heat at the nanoparticle surface. However, optimum heat generation is dictated by the particle size and shape, requiring surface passivation with surfactants such as oleic acid (OA) during the synthesis. Unfortunately, the utilization of such ligands hinders induction heating catalysis applications by limiting the available surface area (SA) and acting as an insulating layer to heat transfer. Consequently, the catalysts require larger alternating magnetic fields (AMF) to generate the required temperatures in the reaction media, while adsorbates are overheated, leading to coking and deactivation. Current solutions for nanoparticle surface treatment entails phase transfer procedures, and ligand replacement, in which neither results in SA enhancement. In contrast, the development of thermal probes that are not susceptible to an AMF, still require concentrated solutions of nanoparticles or provide indirect surface temperature measurements with thermo-sensitive molecules that are unable to withstand high temperatures (>70 °C) without degradation. This work aims to address the challenges in developing tunable induction heating catalysts by surface functionalization, allowing the determination of the surface temperature of these nanoparticles *in situ*, and increasing the available SA with ligand displacement.

In this work, morphologically-controlled Fe_3O_4 particles, prepared via colloidal decomposition, were treated with pyridine and tetramethylammonium hydroxide (TMAOH) to determine the extent of surfactant removal. The effect of ligand binding on different Fe_3O_4 morphologies was investigated with Inelastic Neutron Scattering (INS) to facilitate the distinction between oleate ions and OA, and avoid the interference of IR absorption by the nanoparticle on signal analysis and surfactant detection. Through the understanding of the OA binding, improved heat generation and transfer to adsorbates were possible, resulting in a 4x increase in surface area and a 2x increase in heat generation, measured in terms of Specific Loss Power (SLP). Simultaneously, a europium doped sodium yttrium fluoride ($NaYF_4$: Eu^{3+}) shell was deposited on Fe_3O_4 nanoparticles of similar size and shape, to determine the surface temperature based on the change in luminescent intensity/lifetime under an applied AMF. The luminescent probe allows high-resolution thermal monitoring without the need for concentrated nanoparticle solutions or macroscopic temperature increase. This method is used to standardize surface temperature to the nanoparticle structural properties and applied field, ultimately providing control of the selectivity of induction heating catalysis. Consequently, dehydrogenation of alcohols on Fe_3O_4 resulted in aldehyde selectivity and a twofold yield on cubic nanostructures compared to spheres, while non-morphologically controlled particles resulted in ester formation.

601382 Synthesis of Ternary Metal Phosphide Nanocrystals for Optoelectronics

Abstract:

Our growing energy usage has created an urgent need for sustainable sources of electricity and devices including photovoltaics and lighting. Successful development of such technology relies on the discovery of scalable photonic materials that can compete with current state-of-the-art III-V (GaN, GaAs) semiconductors, but have low toxicity and processing costs. Comprising of abundant, non-toxic, and recyclable elements, we present II-IV-V_2 systems (II-IV-V_2, II: Zn, IV: Ge, Sn, V: P) as candidates for such applications. Ternary pnictides offer the potential of tuning the band gap via cation disorder at fixed compositions and nearly constant lattice parameters; ternary chemistry also enables self-doping due to off-stoichiometry, which can be used to control carrier concentrations. Realizing the potential of II-IV-V_2 materials, however, has been difficult to accomplish. Syntheses have typically led to impure materials due to the difficulty of controlling the presence of defect phases and anti-site defects, leading to uncontrollable optoelectronic properties. In this work, we present colloidal partial cation exchange reactions from II-V_2 to II-IV-V_2 nanoparticles as a means of controlling crystal growth and mitigating the presence of defects. The relationship between structure and optoelectronic properties are explored via x-ray diffraction, x-ray absorption spectroscopy, transmission electron microscopy, and optical spectroscopy.

609028 Cycled Organophosphate (DMMP) Decomposition over Plasmonic Copper TiO_2 Aerogels

Abstract:

Degradation of organophosphates for remediation of environmental pollutants (ex: diazinon, chlorpyrifos) and protection from chemical warfare agents (CWAs, ex: soman, sarin) has spurred research into the development of sorbent and degradative materials. We couple plasmonic copper nanoparticles with mesoporous nanoparticulate metal oxide aerogel architectures (TiO_2, CeO_2) and demonstrate thermal and photo-driven degradation of CWA simulant Dimethyl Methylphosphonate (DMMP). While TiO_2 and CeO_2 adsorb DMMP, thermal and photo-driven degradation functionality is increased by depositing Cu nanoparticles onto the oxide network. Degradation pathways, rates, active lifetime, and product distributions differ tremendously based on the illumination and humidity conditions. Long term stability for DMMP degradation is monitored via *in situ* DRIFTS over the course of 60 hours under constant dilute DMMP feed. The degree and reversibility of DMMP binding to the composite aerogels was characterized via in situ DRIFTS after multiple regeneration cycles and by *ex situ* X-ray photoelectron spectroscopy (XPS) pre- and post-DMMP exposure.

607060 Mechanical and Rheological Characterization of Lunar and Martian Regolith-Based Cement for in-Situ Extraterrestrial Construction

Abstract:

Building infrastructure on Earth's moon or Mars is an engineering challenge. For instance, on the Moon, the environment is subject to thermal cycling, strong vacuum, and low gravity. These conditions present challenges for the traditional methods used for manufacturing building materials on Earth. Moreover, there are significant advantages in in-situ resource utilization of Lunar/Martian regolith for construction there, and hence, minimal upload of Earth resources. Geopolymers, a new class of cement-binding materials, are gaining attention for extraterrestrial construction applications due to their physical and thermal properties. This talk will focus on alkali-activated extraterrestrial soil simulants formulated akin to geopolymer cement. The lava flow-based Lunar mare simulants BP-1 and JSC-1A were formulated and cast into structural elements. The resulting materials were tested for compressive strength. The cast structural elements offer compressive strengths comparable with and surpassing conventional cement. Rheological studies of the effect of particle size, concentration, and curing temperature and time were used to determine the resulting workability and strength properties of the regolith-based cement. The results provide valuable information for manufacturing extraterrestrial infrastructure including but not limited to landing pads. The aim of these investigations is critically aligned with NASA's Artemis program to return to the Moon to stay by 2024.

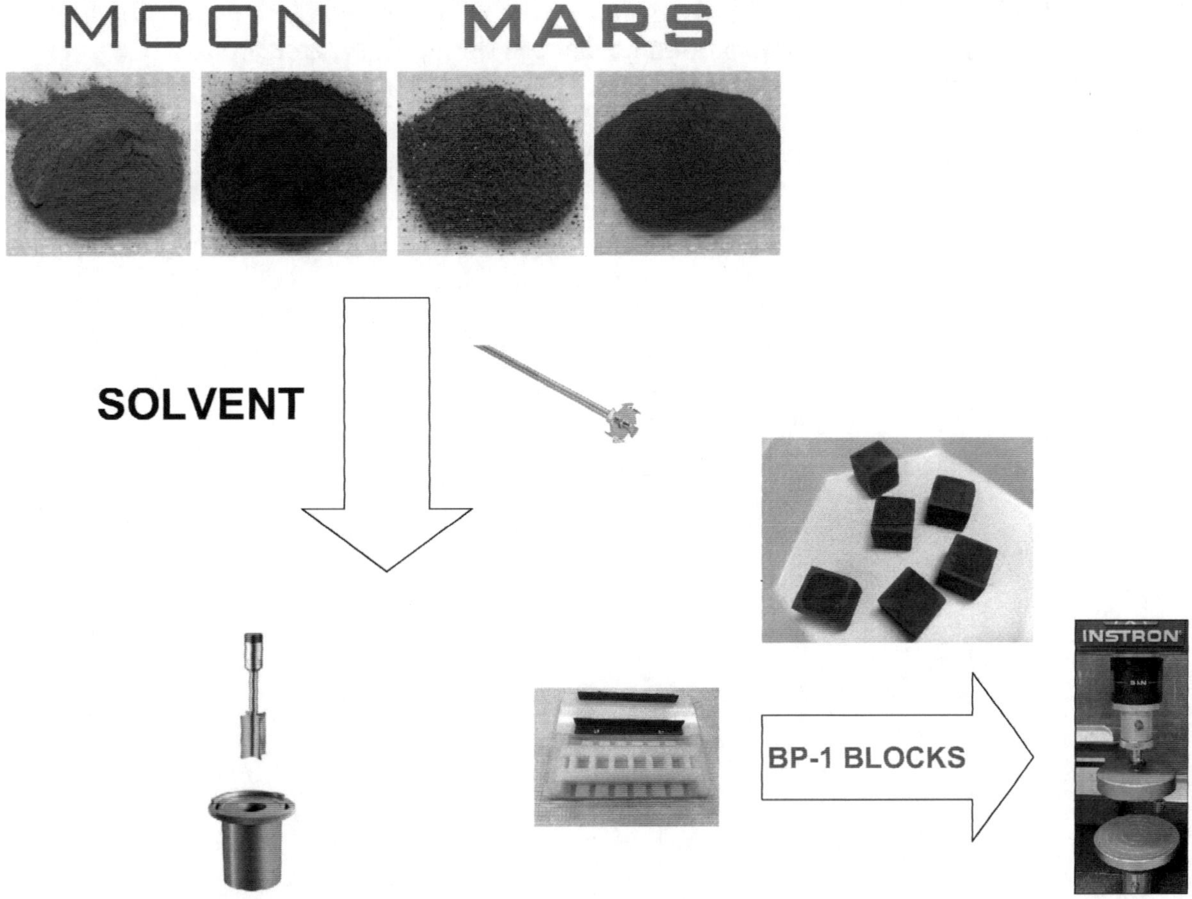

598448 One-Step Fabrication of Amphiphilic Graphene Quantum Dots As pH-Responsive Pickering Emulsifier

Abstract:

Pickering emulsions have attracted considerable interest, which are stabilized by inorganic nanoparticles. Carbon-based nanoparticles have low cytotoxicity, chemical inertness, cost-effectiveness, and biocompatibility. In this work, we developed a facile one-step bottom-up method to synthesize modified graphene quantum dots (C-GQDs) surfactant with controllable wettability. Pristine GQDs fabricated by bottom-up method using citric acid are hydrophilic due to a large of oxygen-containing groups. By introducing hydrophobic functional groups, such as alkane chain, improve hydrophobicity. The results from attenuated Fourier transform infrared spectroscopy (FTIR) and X-ray photoelectron spectroscopy (XPS) indicate that dodecyl alkane chain was introduced on the GQDs surface. The different values of water contact angle for two sides of GQDs (WCA) indicate. The sizes of nanoparticles were investigated by Transmission Electron Microscope (TEM) and Atomic Force Microscope (AFM). These results indicate these C-GQDs lateral size is about 20 nm, the C-GQDs are few-layer. And C-GQDs could form stable Pickering emulsion under neutral conditions. Meanwhile the emulsion could be rapid responsive and reversible by pH due to the suspend carboxyl groups. When added little acid solution in to emulsion, emulsion could be demulsified without shaking. And then added based solution, the solution could be re-emulsified. These results indicate great potential for application in enhanced oil recovery, emulsion polymerization, sensor and catalyst.

605628 Characterizing Changes in Powder and Surface Adhesion Using the Enhanced Centrifuge Method

Abstract:

Powder technology plays a significant role across industries including food, pharmaceuticals and defense. Examples of powder processes include milling, blending, spreading, and compaction. The behavior of the powders under these processes is impacted by the adhesion of the individual particles which comprise a powder. Problems such as poor flowability, dust hazards, and equipment wear arise due to uncontrolled particle adhesion and can lead to production chllenges. Computational models have been developed to predict the behavior of highly idealized powders (*i.e.* powders comprised of simple geometries such as spheres) under various processes but are limited in their ability to model and optimize the manufacturing and handling of powders comprised of many complex particles.

This work focuses on developing an experimental and modeling framework that maps particle-scale and surface properties onto experimentally-validated 'effective' Hamaker constant distributions that describe van der Waals adhesion forces between particles in powders. These distributions represent an engineering approach that allows powders comprised of particles of complex shape and roughness, which are challenging to model, to be described as if they were perfect, smooth spheres, which are comparatively simple to model. The complexity associated with the shape and size distributions of the individual particles is captured by the 'effective' Hamaker constants. This prior work is extended here to investigate the effects of topographical changes on the adhesion force distribution of powders against different surfaces and to characterize the powder behavior and resulting effective Hamaker constant distributions. These size-dependent effective Hamaker constant distributions provide a quantifiable measure of the change in the powder and surface adhesion that reflects the size, shape, and topographical features on the powder and surface with which it interacts.

607392 Characterization of Polymeric Powders through Capillary Flow for Additive-Manufacturing Techniques

Abstract:

Binder jetting is an additive-manufacturing (AM) method that deposits a liquid binding agent onto a powder surface in a layer-by-layer fashion to form a solid part from a computer-aided-design (CAD) file. Powders that are utilized in AM include polymers, sands, and metals. The production of parts requires input materials to have quantifiably repeatable characteristics and properties, which then requires standardized methods of characterization for powders. In this investigation, we use a non-invasive frugal imaging technique to characterize fluid flow in a packed powder bed to investigate fluid-powder interactions. Laser- diffraction particle size and sieve analysis are used for size and shape analyses, as well as scanning electron microscopy (SEM) to investigate the surface topography. The results of this study demonstrate how particle size, size distribution, particle shape, surface roughness, and surface chemistry affect the packing density and therefore the binder-jetting process.

This study investigates the interactions of the polymer powders with fluids at the fundamental level. Results will aid in the understanding of other AM methods (e.g., cold spraying, drop on demand (DOD), selective laser sintering (SLS), electron beam melting (EBM)), which use polymer powders. Since this study uses a frugal technique, which requires only a mobile phone and a flow cell to analyze the fluid-powder interactions, characterization of the properties of any polymer powder is readily available to all industries.

597715 Comparing Dynamic and Quasi-Static Powder Flow Testing Methods

Abstract:

Granular materials are ubiquitous in our daily life. A special class of them is cohesive powders consisting of fine particles that may flow when shaken or tilted, but could cake when left at rest or being compressed during storage.

In order to measure the complex flow properties of these powders, a variety of characterization techniques have been proposed: the angle of repose, the tapped density, the dynamic rotating drum, rotating blade systems or shear testing procedure. These techniques are based on different geometries with very different flow fields under different confining stresses, and make the cross link between them non-trivial. In this study, we attempt to bridge the gap between the static angle of repose test (GranuHeap), the dynamic rotating drum measurement (GranuDrum), the quasi-static packing dynamics measurement (GranuPack) [1] and the quasi-static ring shear test (RST-01) using both free flowing and cohesive limestone powders. Starting from the dynamic flowing angle in the rotating drum and going slower, one can extrapolate to the limit of zero rotation rate and compare it with the repose angle. Moreover, when stretching the ring shear test to its lowest confining stress limit, the bulk friction angle coincides with the drum flowing angle in the zero rotation rate limit [2]. Moreover, the analysis of packing dynamics gives additional insight on the powder behaviour.

This connection between different tests opens new perspectives in the field of powder flow characterization for measurement interpretation and also for a better link with powder behaviour in industrial process lines.

[1] Measuring the flowing properties of powders and grains, Powder Technology **224**, 19–27 (2012)

[2] Stretching the limits of dynamic and quasi-static flow testing on cohesive limestone powders, H. Shi, G. Lumay, & Stefan Luding, Powder Technology **367**, 183–191 (2020)

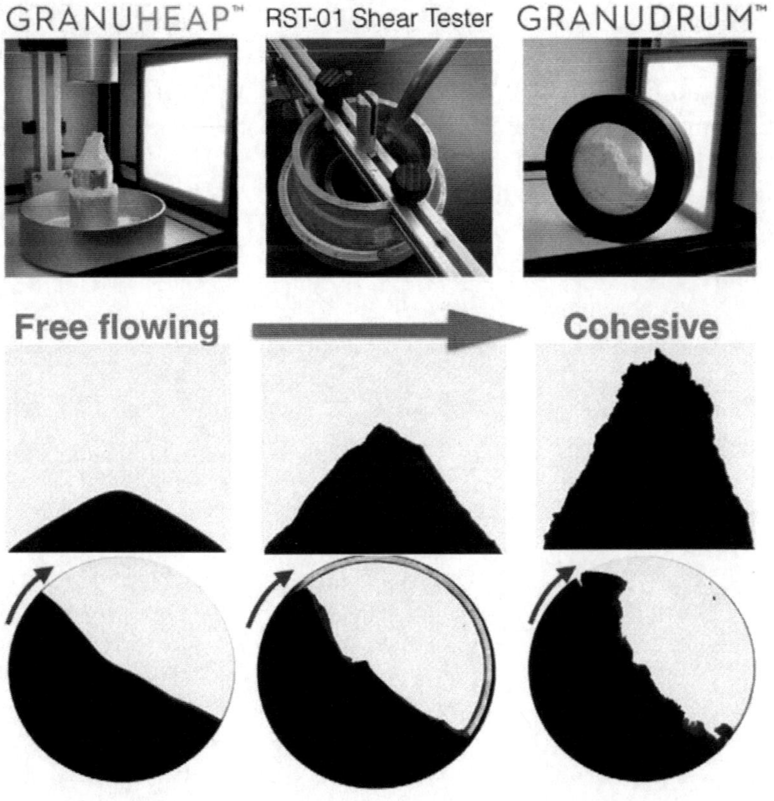

608721 Powder Characterization at the Edge of Sintering – High- and Low-Temperature Ring Shear Testing

Abstract:

ABSTRACT While the behaviour and mechanical coefficients of solids under the influence of temperature is well understood, how this translates to their behaviour when in the form of particulates is not studied in great depth.[1]

In this Study, we will show both cryogenic as well as high-temperature measurements exploring the different ways that temperature influences the behaviour of bulk solids using dedicated measurement equipment.

From the onset of melt behaviour in partially crystalline polymers like Polyamide to the effect of the freezing of surface water in flour and other biological samples, the embrittlement of resins at low temperatures to the effect of drawing surface water from the atmosphere in cement.

Special consideration will be given to the functional mechanisms that are occurring at the edge of sintering temperature and the normal-stress dependent evolution of powders in or near a sintering regime.

All of these effects have their distinct signatures upon the flow behaviour and of course, play a major role in their caking behaviour. Both the data, their influence upon silo and hopper design as well as interpretations of the underlying mechanisms will be shown.

REFERENCES

[1] Tomasetta, Igino, Diego Barletta, and Massimo Poletto. "Correlation of powder flow properties to interparticle interactions at ambient and high temperatures." Particuology 12 (2014): 90-99.

609041 Granular Material Characterisation at the Macroscopic Scale for the Validation and the Calibration of Numerical Models

Abstract:

Numerical simulation is a useful tool for studying the behaviour of powders. Nowadays, very computing efficient DEM codes are available and allows to simulate powder processes at large scale. However, finding suitable values for the microscopic parameters (coefficients of friction, coefficients of restitution, cohesive interactions, ...) to reproduce granular materials behaviour is not trivial, in particular when dealing with real powders made of grains having complex characteristics. Moreover, in many simulation techniques, the input parameters are not directly linked with physical characteristics of the particles.

Measurements can be performed at the scale of the grains with AFM techniques for example to obtain a microscopic characterization. Even if this bottom-up method makes sense conceptually, theorical background linking micro model parameters to macroscopic behaviour is still missing.

We propose a top-bottom method based on a macroscopic characterization workflow composed of three measurements: the packing fraction dynamics (GranuPack [1]), the flowability and cohesiveness (GranuDrum [1]) and the ability of the powder to create electrostatic charges during a flow (GranuCharge [2]). These measurements are performed in well-known geometries (tubes or rotating drum) and thus can be easily simulated. Therefore, the experimental results can be directly compared with the simulations to make a validation of the selected parameters. After a description of the measurement workflow, case studies are presented.

609275 Size and Shape Characterization of Plate-like Particles

Abstract:

Particle size and shape are important in crystallization processes as they dictate downstream properties like the flowability and bioavailability of the crystallized product. Most pharmaceutical products crystallize as nonequant particles (i.e. needles, platelets), which are more difficult to process than equant particles.[1] To manipulate the crystal shape, a quantitative size and shape characterization device is needed. Dual projection imaging has been used in the past to successfully characterize and manipulate the particle size and shape distribution (PSSD) of an ensemble of needles. [2,3] Despite the relevance of platelets in pharmaceutical industries, there is a severe scarcity of techniques to reliably characterize their size and shape. This serves as a motivation for this work. Initial simulation studies that mimic a single and dual projection imaging technique shows that platelets are more challenging to characterize, since they exhibit a high rotational variance, as illustrated in Figure 1. Using single projection onto two different planes, two PSSDs can be obtained. The two PSSDs thus obtained are significantly different from each other and neither of them is a good approximation of the true 3D PSSD. Dual projection mitigates to some extent the effects related to particle orientation, but preliminary studies indicate that it is also not sufficient for a quantitative characterization. Since platelet dimensions (with $L_1 \geq L_2 \geq L_3$) cannot be measured accurately with either mono or dual projection imaging, new techniques have to be developed. This work will systematically characterize the accuracy of the state-of-the-art dual projection imaging method and the newly proposed methods for platelets at different conditions.

A simulation tool, henceforth referred to as the "Virtual Test Bench" (VTB),[2] was used to quantify the accuracy of the different measurement methods. The VTB imitates the dual projection imaging device.[3] The VTB samples 5000 particles from a given population, places them within a virtual flow channel, calculates their two projections, and processes them to obtain the characteristic dimensions of the imaged particles. The measured population can then be compared to the true population to identify weaknesses of the measurement technique. Additionally, a population balance solver is also coupled with the VTB to observe the evolution of the measured and the true populations of crystals in a crystallization process. Based on this, the effectiveness of the methods for estimating kinetic model parameters and for manipulating the crystal shape is evaluated.

A preliminary case study revealed that the average relative error on the particle thickness L_3 and on the particle volume reaches up to 200 %. Additionally, the errors heavily depend on the alignment of the particle with respect to the camera and their aspect ratios, both of which are impossible to control by a user.[4] Therefore, two approaches are tested *in silico* in an effort to reduce the measurement error. One involves adding an additional projection to provide more shape information and the other involves building a mathematical model that maps the raw images to the particle dimensions. The former approach is successful in reducing the error on the platelet volume. However, the accuracy in characterizing the platelet thickness is only marginally improved. The latter approach trains a data-driven model using data generated using the VTB. Several regression techniques are optimized, including support vector machines, regression trees, and artificial neural networks.[5] The model input features are extracted from the 2D contours and the 3D reconstructed volume of individual particles. The final model targets are the three characteristic dimensions of the platelet. The training data includes over 450 000 particles with varying size, shape, aspect ratios, and alignment with respect to the camera. Average relative errors of 2 %, 7 %, and 20 % on L_1, L_2, and L_3, respectively, are achieved on individual particles by utilizing an artificial neural network-based model. To evaluate the effectiveness of the proposed enhancements, populations obtained through a simulated crystal growth and dissolution process are virtually measured using the VTB. The true and measured time-resolved cross moments of the distributions are compared. The study highlights the superior performance of the dual projection imaging device coupled with the data-driven model when compared to the state-of-the art technique to characterize the average dimensions and distribution width.[4]

In conclusion, the state-of-the-art dual projection imaging method is inadequate for quantifying platelet dimensions, so two potential improvements are proposed and tested in a simulation framework. A mathematical approach using a data-driven model proves to be successful. This model significantly improves the measurement of platelet dimensions and paves way for developing processes aimed at manipulating the shape of such particles toward more desirable ones that eventually enhances their downstream processability.

References

[1] Megarry, A. J.; Swainson, S. M. E.; Roberts, R. J.; Reynolds, G. K. A big data approach to pharmaceutical flow properties. *Int. J. Pharm.* **2019**, 555, 337–345.
[2] Schorsch, S.; Ochsenbein, D. R.; Vetter, T.; Morari, M.; Mazzotti, M. High accuracy online measurement of multidimensional particle size distributions during crystallization. *Chem. Eng. Sci.* **2014**, 105, 155–168.
[3] Rajagopalan, A. K.; Schneeberger, J.; Salvatori, F.; Bötschi, S.; Ochsenbein, D. R.; Oswald, M. R.; Pollefeys, M.; Mazzotti, M. A comprehensive shape analysis pipeline for stereoscopic measurements of particulate populations in suspension. *Powder Technol.* **2017**, 321, 479–493.
[4] Jaeggi, A.; Rajagopalan, A. K.; Mazzotti, M. How to Characterize Ensembles of Plate-like Crystals in Suspension? **2020** (In Preparation)
[5] Kuhn, M.; Johnson, K. *Applied Predictive Modeling*; Springer New York: New York, NY, **2013**.

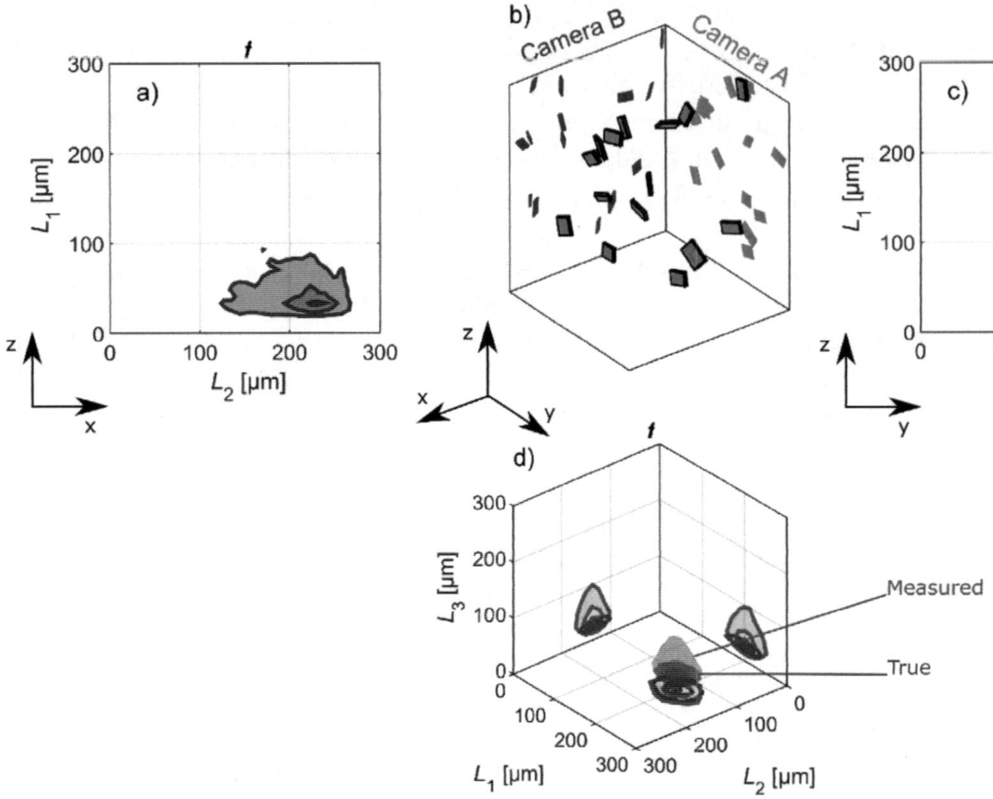

Figure 1: A virtual flow cell (b) with plate-like particles (gray) with projections along xz (blue) and yz (red) plane as seen by the two cameras. 2D PSSD measured by the cameras on a) xz and c) yz plane. d) 3D PSSD measured *in silico* using a state-of-the-art dual projection device (purple) and true 3D PSSD (gray).

602024 Emissions Control Technologies for RTI's Non-Aqueous Solvent for Carbon Capture

Abstract:

Advanced water-lean solvents (WLS) for post-combustion CO_2 capture have been gaining interest due to their ability to reduce the parasitic penalty from energy needed for solvent regeneration. RTI International, with funding from the US Department of Energy, has been involved in development of its non-aqueous solvent (NAS), eCO_2Sol^{TM}. The solvent has demonstrated specific reboiler duties as low as 2.3 GJ/t-CO_2 at pilot plant scale CO_2 capture system at Tiller, Norway. Commercial implementation of these novel CO_2 capture technologies hinges on successful control of amine emissions. Vapor phase emissions can be controlled through use of water wash that scrub the amine components out of the treated gas as it exits the absorber column. While significant research work has been done to characterize and control aerosol-based emissions from aqueous solvents, similar information is not available for emissions from use of these novel water-lean solvents.

RTI has been conducting studies from fundamental and operational aspects to reduce the overall amine emissions from the WLS systems, specifically RTI's non-aqueous solvent eCO_2Sol^{TM}. The work has been carried out at 4-6 kW equivalent using the RTI's bench-scale gas absorption system (BsGAS). It consists of a CO_2 absorber column with intercoolers and water wash section, and a solvent regenerator column with a thermosiphon reboiler and inter-stage heaters. RTI's BsGAS was modified to add the aerosol generation system and monitoring equipment to determine the aerosol characteristics during the eCO_2Sol CO_2 capture process. The system is now capable of producing aerosols with the peak diameter and concentration of 50 mm and 1.2 $x10^7$ cm^{-3}, similar to those observed in the actual coal-fired power plant flue gas at the inlet of the CO_2 capture system.

RTI has conducted over 1,300 hours of parametric testing to evaluate the impact of the aerosols and operating conditions during the CO_2 capture with NAS on the overall amine emissions in the treated flue gas. A simulated flue gas, with 15% CO_2, 2.3-4.2% H_2O and balance air, was used in the study. SO_3 particulates were generated by reacting SO_2 with O_2 in air over a silica supported vanadium pentoxide (V_2O_5/SiO_2) catalyst at 450°C. The generated SO_3 particulates were injected into the simulated flue gas fed to the absorber. The SO_3 particulates mimic the presence of SO_X, fly ash and other particulates in the flue gas exiting the direct contact cooler and provides nucleation sites for growth of aerosols in the CO_2 capture process. A Scanning Mobility Particle Sizer and an Aerodynamic Particle Sizer were used to monitor the aerosol particles size and number distribution at various locations in the CO_2 capture system. The amine emissions at the absorber outlet and water wash outlet were monitored using a MKS MultiGas™ FTIR spectroscopy gas analyzer. Initial testing found that the CO_2 capture rate and water wash temperature were the two most dominant process factors on vapor emissions. Further testing fixed those two process variables to evaluate other variables including liquid/gas ratio (L/G, mass basis), intercooling profile (extent of intercooling at the top and bottom intercooler), temperature of the lean solvent to the absorber, the water wash L/G, temperature of the water wash return stream, and the inlet saturation temperature of the flue gas (which dictates both the extent of water in the flue gas, and the water wash temperature required to maintain water balance in the system). The CO_2 capture rate was maintained close to 90% capture rate.

Results from parametric testing suggested that the presence of the aerosols in the flue gas could increase the overall emissions by 10X compared to the baseline emissions from NAS's vapor pressure. The temperature difference between the temperature bulge seen in the absorber column and the water wash temperature impacts the particle growth, with the aerosol emissions increasing with the increase in the temperature difference. Most of the aerosols did not grow substantially in the system, and the particle concentrations remained nearly constant between the absorber inlet and wash outlet. Only a small portion of the particles were found to grow significantly. A second water-lean solvent was also tested and elicited insights into how to minimize vapor and aerosol emissions through solvent modification.

Results from parametric testing have established the emission baseline from water-lean solvent processes. The BsGAS has been modified to add in vapor and aerosol control technologies such as a water wash acidifier, an amine sorbent removal and regeneration system, high efficiency demisters, and a lime-coated filter which mimics a filter baghouse. Testing of the control technologies has shown their ability to significantly reduce the emissions from the water-lean solvents. Results from emission studies post-BsGAS modifications will be presented to show the effectiveness of the emission control strategies to minimize emissions from the CO_2 capture process using water-lean solvents.

606548 Optimization and Transport of CO_2 from Refineries in Southeast Texas for Offshore Subsea Storage

Abstract:

Behind fossil fuel-based power generation plants, petroleum Refineries are the second largest stationary source of CO_2emissions. Refineries often undergo process upgrades and plant expansions to maximize the crude throughput, thus maximizing profits. This in turn causes the amounts of CO_2 emissions from the refineries to vary annually, although predictably. For instance, process upgrades generally result in lower CO_2 emissions (~10%) based on total number of barrels of throughput. In addition, CO_2 emissions increase by 85 – 90% when the plant capacity is doubled. Due to these constant changes in the refineries, estimating the amount of carbon dioxide to predict the size of the compressor and the recompression station required for the purposes of Carbon Capture and Storage (CCS) becomes complicated.

In this study, the Petroleum Refineries present in the Beaumont and Port Arthur, Texas area were analyzed for CO_2output using data available from EPA's Facility Level Information on GreenHouse Gases Tool (FLIGHT). Refinery Capacity and actual production was determined from quarterly and annual reports from the respective refineries. The petroleum refineries considered for this study include Valero Refinery (Port Arthur, TX), Total Petrochemicals & Refining USA, Inc. (Port Arthur, TX), ExxonMobil Beaumont Refinery (Beaumont, TX), and, Motiva Enterprises, LLC (Port Arthur, TX).

Among the six different types of CO_2 emission sources found within refineries, Stationary Fuel Combustion Sources, Catalytic Cracking and Reforming Units, and Sulfur Recovery Units accounted for about 80 – 85% of the total carbon dioxide emissions. The CO_2 emissions produced from these units was found to have the highest correlation with crude throughput. These findings were used to size compressors and a design CO_2 gathering network for delivery of $CO_2$from these facilities to offshore, subsea locations in the Gulf of Mexico.

607589 Dynamic Modeling, Optimization, and Control Studies of a Moving Bed Process for CO_2 Capture Using a Micro-Encapsulated Solvent

Abstract:

A number of potential solvents for CO_2 capture exhibit undesirable characteristics like high viscosity, solids formation thus making it difficult to use them in conventional towers [1]. Micro encapsulation of carbon sorbents (MECS) can help in overcoming these issues seen in highly viscous and phase change materials. Micro-encapsulation of solvents is a technology proposed [2] for post combustion carbon capture where the solvent of interest is encapsulated inside a polymer. Microencapsulation of carbon sorbents (MECS) involves fabrication of microcapsules containing a solvent with CO_2 capture capabilities. Even though capsules are manufactured with a known solvent concentration, water is known to permeate some capsule materials. H_2O gain or loss due to diffusion into or out of the microcapsules can significantly affect the solvent concentration and is one of the key operational challenges for reliable and efficient operation of MECS technology. While there are a few studies [2-3] that have investigated different reactor configurations for MECS, no studies have been conducted with the intent to understand and mitigate the issue of time-varying water concentration. Extracting solvent from microcapsules to measure the concentration may require collecting huge number of microcapsules from the reactor. The ambient conditions can further change the water content and may not represent the true concentration value in the reactor. Therefore one must rely on other measurable process variables to get an estimate of capsule concentration.

The present work is focused on dynamic modeling, optimization, and control of a moving bed reactor for MECS, where controlling the desired solvent concentration is one of the key control objectives. Moving bed reactors have a strong potential for solid-sorbent based CO_2 capture [4-5]. The counter-current contact between gas and capsules can help to achieve larger CO_2 driving forces in the absorber and regenerator and so is examined in this work. Also, the heat recovery from the capsules leaving the regenerator can help in pre-heating the capsules flowing from absorber to regenerator. Overall, this can reduce the regeneration energy requirements of the process. A 1-D multi-scale moving bed absorber and regenerator using a pre-existing microcapsule model [4] is developed in Aspen Custom Modeler (ACM) for MECS using Na_2CO_3 as the encapsulated solvent. The capsule level model used in the moving bed reactor is validated with experimental data in our previous study [4]. A steady-state optimization of the moving bed configuration that minimizes the Equivalent Annual Operating Cost (EAOC) by considering key operating and design variables is performed. Traditional and model-based advanced controllers are developed for controlling the capture percentage, and solvent concentration within the capsule in the face of input disturbances. An ARMAX type model based on information about measured variables from the rigorous process model will be exploited to estimate solvent concentration and to provide satisfactory control performance.

References

1. Stolaroff, J. K. et al., Microencapsulation of advanced solvents for carbon capture. Faraday Discussions. DOI: 10.1039/c6fd00049e (2016).
2. Vericella, J. J. et al. "Encapsulated liquid sorbents for carbon dioxide capture", Nature Communications, 6:6124, 1-7, 2015.

3. Hornbostel, et al, Packed and fluidized bed absorber modeling for carbon capture with micro-encapsulated sodium carbonate solution, Applied Energy 1192-1204, 2019.
4. Kotamreddy, et al., Process Modeling and Techno-Economic Analysis of a CO_2 Capture Process Using Fixed Bed Reactors with a Microencapsulated Solvent, Energy & Fuels 2019 33 (8), 7534-7549 DOI: 10.1021/acs.energyfuels.9b01255.
5. Knaebel, K.S., Temperature swing adsorption system. 2009, Google Patents.
6. Mondino, Giorgia et al. "Effect of Gas Recycling on the Performance of a Moving Bed Temperature-Swing (MBTSA) Process for CO_2 Capture in a Coal Fired Power Plant Context." Energies, 10(6), 745. 2017.

608069 Enhanced CO_2 Uptake and Selectivity upon Incorporation of Task-Specific Ionic Liquids into Highly Porous MOF-177 at Post-Combustion Capture Condition

Abstract:

Recently, Metal-organic-frameworks (MOFs) have gained tremendous attraction for application in CO_2 capture and storage due to an exceptional surface area, pore volume and for their tunability. Although some of the MOFs have shown excellent CO_2 uptake at high pressure, lower uptake at post combustion condition (partial pressure of CO_2 is 0.15 bar) is a major hindrance for widespread application in CO_2 capture operations. Researchers have deployed various strategies to enhance CO_2 uptake such as grafting amines functionalities covalently to the ligands, impregnating amines or ionic liquids (ILs). Post-synthesis modification of pores of solid sorbents by impregnation of amino functionalized ionic liquid also known as task-specific ionic liquid (TSILs) to enhance CO_2 capture capacity has shown promising results.

In this study, 1-ethyl-3-methylimidazolium [Emim] cations with Glycine [Gly] and Alanine [Ala] as reactive Amino Acid (AA) anions forming [Emim][Gly] and [Emim][Ala] as TSILs were immobilized into the pores of MOF-177 through the incipient wetness technique. TSIL@MOF-177 composites sorbents were prepared for three different loadings (10, 20 and 30) wt.% of TSIL and the developed composites were characterized and investigated from the engineering perspective in terms of CO_2 capture capacity, selectivity, and enthalpy of adsorption. CO_2 adsorption isotherms were measured at three different temperatures (303, 313 and 323 K) in the pressure range varying from (0.1 to 10) bar. In addition, N_2 adsorption isotherms were also measured for all composites and the pristine MOF-177 from which CO_2/N_2 ideal selectivities were calculated. In addition, experimental adsorption isotherms were modelled using the Freundlich model and the isosteric heats of adsorption were calculated and compared to those of the pristine MOF-177.

Impregnation of TSILs into the highly porous metal organic frameworks (MOF-177) had a remarkable impact on CO_2 capture and selectivity. The highest CO_2 uptake achieved by [Emim][Gly]@MOF-177 composite was 0.45 mmol/g-solid and 0.42 mmol/g-solid at 0.2 bar, 303 K for [Emim][Ala]@MOF-177 composite which were 3 and 2.8 times higher than the pristine MOF-177 at the same conditions, respectively. It was also observed that the optimum loading of TSIL was 20 wt.% as further increase to a loading to 30 wt.% did not results in any incremental CO_2 uptake. CO_2 selectivity over N_2 was also enhanced from 5 for the pristine MOF-177 to 13 and 11 for [Emim][Gly] and [Emim][Ala], respectively, at 0.2 bar and 313 K. This significant improvement of both CO_2 uptake and selectivity can be attributed to the chemisorption interaction between CO_2 and $-NH_2$ functional group present in the amino acid anion in TSIL which was also reflected in the increase in the isosteric heat of adsorption (Q_{st}) values from the value of the pristine MOF. Significantly high CO_2 uptake and CO_2/N_2 selectivity at the low pressure region (<1.0 bar) makes these functionalized MOFs promising candidates for the post-combustion CO_2 capture operation.

600113 An Adhesion-Based Anti-Corrosion Strategy for Wearable Electrochemical Sensing and System Integration

Abstract:

The large-scale deployment of biomonitoring sensors through integration in wearable consumer electronics enables population-level health and wellness monitoring, and thus can play a critical role in transforming personalized medicine. Currently, physical sensors have been widely incorporated within commercialized wearable platforms, but they cannot capture biomarker information that lie in molecular levels and provide highly specific information about the body's dynamic chemistry. Electrochemical biosensing interfaces can be positioned to target biomarker molecules in a wearable format, as they can be constructed on miniaturized/flexible footprints/substrates and can convert (bio)chemical signals into electrical signals measurable by electronic readouts. The contact pads on the printed circuit board (PCB), which is a core component of wearable consumer electronics, can directly serve as electrodes and as a substrate for electrochemical sensors. However, we found that the commercialized PCB metal contact pads (*e.g.*, gold) were prone to corrosion and unsuitable for electrochemical modifications in solution environments. Additionally, electrochemical biosensors often are not reusable and need to be replaced frequently, while PCB units are not positioned for disposable applications.

To this end, we devise an adhesion-based anti-corrosion strategy for PCB-interfaced electrochemical sensing, which enables seamless sensor-system integration. The strategy is based on decoupling the sensing interface and PCB metal contact pads by an intermediary ultrathin (~50 μm), vertical conductive, and double-sided adhesive film. This strategy has dual benefits of making electrochemical sensors disposable and insulating the contact pads from sounding solutions. We first establish the suitability of the adhesion film to serve as an electrode and as an interconnection film simultaneously. One side of the double-sided adhesive film was patterned with noble metal electrodes, followed by the modifications with recognition layers, such as permselective membrane and enzymes, to render biosensing functionality. The electrochemical biosensors were transferred and adhered to the PCB contact pads with the aid of water-soluble tape. As compared with bare PCB contact pads, this fabrication strategy allows for realizing a corrosion-resistant and stable electrochemical biosensing interface on PCB for operation in biofluid (*e.g.*, sweat). Amperometry and open circuit potential characterization experiments validated the electrochemical corrosion resistance properties in saline solution environments. The excellent anti-corrosion attributes of the modified film could be ascribed to: (1) the polymer glues on both sides of the adhesive film acting as corrosion-protective layers; (2) bypassing the need for oxidation prone-adhesion metal (*e.g.*, chromium), achieved by e-beam evaporation (at the noble metal patterning step); and (3) the electrodeposited permselective membrane (here, phenylenediamine), which has been reported to presented anti-corrosion properties. We specifically tailored the enzymatic layer of the electrochemical interface to target biomarker molecules such as glucose, lactate, and choline as example of informative metabolites and nutrients. By modifying different (bio)chemical layers various other molecules can be targeted. In this way, our strategy serves as a foundation for electrochemical sensor development and integration with wearable consumer electronics.

607021 Rapid Screening Platform for the Isotopic Determination of Radionuclides in Water

Abstract:

Traditional approaches for the detection of trace-level radioisotopes in water require lengthy offsite sample preparations and do not lend themselves to rapid screening. Therefore, a novel sensor platform that combines onsite purification, concentration, and isotopic screening with a fieldable detection system will be an invaluable tool.

In this presentation, we will discuss findings on the development of a sensor platform based on reactive polymer film coatings for the rapid isotopic screening of waterborne radionuclides. Nanothin coatings were prepared by grafting uranium-selective polymers from polyethersulfone (PES) surfaces via UV-initiated polymerization, and by introducing uranium-selective functional groups to polyacrylonitrile (PAN) surfaces by chemical reaction. Ellipsometry was used to study film growth kinetics on PES films. X-ray photoelectron spectroscopy of modified PAN films revealed conversion of nitrile groups to amidoxime groups to be as high as 40% and showed that the extent and depth of reaction could be varied precisely. Static uptake experiments with solutions of depleted uranium spiked with ^{233}U were conducted to determine uranium binding capacities and kinetics of the modified polymer films at different pH values from 4 to 8. Sorption isotherm data were fitted to the Langmuir model, and the highest sorption capacities were obtained at pH 6 for modified PAN (M-PAN) and PES (M-PES) films. Capacities at pH 4 and 8 were lower and could be explained by differences in sorption mechanisms. Uranium batch uptake kinetics followed a pseudo-second order rate model. Alpha spectroscopy pulse height spectra were analyzed to study the role of selective layer film thickness on peak energy resolution. Excellent resolution with full width at half maximum values from 29 to 41 keV were recorded for M-PAN film and from 26 to 45 keV for M-PES film. Whereas uranium uptake increased with selective layer film thickness and varied with polymer chemistry/extent of modification, the peak energy resolution was independent of layer thickness and polymer chemistry within the experimental measurement uncertainties.

Results from this work were used to modify PES and PAN ultrafiltration membranes with the same experimental conditions to develop uranium selective membranes. High-throughput uptake experiments were performed to study the effect of residence time on detection efficiency and isotopic resolution. This membrane-based detection method enables the rapid, fieldable screening analysis of radionuclides in water for environmental monitoring studies and nuclear forensics investigations.

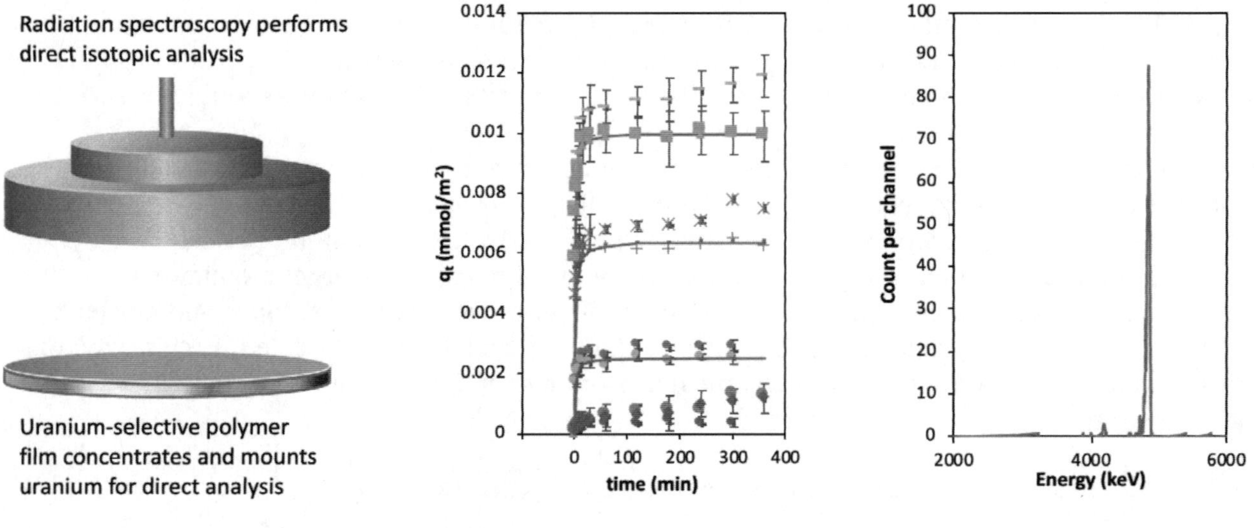

600875 Electrochemical Detection of Environmental Pollutants

Abstract:

Many small molecule pollutants interact with human hormonal systems. These compounds, known as endocrine disruptors, are ubiquitous, as they are found in plastics, medications, and pesticides. Such chemicals are especially problematic because they have been linked to severe health problems, including cancer, infertility, and diabetes. We have developed two complementary electrochemical methods to rapidly detect and quantify these compounds. One technique involves the specific monitoring of Bisphenol A (BPA) using a DNA aptamer immobilized on an electrode. The second method involves the measurement of the biological activity of chemically dissimilar endocrine disruptors using a human hormone receptor as a component of an electrochemical sandwich assay. These electrochemical platforms require no specialized skills to implement and have enabled very sensitive detection (sub-ppb levels) of these environmental pollutants from complex solutions, such as baby formula.

604144 Fabrication of High-Resolution Graphene-Based Flexible Electronics Via Polymer Casting and Microfluidic Approaches

Abstract:

Graphene-based flexible electronic devices for sensor applications has been receiving growing interest due to unique properties of graphene.[1] Various fabrication methods, such as photolithography, printing and chemical vapor deposition (CVD), has been used to produce high resolution and low feature size graphene-based devices.[2, 3] However, most of these methods are complex and require multiple processing steps (i.e. stamping, vacuum drying and etching) as well as additional post treatments (i.e. high temperature (300-1000 °C) baking or laser annealing) that can thermally or chemically degrade polymer-based substrates limiting substrate material selection.[2, 3] In addition, these methods can only be applied to 2D planar substrates and are not feasible to fabricate 3D circuits and some of them, such as sticky/adhesive tape peeling method, require high amounts of graphene consumption.[4] Despite the recent progress in the field, there is still a need to develop novel fabrication approaches that can address the current limitations of the existing methods.

In this study, we present two flexible electronics device fabrication methods based on simple polymer casting and microfluidics approaches.[5, 6] The first method focuses on simple polymer casting based graphene transfer at room temperature that does not require any post-processing. Briefly, our method consists of two main steps; (i) the formation of graphene patterns/films on substrates/molds via conventional methods such as CVD, channel filling or ink-jet printing and (ii) direct casting of target substrate polymer solution on the substrates/molds with graphene micropatterns and direct graphene transfer to the target substrate via peeling off upon drying and film formation. This method simply relies on the differences in the surface energies and adhesive forces between the graphene/mold and graphene/target polymer substrate.[5] Alternatively, the second method concentrates on a novel microfluidic approach to fabricate high resolution and low feature size graphene-based circuits/devices.[6] This method involves controlled and selective filling of microchannels on substrate surfaces with a conductive binder-free graphene nanoplatelets (GNP) solution. The ethanol-thermal reaction of GNP solution at low temperatures (~75 °C) prior to microchannel filling (as called "pre-heating") potentially reduces and homogenize GNP, which in turn enhances conductivity and reduces sheet resistance (down to ~0.05 kΩ sq^{-1}). Therefore, harsh post-processing of GNP is eliminated, and room temperature fabrication is enabled, which allows the use of degradable substrates, including biodegradable and natural polymers. This simple process is followed by selective administration of conductive GNP solution to the predetermined microfluidic channels on flexible substrates in a controlled manner via a syringe pump to create high resolution and low feature size graphene circuits/devices. This microfluidic approach can also be used to create 3D circuits with different geometries, using combined approaches with 3D printing, which is difficult to obtain with other methods (i.e. inkjet, screen or aerosol printing) that work mainly with 2D planar substrates. [6]

We demonstrated that using these fabrication methods, we obtained high resolution and low feature size graphene circuits with different geometries, dimensions and 3D superficial microstructures. We investigated the effect of temperature and microfluidic parameters (such as concentration, volume and flow rate) as well as the type of substrate materials on the graphene structure and formed micropatterns. The XPS, Raman, SEM and TEM results validated the structural changes of graphene along with confirming the stable presence of graphene patterns on the target substrates.[5, 6] The feature sizes of the graphene patterns could range from a few micrometers (down to ~15 μm in width and ~5 μm in depth) to a few millimeters with very small amounts of GNP used (~2.5 mg of graphene to obtain ~0.1 kΩ sq^{-1} of sheet resistance for 1 cm^2). [5, 6] The generated graphene patterns demonstrated significant stability after multiple washing and bending cycles. These methods were also implemented using other conductive inks, such as conductive graphene-silver based composite ink.[5, 6] Room temperature processing also provided precise control of 3D microstructural and mechanical properties (such as film porosity, pore size, elasticity etc.) of the target substrate materials, as demonstrated by SEM images.[5, 6] Using these methods, we successfully fabricated pressure sensors and near field communication antennas. [5, 6] Our results demonstrated that the devices worked with high sensitivity and stability.

In conclusion, the circuits/devices fabricated using this method can easily be implemented as wearable electronic, sensors and batteries, robotic components or motion detectors. In a broader sense, these simple, environmentally friendly and low-cost methods could pave the way for the fabrication of 2D and 3D electronic circuits/devices on different degradable flexible substrates using various graphene-based conductive inks.

References:

1. Jang, H.; Park, Y. J.; Chen, X.; Das, T.; Kim, M. S.; Ahn, J. H., Graphene-Based Flexible and Stretchable Electronics. *Advanced Materials* **2016,** *28* (22), 4184-4202.
2. Das, S. R.; Nian, Q.; Cargill, A. A.; Hondred, J. A.; Ding, S.; Saei, M.; Cheng, G. J.; Claussen, J. C., 3D nanostructured inkjet printed graphene via UV-pulsed laser irradiation enables paper-based electronics and electrochemical devices. *Nanoscale* **2016,** *8* (35), 15870-15879.
3. Das, S. R.; Srinivasan, S.; Stromberg, L. R.; He, Q.; Garland, N.; Straszheim, W. E.; Ajayan, P. M.; Balasubramanian, G.; Claussen, J. C., Superhydrophobic inkjet printed flexible graphene circuits via direct-pulsed laser writing. *Nanoscale* **2017,** *9* (48), 19058-19065.
4. Oren, S.; Ceylan, H.; Schnable Patrick, S.; Dong, L., High-Resolution Patterning and Transferring of Graphene-Based Nanomaterials onto Tape toward Roll-to-Roll Production of Tape-Based Wearable Sensors. *Advanced Materials Technologies* **2017,** *2* (12), 1700223.
5. Uz, M.; Jackson, K.; Donta, M. S.; Jung, J.; Lentner, M. T.; Hondred, J. A.; Claussen, J. C.; Mallapragada, S. K., Fabrication of High-resolution Graphene-based Flexible Electronics via Polymer Casting. *Scientific Reports* **2019,** *9* (1), 10595.
6. Uz, M.; Lentner, M. T.; Jackson, K.; Donta, M. S.; Jung, J.; Hondred, J.; Mach, E.; Claussen, J.; Mallapragada, S. K., Fabrication of Two-Dimensional and Three-Dimensional High-Resolution Binder-Free Graphene Circuits Using a Microfluidic Approach for Sensor Applications. *ACS Applied Materials & Interfaces* **2020,** *12* (11), 13529-13539.

601014 Reduced Order and Higher Order Models for Bubbling Fluidized Bed Reactors

Abstract:

Fluidized bed reactors offer some of the most complicated hydrodynamics that a chemical engineer could experience. For fast reactions, being able to capture the bubble hydrodynamics is paramount in developing a model useful for scale-up. D. Kunii and Octave Levenspiel first addressed this with their bubbling fluidized bed model where the gas flow through the bed is modeled through uniformly sized bubbles, surrounded by clouds and followed by wakes, rising through an emulsion of downward moving solids. Interchange of gas occurs continuously among bubble, cloud-wake, and emulsion regions. The model contains one key parameter, the bubble size, and all internal flow and interchanges in the bed are derived from it. It was a simple useful model providing the bubble size was captured.

Today, such models involve using computational fluid dynamics. Inherent knowledge of the bubble properties is not required. However, the spatial resolution of the computational domain is a trade-off between the accuracy of the bubble properties and the simulation time. So, which is better or more appropriately, which model should be used and when. This talk will focus on the merits, compromises, and limitations of these models.

601062 DEM-CFD Analyses of Flexible Fiber Fluidization

Abstract:

Gas-fluidized beds of flexible fibers are investigated using the coupled approach of Discrete Element Method (DEM) and Computational Fluid Dynamics (CFD). Gas-fiber interactions are modeled by calculating the interaction force for each constituent element in the fiber, rather than the resultant force on the whole fiber, as is typically done in DEM-CFD simulations. This treatment enables improved fluid field resolution. Pressure drop and fiber orientation results for rigid fibers with various aspect ratios are in good agreement with previous experimental and simulation results. Novel results consider the effect of fiber flexibility. Larger gas-fiber interaction forces (before the bed is fluidized) and smaller minimum fluidization velocities are observed for more flexible fiber beds, which have smaller porosities. Both the fiber aspect ratio and the fiber flexibility impact on the solids mixing rate by affecting the minimum fluidization velocity.

601110 Nonmechanical Control of Solids Flows in Chemical Looping Systems

Abstract:

Nearly all chemical looping systems being developed or proposed involve substantial flows of solids around the system. Typically, the temperatures in chemical looping systems are too high to use standard mechanical valves for control of the solids. Therefore, nonmechanical means are employed to control the solids flows around the systems. The nonmechanical configuration used to control the solids depends on the type of system used as well as the particle size distribution of the solids.

In this presentation, the different types of chemical looping systems are evaluated with respect to the different nonmechanical means employed.

598962 Daizo Kunii, Kunii-Levenspiel and the Historical Meaning of His Work

Abstract:

For decades “Fluidization Engineering” (1st ed.: 1969; 2nd ed.: 1991) has been the leading monograph/textbook worldwide for students, researchers and engineers. Its first edition was based on Daizo Kunii (1923-2019)’s book “Fluidization Methods” published in 1962 in Japanese, which was later found by his lifelong friend Octave Levenspiel (1926-2017) and they worked together to write their comprehensive book.

Daizo was a person of strong mission shouldering the responsibility of constructing the edifice of modern chemical engineering in the post war Japan.

In this presentation I am revisiting the background of their mission, the value of their work in the history of fluidization science and engineering and the lessons we should absorb from them.

600877 Chemical Looping Gasification, Reforming and Chemical Syntheses

Abstract:

Chemical looping technology is a manifestation of the interplay among such key elements of metal oxide reaction engineering and particle science and technology as particle synthesis, reactivity and mechanical properties, flow stability and contact mechanics, gas-solid reaction engineering and particulates system engineering. This presentation will describe the fundamental and applied features of modern chemical looping technology in the context of the circulating fluidized bed platform that utilizes fossil and other carbonaceous feedstock. It will discuss the reaction chemistry, ionic diffusion mechanisms, metal oxide synthesis and thermodynamics, reactor design, and system engineering along with energy conversion efficiency and economics of the chemical looping processes for, particularly, partial and selective oxidation for syngas and chemicals production. The Ohio States University has developed a number of advanced chemical looping gasification and reforming processes which will be highlighted in this presentation. Specifically, using CO_2 as a feedstock and its reactor modularization strategy in a chemical looping system adopted based on a non-linearity relationship in the variation of CO_2 and H_2O inputs with the syngas yields for gasification and reforming processes are illustrated from both the experimental and thermodynamic simulation viewpoints. The modularization strategy can directly be applied to scaling up processes when multiple reactors are needed leading to a considerable saving of the carbonaceous feedstock required to achieve a given syngas production.

601069 Revisiting the Nature of Turbulent Fluidization

Abstract:

Looking back to the history of the gas-solids fluidization, the term "turbulent fluidization" has been evolved from describing a characteristic phenomenon to defining a distinct flow regime with the increasing understanding of the nature in gas-solid fluidization system. The discussion on the existence of the turbulent fluidization regime started in the 1970s after Kehoe and Davidson (1970) first came up with the concept. The demarcation of the turbulent flow regime was first proposed by Yerushalmi et. al (1978) and then widely studied and accepted that the transition to turbulent fluidized bed (TFB) is signified by the maximum pressure fluctuations. Getting into the 1990s, more studies have been expanded to the general hydrodynamics, as well as on gas-solids mixing and heat/mass transfers. In the industry, however, the application of the turbulent fluidized bed reactors actually pre-dated the inception of its "official name", recognizing its practical benefits with its high solids holdup, excellent gas-solids contacting, and efficient heat/mass transfers, as early as the 1950s. On the other hand, the academic studies provided more theoretical understanding, that in turn benefited the industry, such as leading to the switch of the Sasol Synthol reactor to turbulent fluidized bed operation. The research and development on the turbulent flow regime and the turbulent fluidized reactors not only manifests the significance of critical thinking but also inspires more young scholars to search for novel variations of fluidized beds for newer applications.

At the 50-year point, it is perhaps a good time to revisit the nature of turbulent fluidization with some possible new interpretations. Different transition velocities to TFB regime based on various measuring techniques and criteria especially developed after 2000 will be reviewed. The flow mechanisms and the interactions between the dilute bubble/void phase and the dense particulate phase will be discussed based on both the time average data such as the overall pressure drops or the solids holdups and the instantaneous signals such as the time series analysis of signals or images. The two-phase flow structure which is well accepted in bubbling flow regime will be examined and extended in the turbulent regime to reveal the link between the conventional flow regimes. More quantitative analysis on the dynamic behavior of both the gas bubbles/voids and particles will be reviewed and the spatial heterogeneity of flow structure will be discussed.

602205 A Hysteretic Nonlinear Elastoplastic Contact Model for Discrete Element Modeling of Milled Woody Biomass

Abstract:

Abstract: Milled woody biomass, such as pine residuals, features irregular-shaped flexible particles. According to the axial loading-unloading experimental tests, milled pine particles manifest a strong nonlinearity in their bulk stress-strain behavior, which is challenging to simulate with discrete element methods (DEM) with simple contact models. This work presents the development of a hysteretic nonlinear elastoplastic contact model for spherical DEM. The formulation of the contact model is conceptualized with the bulk stress-strain relations measured in the physical tests. In particular, a hysteresis of plasticity between each two DEM particles in contact is proposed to capture the plastic deformation of a bulk of milled pine particles subjected to repeated compression. The contact model is implemented to an open-source DEM package, LIGGGHTS, and calibrated through simulations of a cyclic loading-unloading test to match the equivalent physical test results.

600882 Discrete Element Modeling of Irregular-Shaped Switchgrass Particles for Integrated Process Optimization

Abstract:

Switchgrass is a perennial herbaceous plant that is regarded as a biomass energy crop for its high adaptability and yield potential. Mechanical preprocessing and handling of switchgrass are challenging due to the erratic mechanical and flow behavior originating from its intrinsic particulate properties. In this work, we present our recent efforts in developing and validating a discrete element model (DEM) to simulate and predict the bulk behavior of switchgrass particles in support of integrated process optimization. The DEM model can capture three distinct particle characteristics of switchgrass, i.e., fibrous particle shapes, a wide range of particle size (both width and length) distribution, and high particle deformability. Particle shape templates were designed and generated based on results from image analysis of real switchgrass particles. Contact parameters of the numerical model were calibrated using data of a uniaxial compression test and Schulze ring shear test on both chopped and grinded switchgrass specimens. This numerical model was used to simulate and predict the bulk properties of switchgrass particles with a wide range of mean particle sizes, particle size distributions, and moisture contents. Furthermore, preprocessing tests of switchgrass bales were conducted using the Process Demonstration Unit at Idaho National Laboratory. The generated data were used to validate the developed DEM model. Validation shows the DEM model predicts the bulk densities of switchgrass with a high accuracy. Finally, the simulation results were compiled and integrated as mathematical regression models correlating particle bulk density to the particle size parameters as well as the moisture content, which is used in a system model developed to optimize the system performance.

602438 A Multi-Regime Continuum Approach for Modeling Flow of Granular Pine Chip

Abstract:

The flowability issues in biorefineries are due to poor understanding of the flow behavior of biomass granulate material, which leads to inappropriate equipment design and operation. With overall objectives of providing design tools and operational envelopes of each handling equipment for biomass-derived feedstock (ground pine chip), this talk will present the computational modeling efforts to address these issues by combining the discrete element method (DEM) at the particle scale and the Eulerian mesh-based finite element method (FEM) at the continuum scale. At the particle scale, we evaluate and identify the most suitable DEM model to efficiently and accurately realize the shear behavior of ground pine chip with complex particle size & shape distribution and particle compressibility. X-ray CT scan data and sieve analysis are used to generated DEM particle morphology, axial shear experiments are then used to calibrate and validate the contact model parameters. The high-fidelity DEM model with calibrated parameters is used to generate ground pine bulk flow behavior (stress-strain rate relation), and then up-scaled to evaluate and formulate the constitutive model. At the continuum scale, the critical state theory based NorSand model, incapable of modeling the nonlinear elasticity, the pressure- and void ratio-dependent shear flow, softening and hardening accompanied by volume compaction and dilation, is evaluated and identified as the most suitable constitutive model to capture milled biomass flow at the quasi-static condition (solid-like). We enrich the NorSand model with dynamic modeling capability by incorporating a rheology model, in which the shear strength of granular material is dependent on shear rate and pressure (fluid-like). We calibrate the constitutive model parameters controlling quasi-static behavior from cyclic uniaxial compression and Schulze ring shear test, and we obtain the constitutive model parameters controlling fluid-like behavior from inclined plate flow experiments and DEM generated stress-strain rate relationship. After implementing the extended model with the capability to model multi-regime flow from quasi-static to dynamic into the FEM package, it is utilized to model the hopper flow with the same geometry and initial conditions as the experiments at the pilot-scale. The predicted mass flow rate with the variation of hopper inclination angle and of the opening size is compared to experimental data to validate the model. We further perform sensitivity analysis by using the new multi-regime model and its calibrated parameters to search for the optimal design of hopper at the industry-scale.

609983 A Kinetic Approach to Simulating Spherocylindrical Particles

Abstract:

The characteristics of granular flows exhibit varying behavior dependent on the shape of the individual particles. In the case of non-spherical particles, analytical closures for determining the characteristics of granular flows such as the equation of state is mostly unknown. Given the relative difficulty in physically measuring granular flows non-intrusively, we look to numerical methods to understand more complex granular flows. Unfortunately, deterministic discrete element methods are restricted to relatively small-scale systems to limit the number of contact forces that must be resolved to a computationally manageable size. To extend numerical solutions to large scale granular flows, particles can be treated as an ensemble, and the particle interactions can be resolved stochastically, akin to the direct simulation Monte Carlo (DSMC) method. Here, we extract various quantities needed to formulate a DSMC-like method for spherocylindrical particles. Pairwise collisions between spherocylinders are deterministically simulated using a soft-sphere discrete element method and consolidated into probability density and scattering functions. Collisional cross sections and their dependence on the initial conditions of the particle pair, e.g. orientation and rotational energy, are also measured from the discrete element simulations. Granular gases composed of spherocylinders are simulated to determine bulk characteristics such as the collision frequency and rotational relaxation time. The measured properties from discrete element simulations are then incorporated into a direct simulation Monte Carlo solver. The DSMC simulations results are compared to deterministic discrete element simulations and the accuracy and computational costs of the DSMC approach are discussed.

600881 Superdem-CFD: Open Source Parallel Solver for Non-Spherical Particle-Fluid Fluidization Systems

Abstract:

Understanding the fluidization behavior of non-spherical particles is important for fluidized bed design and scale-up. In this study, a SuperDEM-CFD coupled solver in open-source MFiX for non-spherical particle fluidization systems was developed and validated. The non-spherical particle was modeled using a superquadric method, most of the shapes can be easily represented by only varying five parameters. A new interpolation method (called DPVM-Satellites) for non-spherical particles was developed to distribute the particle volume into the neighboring fluid cells. A general algorithm to calculate the projected area perpendicular to the flow was developed. Several non-spherical drag models that consider both particle orientation and fluid voidage effects were compared. The solver was validated by comparing the simulation results with various fluidization experiments of non-spherical particles, including cylinder, elongated cylinder, cubic, etc. The model can correctly simulate the pressure drop, particle height distribution, particle orientation distribution for different particle shapes. Besides, large-scale systems with millions of non-spherical particles were parallel simulated to demonstrate the ability of the solver for industrial-relevant flows simulation.

610060 Computational Fluid-Dynamics Modeling of Biomass-Feedstock Flow Using a Non-Local Granular-Fluidity Constitutive Model

Abstract:

Upsets in milling, material handling, and reactor feeding systems significantly contribute to the downtime in biorefineries. We are working towards reducing these upsets by developing constitutive relations derived from the physical attributes of biomass feedstocks to couple with Computational Fluid Dynamics (CFD). In particular, here we present work on using the Non-Local Granular Fluidity (NLGF) constitutive model to represent dynamic flow of bulk biomass feedstocks. Whereas other constitutive models, such as Drucker-Prager-Cap and inertial rheology, are generally capable of resolving bulk flow in many flow regions, they are not capable of representing the well-known nonlocal phenomena in granular flow. Nonlocal phenomena emerge in regions where flow features occur on length scales that are near the scale of the system geometry, e.g., the jamming of hopper outlets and the stop-height of piles on inclines. The NLGF model handles the nonlocality of flow by introducing an intrinsic diffusive length scale into the granular rheology model. The developed CFD model, based on NLGF, was used to simulate biomass feedstock flow in feed-handling equipment and into reactors. Toward this goal, we extend the existing software packages and algorithms to accommodate the computational modeling work. Further, we validate the NLGF model-based CFD simulations of biomass feedstock flow by comparing simulation results against lab-scale flow and deformation experiments. Specifically, the flow of particulate pine residues in three systems will be presented: inclined plane setup, ring-shear tester, and a wedge hopper. This experimentally validated model will be used for providing actionable guidance for operating and optimizing lab to pilot scale biomass-conversion feeding systems.

603140 Computational Modeling of a Biomass Screw-Feeder with Compressible Non-Newtonian Rheology

Abstract:

Lignocellulosic biomass, such as agricultural waste and forestry byproducts, can be a carbon-neutral source of energy and materials, providing an alternative to fossil sources. Handling of such feedstocks with high variability is critical to efficient biochemical or thermochemical conversion. Conveying of biomass into high pressure pretreatment reactors using screw-feeders is one of the important feedstock handling steps. Mechanical failure and blockage in screw-feeders are among the difficulties that biorefineries face during operation. The complex properties of biomass feedstock, i.e. moisture content, insoluble solids concentration, and non-ideal physical properties, make the blockage prediction very challenging.

The focus of this study is to develop a computational-fluid-dynamics (CFD) model that can predict the torque requirements in these screw-feeders by including the complex rheological and compressive characteristics of biomass feedstock. As a result, we can improve the mechanical design and avoid operating conditions that lead to blockage and failure. The screw-feeder studied in this work is 1.7 m long and is enclosed inside a grooved housing with tight clearance ~ 1 mm. Both the screw and the housing are tapered with outer diameter decreasing from 0.53 m to 0.4 m. A conical back pressure damper at the housing exit assures that the biomass is compressed to the required pressure before moving from the feeder to the pretreatment reactor. Our model is validated against torque measurements from experiments that are performed for various screw speeds and feeding rates.

In order to model this screw-feeder, we have developed a customized solver in OpenFOAM [1], an open-source CFD library. The biomass feedstock is modeled as a single-phase compressible mixture whose density varies with pressure, based on a feedstock dependent equation-of-state, derived from a recent study by Duncan et al. [2]. This is a reasonable approximation because the biomass particles are densely packed in the screw-feeder during compression. The transport equations for mass and momentum are solved using a pressure-based low Mach algorithm. The rheology of the biomass is considered using two different non-Newtonian viscosity models; density-dependent Cross model [3] and a Bingham model [4] with density-dependent yield stress. The parameters for these models are obtained from rheology experiments that characterize yield-stress and strain-rates on a small-scale screw extruder. The screw-feeder torque is calculated for different rotation speeds using both of the proposed constitutive models. Finally, our modeling results are compared with our experimental torque data for model validation.

References:

[1] Weller, Henry G., Gavin Tabor, Hrvoje Jasak, and Christer Fureby. "A tensorial approach to computational continuum mechanics using object-oriented techniques." Computers in physics 12, no. 6 (1998): 620-631.

[2] Duncan, Joshua C., Anaram Shahravan, Joseph R. Samaniuk, Thatcher W. Root, Michael D. Graham, Daniel J. Klingenberg, C. Tim Scott, Keith J. Bourne, and Roland Gleisner. "Pressure-driven flow of lignocellulosic biomass: A compressible Bingham fluid." Journal of Rheology 62, no. 3 (2018): 801-815.

[3] Cross, M. M. (1965). Rheoogy of non-Newtonian fluids: A new flow equation for pseudoplastic systems. *Journal of Colloid Science*, 20(5), 417–437.

[4] Bingham, E.C. (1916). "An Investigation of the Laws of Plastic Flow". *Bulletin of the Bureau of Standards*. **13** (2): 309–353.

607327 Flow Characterization of Fibrous Woody Biomass Using Experiments and Modeling

Abstract:

Critical problems often arise when feeding biomass in large scale processes. Frequently, solid feeding devices, such as screw feeders and hoppers, become blocked, and fail to provide uniform and continuous flow of the feed material. Reliable, efficient, and economical feeding of lignocellulosic biomass into pyrolysis reactors makes biofuels production one step closer to the goal of replacing fossil fuels. In this work, transporting biomass solids in compression-screw feeders is the primary focus. In order to mimic the flow properties of compressed biomass inside a compression-screw feeder, rheological properties of compressed lignocellulosic biomass were investigated. Then, the experimental results were validated by Computational Fluid Dynamics (CFD) simulations using OpenFOAM software package for a simple pipe flow case. To conduct the experiments, woodchips were fed into a twin-screw compounder in order to obtain compression conditions experienced in a compression-screw feeder in an industrial scale process. Then, changes in flow behavior were measured when varying the material and operating conditions, including screw speed, moisture content, and particle size distribution. After solving the equations of continuity and conservation of momentum for the microcompounder, screw speed and force were converted to shear rate and viscosity, respectively. We found that at higher screw speeds and moisture contents, the material flows easier. In addition, the viscosity of biomass in the microcompounder falls within the shear thinning region of a Cross model fit. Moreover, the impact of particle size distribution on the flow behavior of compressed biomass is insignificant. For the modeling and simulation efforts, biomass was modeled as a compressible fluid with a pressure-dependent density and density-dependent yield stress. Density-dependent Bingham and Cross models developed were validated for biomass pipe flow under pressure. Rheological values obtained from pipe flow simulations agreed with experimental data obtained from literature showing that the models developed are able to capture the flow behavior of compressed biomass under pressure. The next phase of this project will focus on performing computational simulations of the industrial scale compression-screw feeder and refining the model parameters and validating the simulation results against pilot-scale experimental data.

599338 True Bridging Liquid-Solid Ratio (TBSR): Redefining a Critical Process Parameter in Spherical Agglomeration

Abstract:

Spherical agglomeration is a particulate process which allows the downstream processing of high value products to be improved through size enlargement. In the case of pharmaceuticals, spherical agglomeration is particularly well suited for problematic active pharmaceutical ingredients, especially those with needle-like morphologies. In spherical agglomeration, crystals are agglomerated through the use of a partially-immiscible bridging liquid into dense spherical agglomerates with superior mechanical properties. This includes their flowability, compressibility, and tabletability, making the product highly suitable for direct compression without subsequent processing steps. As such, downstream processing becomes less energy, time and cost intensive, a current drive within the industry.

Currently, the mechanistic understanding of the process is limited, and few studies specifically investigate and elucidate these mechanisms to further a wider understanding as a whole (Pitt et al., 2018). However, it is recognised that agglomeration is not only successful, but highly controllable if the ternary solvent system is within the immiscible region. Here, two distinct immiscible phases form. As the bridging liquid preferentially wets the particles of interest, agglomeration in these immiscible systems is highly successful. Conversely, the current literature provides a detailed account of the influence of process and formulation parameters on the final agglomerate properties.

One critical process parameter is the bridging liquid to solid ratio (BSR), which quantifies volumetrically the amount of both components within the process. This ratio must be accurately controlled as, ultimately, it directly dictates the performance of the process. Many studies recognise and define an optimal range for this parameter, within which dense, spherical agglomerates are formed. Outside this range, agglomeration is either inefficient as there is not enough bridging liquid, or, a paste-like substance is formed. Each of these scenarios is undesirable. Unfortunately, the critical range only is specific for each bulk solution-bridging liquid-solid system. If any of these three components are changed, the critical range no longer holds. This is, in part, because BSR fails to account for the influence of bridging liquid miscibility within the bulk solution.

Here, a new definition is introduced to standardise bridging liquid volume reporting as a dimensionless parameter: the true bridging liquid-solid ratio (TBSR), which incorporates the true volume of the bridging liquid rich phase. To evaluate solvent miscibility, the ternary phase diagram is pivotal. Through a combined approach of experimental and simulation work, these diagrams have been determined with a high degree of accuracy and agreement. The diagrams were used to evaluate bridging liquid miscibility in a variety of hypothetical agglomeration systems. As a result, a relationship BSR and TBSR has been identified. Critically, findings based upon these theoretical systems supported the rationale and need for the definition as a whole.

With the relationship between these parameters known, a target true bridging liquid-solid ratio can be easily identified. This includes calculation of the bridging liquid volume required to surpass solvent miscibility and form an immiscible agglomeration system. Experimental validation using water-acetone-bridging liquid systems produced highly similar agglomeration profiles for three different bridging liquids with varying miscibility with water-acetone mixtures. Changes in the bulk solution composition, or the bridging liquid, produced only small changes in these profiles. Interestingly, the percentage of solids not agglomerated remained consistent across all conditions. This demonstrates the suitability of the TBSR as a means of standardising bridging liquid volume reporting, whilst accounting for bridging liquid miscibility. As a dimensionless parameter, further mechanistic studies into spherical agglomeration should make full use of this new definition, especially those which are focussed on the wetting and nucleation rate processes. Ultimately, such studies will form the basis of regime maps for the development of robust spherical agglomeration processes.

595728 Role of Initiator & Surfactant in Multi-Stage Core-Shell Polymer Aqueous Agglomeration & Powder Properties

Abstract:

A multi-factor analysis study was performed on the effects of surfactant choice—sodium c14-16 alpha olefin sulfonate vs. sodium lauryl sulfate—and initiator choice—potassium peroxydisulfate (KPS) vs. tert-butyl hydroperoxide (tBHP)—on the temperature of agglomeration and product bulk density of two variants of acrylic processing aids—the K-128P innovation product **PMMA1** and one of our established processing aids **PMMA2**. The goal of the experimental work was to enhance the final product performance of a newly launched acrylic processing aid product designed for rigid PVC applications.

After initial development and customer sampling, challenges were uncovered related to the product powder production, packaging, and product conveying experience at the customer. This series of studies focused on resolving the two overarching parameters: I) Lowering the temperature of agglomeration and II) Increasing the bulk powder density in the final packaged product. These were selected out of a range of desirable scaleup improvements such as high foaming in the wastewater, high dust-loss during drying, high temperature needed for agglomeration and bulk powder density below that expected of our comparable processing aids.

The study concluded that emulsion polymers of the processing aid family respond to changes in surfactant & synthesis initiators in the parameters studied (I & II). As a result, the new product was modified with the optimal composition and improved isolation techniques proved effective at mitigating all of the manufacturing concerns identified during early scaleup, enabling reliable plant-scale production of the innovation product in the year since the changes have been implemented. The product has closed production gaps with our existing portfolio without compromising the innovative properties of the new chemistry.

609943 Improved Properties of Binary Blends of Cohesive Drug Powders for Direct Compression Using Fine Grade Engineered Excipients

Abstract:

Direct compression (DC) is one of the most efficient and economic methods for tablet manufacturing since only blending and compression operations are required, greatly simplifies the process-train. However, the popularity of direct compression is limited by its stringent requirements, such as good flowability, high packing density, and excellent compaction for excipient-active pharmaceutical ingredient (API) blends. When the blend has high drug loading with a poorly flowing API, the flowability and tabletability of blends is generally decreased sharply. In a recent work, it has been shown that surprisingly, fine excipients of 20 mm size that have improved flow and bulk density after silica dry coating, can greatly improve direct compression performance [1]. Unfortunately, it could not work for very fine cohesive APIs such as 10 mm acetaminophen (APAP) at higher drug loading. In another work, fluid energy mill (FEM) was used to simultaneously mill and dry coat microcrystalline cellulose (MCC) with nano-silica to produce different sizes of fine, well flowing MCC excipients [2]. Subsequently, a larger sized dry coated MCC (30 mm) was used in multi-component blends with a greater success in DC tableting [3]. In this work, binary blends of several fine sized dry coated MCC-based excipients are considered with three different fine, cohesive APIs at high drug loadings. Bulk density, FFC, and tablet compactability for all these blends are measured and compared to determine which excipient sizes provide the best overall performance. The results are analyzed for the relative sizes of the excipients and APIs to better understand the interplay between the components of the blends and assessment of the extent of silica transfer from coated excipient powders to API powders and its effect on blend bulk density and flowability. It is expected that using fine excipients with fine APIs may also help in reduced segregation tendency during overall processing. Hence, these results are expected to provide industry relevant guidelines for the use of finer surface modified excipients in DC tableting.

610395 The Effect of Particle Shape in Screw Feeders.

Abstract:

Granular Material 9GM) is the second most manipulated substance in the world and present in a number of industries either as input or final products. At the macro-scale GM exhibits complex behavior which most often does not fit into empirical relations that exists for fluids. This complex macro-scale behavior is a result of the interactions of individual particles at the grain scale, hence a numerical method like DEM is the most feasible to aid in the controlled manipulation of GM. While the DEM has been used since the early 90's to simulate industrial devices such as ball mills and slio's its usage has been limited owing to simplistic particle shape assumptions and low number of particles as a result of its computational expense. While in some cases the assumption of spherical particles and a reduced scale is sufficient, in the case of screw feeders which are used in a number of industries to deliver precise dosages this oversimplification maybe lead to erroneous results. In this paper we study the effect that particle shape has on the spatial distribution of particulate matter in screw feeders as well the degradation of the material which effects product quality and production rates using the polyhedra particle based open-source GPU code Blaze-DEM. We will also discuss our utilization of GPU computing technology in the commercial code XPS that enables us to tackle large scale problems concerning the screw feeding of pharmaceutical powders.

600669 Enhancement of Battery Performance of All-Solid-State Lithium-Ion Battery with Structural Control of Dry Coated Particls

Abstract:

As a promising candidate of next generation secondary battery with both high safety and high energy density, all-solid-state lithium-ion secondary batteries (ASS-LIBs) have attracted much attention. In the ASS-LIBs, inorganic solid electrolytes (SEs) are used instead of the organic liquid electrolytes. SEs are non-flammable materials and stable under higher voltage and broad temperature range. Due to these features, the ASS-LIBs are highly expected to be a next generation secondary battery for EVs. In the ASSLIBs, transfer of lithium ions during charging and discharging only occurs at the interfacial contact surface between an electrode particle (active material) and SE. Therefore, development of particle processing technologies to construct the solid-solid interfacial contacts is important issue. We reported for the first time a dry coating process to produce composite particle of electrode active material and solid electrolytes. By means of the dry coating process, lager electrode active material (core particles) were directly coated with SEs (guest particles) without using any solvents and binders, resulting the solid-solid interfacial contacts between active material and SEs within the dry coated particles. However, infulence of the structure of the dry coated particle on the battery performance has been unknown.

In this study, we have focused on the thickness of the SE layer and influence of the thickness on the battery performance was investigated. Consequently, for enhancement of the performance of the ASSLIB the thickness should be a certain optimum range, where the ion and electron conductivities can be balanced well.

Magnetic Nanoparticle Endocytosis Mechanism: Characterization and Quantification by Magnetic Cytometry

Abhinav Sannidhi[1], Paul W. Todd[2], and Thomas R. Hanley[1]
[1]Department of Chemical Engineering, Auburn University, AL 36849
[2]Magnaquant LLC, 11760 Commonwealth Drive, Louisville, KY 40299, USA

Abstract

Magnetic nanoparticles (MNPs) with varied surface modifications are extensively used in cell labeling, cell purification, cell physiology, endocytosis, biochemical micro assays, endosome research, cell separation, drug targeting, and in vivo diagnostics. The Hyperflux™ velocimeter uses particle tracking velocimetry to measure magnetophoretic mobility, size and other morphological parameters of magnetic particles and labeled cells. Magnetic cytometry by velocimetry records the motion of labeled cells in an isodynamic magnetic field thereby estimating the key parameter, magnetophoretic mobility of labeled cells. Magnetophoretic mobility is proportional to the number of particles ingested per cell. The rapid estimation of magnetophoretic mobility by the instrument and collection of multiple thousand data points facilitates cellular uptake quantification and kinetic studies in less time than any other existing technique. The receptor-independent uptake by cultured CHO (Chinese Hamster Ovary) cells of 100 nm iron oxide nanoparticles with different surface coatings, namely starch, dextran sulfate (DXS) and amino groups was studied to reveal the role of nanoparticle endocytosis mechanisms. Caveolae-mediated and clathrin-coated endocytosis are revealed by using specific mechanism-based inhibitors such as Genistein and Methyl-β-cyclodextrin. Uptake inhibition has been investigated at different inhibitor concentrations during endocytosis by evaluating the cellular survival rate, toxicity, and uptake of nanoparticles with the different surface coatings, which also reveal relevance of surface charge. Characterization of labeled cells and magnetic particles by magnetic velocimetry facilitates the rapid estimation of intrinsic magnetic properties, the optimization of MNP coatings, a better understanding of cell labeling, and the evaluation of the role of different cellular endocytosis mechanisms.

Keywords: magnetophoretic mobility, magnetic nanoparticles, CHO cells, endocytosis mechanism, inhibitors

Introduction

Magnetic nanoparticles with the superparamagnetic nature of the inner iron oxide core and biocompatibility of the outer shell with different surface coatings are widely used in different therapeutic and biomedical applications such as cell separation [1,2], hyperthermia [3], magnetic resonance imaging [4–7] and cellular phagocytosis/endocytosis studies [8–14]. Ongoing research for quantifying cellular uptake and different endocytosis mechanisms of magnetic nanoparticles and non-magnetic nanomaterials (such as gene vectors, liposomes, air contaminants and viruses) employs a variety of methods such as fluorescence spectroscopy [15,16], flow cytometry [17], magnetic bead enzyme-linked immunosorbent assay (ELISA) method [18–20], immunofluorescent microscopy [21,22], inductively coupled plasma (ICP)-based spectroscopic techniques including -optical emission spectrometry (ICP-OES) [23], mass spectrometry (ICP-MS) [24], focused ion beam scanning electron microscopy (FIB-SEM) [25,26], transmission electron microscopy (TEM) [16]. Magnetophoretic mobility measurement of magnetically labeled cells is a facile, cost-effective and more robust method for estimating superparamagnetic iron oxide nanoparticles (SPIONs) uptake. The cellular uptake of SPIONs by Chinese Hamster Ovary cell line (CHO-K1) with the velocimeter instrument has been investigated and optimum instrument threshold and size gating settings were determined previously [27]. The study of roles of endocytosis mechanisms of the magnetic beads with different surface coatings is focused in this research.

Cellular endocytosis is classified into two main categories: Phagocytosis (restricted to specialized cells only) and pinocytosis (occurs in almost all cell types). Pinocytosis is further classified into three different mechanisms, namely, clathrin dependent [28]; clathrin independent and macro-pinocytosis [29,30]. The clathrin independent mechanism is further subdivided into caveolae mediated [31] and caveolae and clathrin independent mechanism [32]. The most commonly used chemical inhibitors such as chlorpromazine, potassium depletion, genistein, methyl-β-cyclodextrin (M β CD) have heterogeneous effects in terms of cytotoxicity and inhibitor specificity and are cell line dependent [33]. The adopted endocytosis and uptake mechanisms of nanoparticles depends on size, shape and intrinsic surface properties of individual particles and plays a major role in characterization and quantification of the labeled cells. The magnetic properties of individual nanoparticles such as magnetophoretic mobility and intrinsic magnetic properties are of much importance in assessing the particle quality and cellular uptake kinetics [9,34,35]. In this study, caveolae mediated and clathrin mediated endocytosis mechanisms were inhibited by using two different chemical inhibitors, genistein and chlorpromazine respectively at different concentrations. Genistein is an isoflavone drug derived from soy products and used in the treatment of many chronic diseases, diabetes, tumors and menopausal symptoms [36,37]. It is a tyrosine-kinase inhibitor and causes local disruption of the actin network at the site of endocytosis [38]. Chlorpromazine, having cationic amphiphilic properties, induces inhibition of clathrin and AP2 transportation at the cell membrane. Chlorpromazine applications include treatment of chemoresistant glioma [39], inhibition of histamine release from human lung fragments [40], cell-cell fusion, clathrin-mediated endocytosis [41], replication of West Nile virus (WNV), Middle east respiratory syndrome-coronavirus (MERS-CoV) [42] and Severe Acute Respiratory Syndrome-coronavirus (SARS-CoV) [43].

The objective of this research is to determine how nanomaterials enter cells (endocytosis pathway mechanism) by characterizing and quantifying cellular uptake, cell toxicity/viability and roles of different endocytosis mechanisms of magnetically labeled mammalian cells by measuring magnetophoretic mobility using a velocimeter. In this study, the cellular uptake mainly depends on two independent parameters, namely zeta potential of the beads (electrostatic interaction between the particle and biological environment), and the inhibitor toxicity for endocytosis. The toxicity constant for the inhibitors were determined using the magnetophoretic mobility data. The major dominant mechanisms of endocytosis of magnetic beads with starch, aminated starch, dextransulfate (DXS) and aminated dextran sulfate coatings by CHO-KI cells were identified.

Theory

Magnetophoretic mobility of the labeled cells is defined as the ratio of the field-induced terminal velocity of the labeled cell, *v*, to the gradient of the magnetic field, $\nabla\left(\frac{B_o^2}{2\mu_o}\right)$, also denoted as S_m.

$$U_m = \frac{v}{\nabla\left(\frac{B_o^2}{2\mu_o}\right)} \tag{1}$$

S.I. units of magnetophoretic mobility are $\frac{m^3}{T\,A\,s}$ (cubic meter per tesla-ampere-second), also denoted as "Rm"- Rembaum, in honor of Alan Rembaum for his pioneering work on magnetophoretic mobility.

Materials and Methods

Magnetic beads

FluidMAG nanoparticles of size 100 nm with different surface coatings such as starch coating (fluidMAG-D) and dextran-sulfate coating (fluidMAG-DXS) were procured from Chemicell (Germany). These beads were modified by aminosilane coating to get aminated starch and aminated DXS nanoparticles.

Mammalian Cells and Chemical Inhibitors

Chinese Hamster Ovary (CHO-K1) Cells (catalogue no: ATCC CCL-61) were obtained from ATCC (Manassas, VA, USA), two different mechanism-based chemical inhibitors namely, Genistein (Catalogue No: TCG0272) and Chlorpromazine Hydrochloride (Catalogue No: TCC2481) procured from VWR (Radnor, PA, USA) were used to focus on each specific type of endocytosis mechanism. The CHO-K1 cell subculture is performed for every 48 hours by growing cells attached to the flask surface following standard procedure.

Hyperflux Velocimeter

The magnetic particle velocimeter, obtained from IKOTECH, LLC, and serviced by Magnaquant (Kentucky, USA) consists of a channel cell with magnetic assembly, video microscopy system, and image processing software. The captured image data are analyzed by the built-in software which generates 26 different characteristics of magnetic particles based on user-selected intensity threshold and size limits.

Results and Discussion

Cellular Uptake with different inhibitors

Cellular uptake in the presence of different inhibitors at different concentrations and their effects on endocytosis were evaluated by measuring magnetophoretic mobilities of the SPION- labeled CHO-K1 cells. The magnetophoretic mobility of the labeled cells is directly proportional to the cellular uptake of SPIONs. The mobility of untreated cells ingesting SPIONs (100 μg/mL) for 24 h is high, as seen tin Table 1.

Table 1 - Average control magnetophoretic mobilities and their standard deviations of CHO-K1 cells labeled with 4 different surface functionalized magnetic nanoparticles for 24 hours at 100 μg/mL.

Particle Type	Average control magnetophoretic mobility (pRm)	SD
Dextran Sulfate (DXS)	4.285	0.063
Aminated DXS	4.182	0.135
Starch	5.678	0.157
Aminated starch	19.372	2.173

Table 1 gives the average control magnetophoretic mobilities of CHO-K1 cells labeled with different functionalized nanoparticles along with their standard deviations. This data shows the reproducibility and accuracy of the method. The average mobilities were estimated by applying data filters- Mobility>0 and Intensity>130 obtained from the tracks.csv spreadsheet. The average magnetophoretic mobility of cells labeled with aminated starch beads is high when compared to other magnetic beads, indicating the higher cellular uptake. The normalized magnetophoretic mobility (expressed as per cent of mobility at zero inhibitor concentration and plotted as "Normal Mobility") of labeled cells with different SPIONs and inhibitors is shown in Figures 1 and 2. A Marquardt-Levenberg least-squares multiparameter fit was used to fit the non-linear exponential curve with the equation $y = A + ((1-A)\exp(-kx))$, where y is the total fraction of uninhibited endocytosis sites, k is the toxicity of the inhibitor and A is the uninhibited fraction of endocytosis events. Genistein with a concentration range of 0 to 300 μM (Figure 1) and chlorpromazine with a concentration range of 0 to 10 μM (Figure 2) were used to inhibit endocytosis of different surface modified SPIONs. The normalized mobilities of cells taking up aminated DXS and starch coated beads and treated with genistein were fitted linearly, as the data did not reach the saturation plateau even beyond the lethal concentration range of 200 μM. The normalized mobilities of dextransulfate and aminated starch beads followed an exponentially decreasing function with the increase in the concentration of genistein. The curve fit parameters are given in Figure 1. The cellular uptake (normalized mobility) is decreased for all the beads as the concentration of genistein is increased.

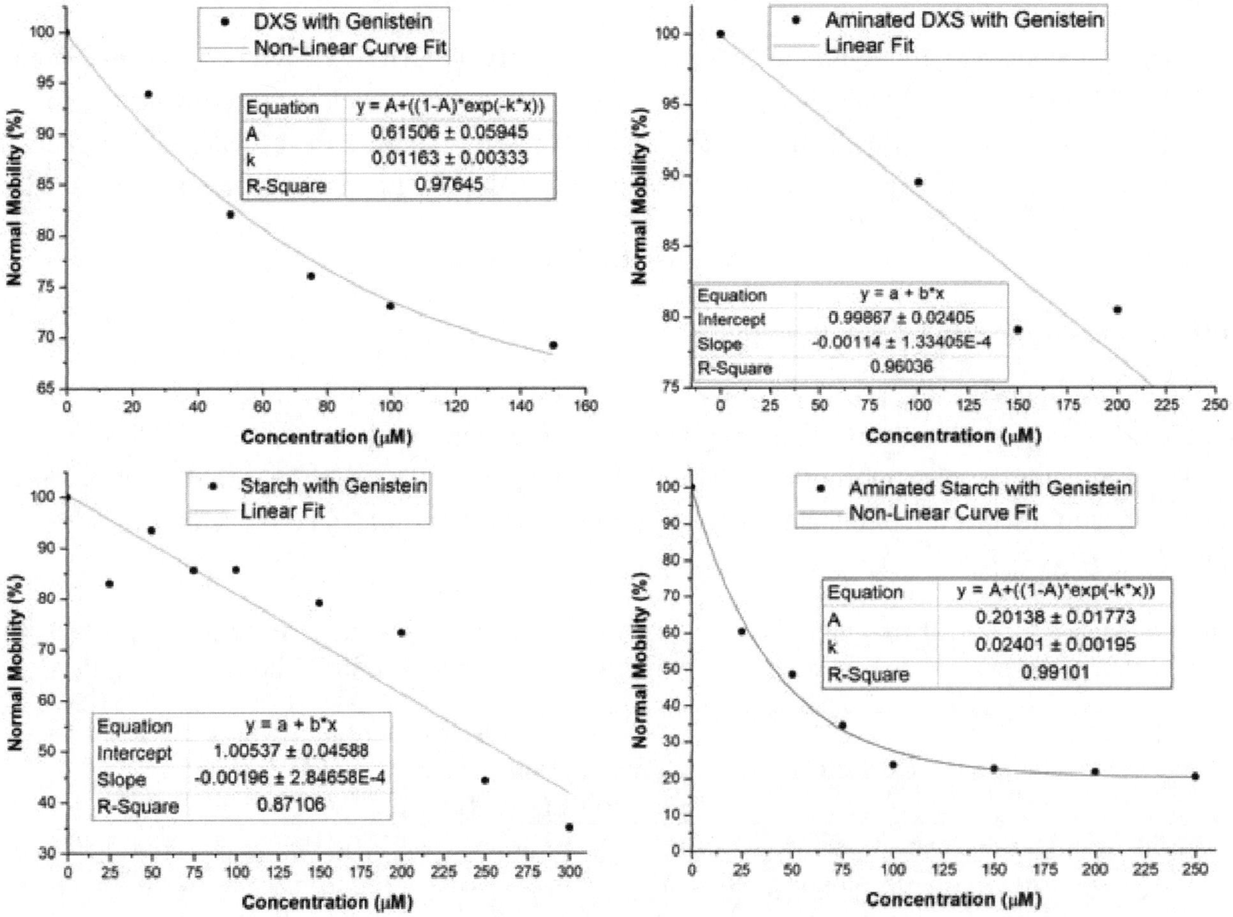

Figure 1 - Normalized mobility change of cells labeled with different SPIONs with respect to genistein concentration.

The normalized mobility (cellular uptake) of labeled cells for all the beads with chlorpromazine drug were exponentially decreasing with the increase in the concentration of chlorpromazine hydrochloride. The curve fit parameters are given in Figure 2, with $R^2 > 0.9$.

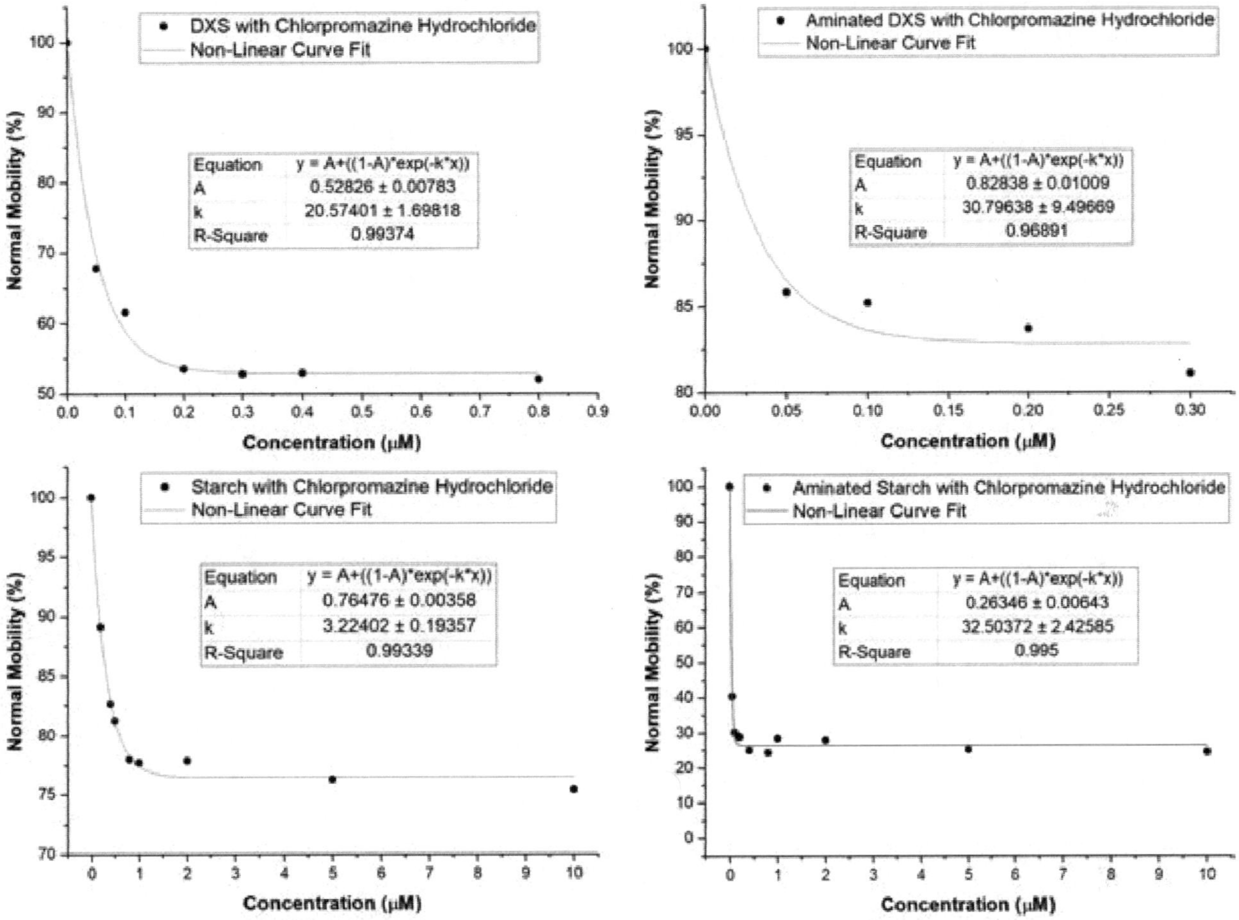

Figure 2 - Normalized mobility change of cells labeled with different SPIONs as a function of chlorpromazine hydrochloride concentration.

Inhibitor toxicity and Zeta Potential

The hydrodynamic size, zeta potential, and inhibitor parameters for different SPIONs are given in Table 2. The inhibitors toxicity was quantified by the toxicity constant (k) derived from the normalized mobility graphs (Figures 1 and 2) as shown in Table 2 and Figure 3. The difference in values of size and zeta potential measured in distilled water (dH_2O) and complete culture media (CM) are due to the presence of balanced salt solution, serum, serum proteins, cationic and anionic species in the cell culture media. The electrostatic interactions of nanoparticles and culture media plays a significant role in protein adsorption and cellular uptake [44]. The better cellular uptake and ligand adsorption can be achieved by proper tuning of surface properties of nanoparticles [45].

The viabilities of the labeled cells with different inhibitors are in the range of 60 to 90%. The viability is decreased w.r.t increase in the concentration of both inhibitors. The cell viability is high and consistent between both experimental methods- MTT Assay and Trypan Blue indicator.

Table 2 - Hydrodynamic size and zeta potential (ZP) measured in distilled water and complete cell culture media along with standard deviations, and inhibitor parameters for different SPIONs.

Particle Type	Size (nm) in dH_2O	Size (nm) in CM	ZP in dH_2O (mV)	ZP in CM (mV)	Inhibitors			
					Genistein		Chlorpromazine	
					k	A	k	A
Dextran Sulfate (DXS)	133.9 ± 1.3	166.5 ± 0.8	-52.93 ± 0.9	-0.98 ± 0.05	0.012	0.615	20.57	0.528
Aminated DXS	164.7 ± 0.6	148.2 ± 0.6	-33.43 ± 0.5	-0.34 ± 0.13	0.001	N/A	30.79	0.828
Starch	108.6 ± 1.4	151.1 ± 0.7	-8.99 ± 0.2	-3.78 ± 0.6	0.002	N/A	3.22	0.765
Aminated Starch	142.7 ± 1.1	128.5 ± 0.5	39.60 ± 0.8	-0.69 ± 0.2	0.024	0.201	32.50	0.264

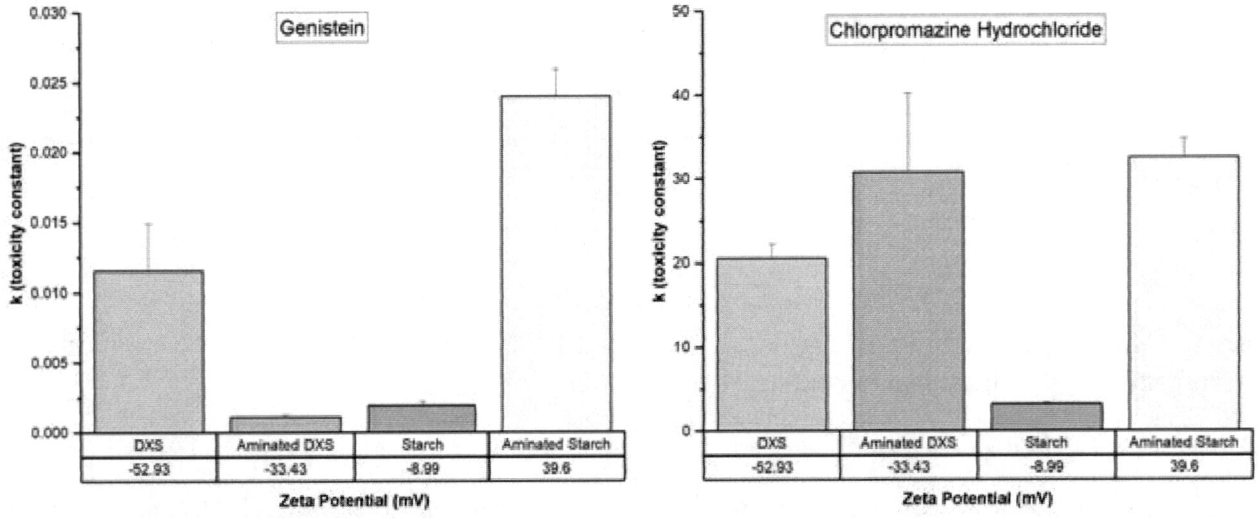

Figure 3 - Inhibitor toxicity constant (k) versus zeta potential (ZP in dH_2O) for different SPIONs.

Genistein showed a low toxicity constant (in the order of 0.01 per micromolar) indicating that high doses are required. Chlorpromazine exhibited higher inhibition toxicity (in the order of 10 per micromolar) and was sensitive to low dosages. For both inhibitors, the toxicity toward the uptake of aminated starch beads was high and that of starch beads was low.

Effect of both inhibitors

The cellular uptake of cells labeled with aminated starch coated beads was quantified by magnetophoretic mobility measurement by treating the aminated starch labeled CHO-K1 cells with both the inhibitors to examine whether both the inhibitors are targeting same or different endocytosis mechanisms. The concentration of chlorpromazine was held constant at 0.3 μM and the genistein concentration was varied from 0 to 200 μM. The decrease in normal mobility values for the genistein and chlorpromazine experiment (red data points in the left panel of Figure 4) when compared to the genistein alone (black data points) experiment reveals that the two inhibitors target and inhibit different endocytosis pathways. The viability data for these experiments shown in right panel of Figure 4 are consistent between methods and decrease with the increase in inhibitor concentrations.

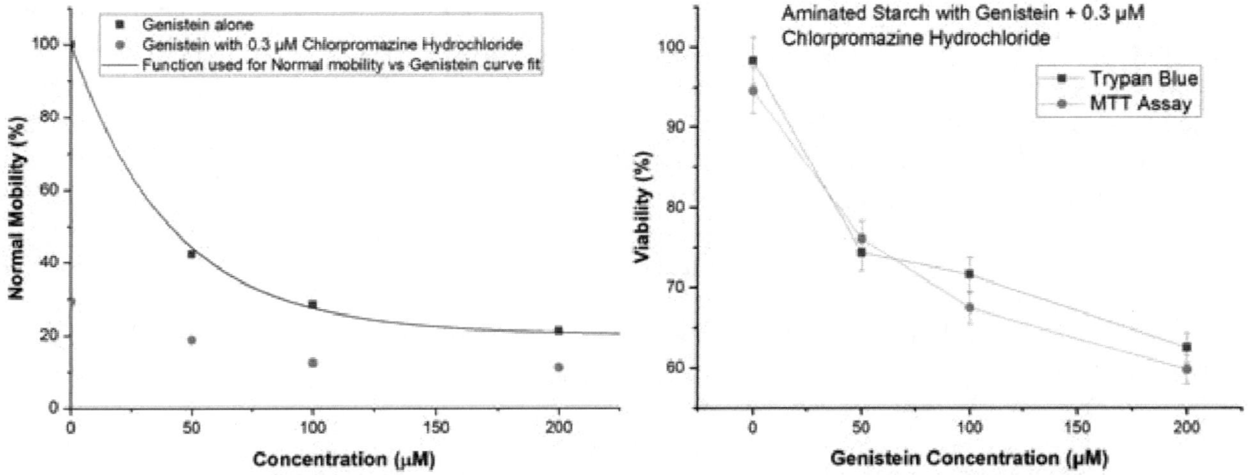

Figure 4 - Normalized mobility of CHO-K1 cells labeled with aminated starch beads and treated with genistein + 0.3 μM chlorpromazine hydrochloride and viability of CHO-K1 cells labeled with aminated starch beads and treated with genistein + 0.3 uM chlorpromazine.

Mechanism of uptake

The magnetic beads with aminated starch and dextran-sulfate coatings exhibited clathrin and caveolae mediated endocytosis as dominant mechanisms. The starch coated beads showed caveolae mediated endocytosis as the major pathway to enter cell membrane. Based on literature [8,46,47], aminated-DXS beads follow a likely clathrin and caveolae independent mechanism as tabulated in Table 3.

Table 3 - Types of endocytosis mechanisms dominant for different SPIONs.

Particle Type	Endocytosis Mechanism	
	Major/ Dominant	Minor/ Low to Moderate
Dextran Sulfate (DXS)	Clathrin & Caveolae mediated	Clathrin & Caveolae independent
Aminated DXS	Clathrin & Caveolae independent	Clathrin dependent & Caveolae mediated
Starch	Caveolae mediated	Clathrin dependent, Clathrin & Caveolae independent
Aminated Starch	Clathrin & Caveolae mediated	Clathrin & Caveolae independent

Conclusions

SPIONs with different surface modifications were studied and utilized in this research to quantify the cellular uptake of magnetically labeled mammalian (CHO-K1) cells. The toxicity and inhibition rate of the mechanism-based inhibitors was quantified by magnetophoretic mobility measurement of thousands of magnetically labeled CHO-K1 cells. The two independent parameters- inhibitor toxicity and zeta potential of magnetic beads accountable for the cellular uptake of SPIONs were characterized and discussed in terms of the types of endocytosis mechanisms dominant for different magnetic bead surfaces. The objective of determining how magnetic beads enter cellular cytoplasm was achieved as given in Table 3. Thus, characterization of labeled cells and magnetic particles by magnetic velocimetry facilitates the optimization of SPIONs coatings, a better understanding of cell labeling, and the evaluation of the role of different cellular endocytosis mechanisms.

Acknowledgments

Research support was provided by Auburn University Research Funding. We thank Dr. Allan E. David and his research group for providing cell culture training and lab facility.

References

[1] M. Zborowski, J.J. Chalmers, Magnetic cell separation, Elsevier, 2011.

[2] P.S. Williams, M. Zborowski, J.J. Chalmers, Flow rate optimization for the quadrupole magnetic cell sorter, Anal. Chem. 71 (1999) 3799–3807. doi:10.1021/ac990284+.

[3] F. Gazeau, M. Lévy, C. Wilhelm, Optimizing magnetic nanoparticle design for nanothermotherapy, Nanomedicine. 3 (2008) 831–844. doi:10.2217/17435889.3.6.831.

[4] D. Eberbeck, F. Wiekhorst, S. Wagner, L. Trahms, How the size distribution of magnetic nanoparticles determines their magnetic particle imaging performance, Appl. Phys. Lett. 98 (2011) 1–4. doi:10.1063/1.3586776.

[5] B. Chertok, B.A. Moffat, A.E. David, F. Yu, C. Bergemann, B.D. Ross, V.C. Yang, Iron oxide nanoparticles as a drug delivery vehicle for MRI monitored magnetic targeting of brain tumors, Biomaterials. 29 (2008) 487–496.

[6] S. Yoffe, T. Leshuk, P. Everett, F. Gu, Superparamagnetic iron oxide nanoparticles (SPIONs): synthesis and surface modification techniques for use with MRI and other biomedical applications., Curr. Pharm. Des. 19 (2013) 493–509.

[7] F. Yu, L. Zhang, Y. Huang, K. Sun, A.E. David, V.C. Yang, The magnetophoretic mobility and superparamagnetism of core-shell iron oxide nanoparticles with dual targeting and imaging functionality, Biomaterials. 31 (2010) 5842–5848.

[8] A.K. Gupta, M. Gupta, Cytotoxicity suppression and cellular uptake enhancement of surface modified magnetic nanoparticles, Biomaterials. 26 (2005) 1565–1573.

[9] C. Zhou, Y.S. Choi, A.E. David, P.W. Todd, T.R. Hanley, Nanomaterial endocytosis: Estimation of particles per cell by magnetic measurement, IEEE Magn. Lett. 9 (2018) 7–11. doi:10.1109/LMAG.2018.2830754.

[10] C. Zhou, Z. Qian, Y.S. Choi, A.E. David, P. Todd, T.R. Hanley, Application of magnetic carriers to two examples of quantitative cell analysis, J. Magn. Magn. Mater. 427 (2017) 25–28. doi:10.1016/j.jmmm.2016.11.009.

[11] C.C. Hanot, Y.S. Choi, T.B. Anani, D. Soundarrajan, A.E. David, Effects of iron-oxide nanoparticle surface chemistry on uptake kinetics and cytotoxicity in CHO-K1 cells, Int. J. Mol. Sci. 17 (2015) 1–15. doi:10.3390/ijms17010054.

[12] G. Jie, L. Yang, Y. Jinghai, Influence of Co-Doping on Synthesis, Structure and Magnetic Properties of Ni Nanoparticles, RARE Met. Mater. Eng. 39 (2010) 328–331.

[13] Y. Jing, N. Mal, P.S. Williams, M. Mayorga, M.S. Penn, J.J. Chalmers, M. Zborowski, Quantitative intracellular magnetic nanoparticle uptake measured by live cell magnetophoresis, FASEB J. 22 (2008) 4239–4247.

[14] S. Prijic, J. Scancar, R. Romih, M. Cemazar, V.B. Bregar, A. Znidarsic, G. Sersa, Increased Cellular Uptake of Biocompatible Superparamagnetic Iron Oxide Nanoparticles into Malignant Cells by an External Magnetic Field, J. Membr. Biol. 236 (2010) 167–179. doi:10.1007/s00232-010-9271-4.

[15] N.Ž. Knežević, V.S.-Y. Lin, A magnetic mesoporous silica nanoparticle-based drug delivery system for photosensitive cooperative treatment of cancer with a mesopore-capping agent and mesopore-loaded drug, Nanoscale. 5 (2013) 1544–1551.

[16] K. Strojan, J. Lojk, V.B. Bregar, P. Veranič, M. Pavlin, Glutathione reduces cytotoxicity of polyethyleneimine coated magnetic nanoparticles in CHO cells, Toxicol. Vitr. 41 (2017) 12–20.

[17] H.J. Hathaway, K.S. Butler, N.L. Adolphi, D.M. Lovato, R. Belfon, D. Fegan, T.C. Monson, J.E. Trujillo, T.E. Tessier, H.C. Bryant, Detection of breast cancer cells using targeted magnetic nanoparticles and ultra-sensitive magnetic field sensors, Breast Cancer Res. 13 (2011) R108.

[18] T. Liao, F. Yuan, H. Yu, Z. Li, An ultrasensitive ELISA method for the detection of procalcitonin based on magnetic beads and enzyme-antibody labeled gold nanoparticles, Anal. Methods. 8 (2016) 1577–1585.

[19] H. Chen, A. Abolmatty, M. Faghri, Microfluidic inverse phase ELISA via manipulation of magnetic beads, Microfluid. Nanofluidics. 10 (2011) 593–605.

[20] M. Kala, K. Bajaj, S. Sinha, Magnetic Bead Enzyme-Linked Immunosorbent Assay (ELISA) Detects Antigen-Specific Binding by Phage-Displayed scFv Antibodies That Are Not Detected with Conventional ELISA, Anal. Biochem. 254 (1997) 263–266. doi:https://doi.org/10.1006/abio.1997.2378.

[21] M. V Kiselev, A.K. Gladilin, N.S. Melik-Nubarov, P.G. Sveshnikov, P. Miethe, A. V Levashov, Determination of cyclosporin A in 20% ethanol by a magnetic beads-based immunofluorescence assay, Anal. Biochem. 269 (1999) 393–398.

[22] Å. Lundkvist, J. Hörling, L. Athlin, A. Rosén, B. Niklasson, Neutralizing human monoclonal antibodies against Puumala virus, causative agent of nephropathia epidemica: a novel method using antigen-coated magnetic beads for specific B cell isolation, J. Gen. Virol. 74 (1993) 1303–1310.

[23] A.H. Nguyen, G.N. Abdelrasoul, D. Lin, H. Maadi, J. Tong, G. Chen, R. Wang, A. Anwar, L.

Shoute, Q. Fang, Polyethylenimine-coated iron oxide magnetic nanoparticles for high efficient gene delivery, Appl. Nanosci. 8 (2018) 811–821.

[24] A. Jeong, H.B. Lim, Magnetophoretic separation ICP-MS immunoassay using Cs-doped multicore magnetic nanoparticles for the determination of salmonella typhimurium, Talanta. 178 (2018) 916–921.

[25] B. Lee, J.S. Yoo, V. Ogay, K.W. Kim, H. Dobberstein, K. Soh, B. Chang, Electron microscopic study of novel threadlike structures on the surfaces of mammalian organs, Microsc. Res. Tech. 70 (2007) 34–43.

[26] S. Ahlberg, A. Antonopulos, J. Diendorf, R. Dringen, M. Epple, R. Flöck, W. Goedecke, C. Graf, N. Haberl, J. Helmlinger, PVP-coated, negatively charged silver nanoparticles: A multi-center study of their physicochemical characteristics, cell culture and in vivo experiments, Beilstein J. Nanotechnol. 5 (2014) 1944–1965.

[27] C. Zhou, Characterization and Quantification of Magnetic Particles and Magnetically Labeled Cells by Magnetic Cytometry, (2017).

[28] S. Xu, B.Z. Olenyuk, C.T. Okamoto, S.F. Hamm-Alvarez, Targeting receptor-mediated endocytotic pathways with nanoparticles: rationale and advances, Adv. Drug Deliv. Rev. 65 (2013) 121–138. doi:10.1016/j.addr.2012.09.041.

[29] G. Sahay, D.Y. Alakhova, A. V Kabanov, Endocytosis of nanomedicines, J. Control. Release. 145 (2010) 182–195.

[30] D. Dutta, J.G. Donaldson, Intended specificity and unintended consequences, (2012) 203–208.

[31] A.L. Kiss, E. Botos, Endocytosis via caveolae: Alternative pathway with distinct cellular compartments to avoid lysosomal degradation, J. Cell. Mol. Med. 13 (2009) 1228–1237. doi:10.1111/j.1582-4934.2009.00754.x.

[32] E.-M. Damm, L. Pelkmans, J. Kartenbeck, A. Mezzacasa, T. Kurzchalia, A. Helenius, Clathrin- and caveolin-1-independent endocytosis: entry of simian virus 40 into cells devoid of caveolae, J. Cell Biol. 168 (2005) 477–488. doi:10.1083/jcb.200407113.

[33] D. Vercauteren, R.E. Vandenbroucke, A.T. Jones, J. Rejman, J. Demeester, S.C. De Smedt, N.N. Sanders, K. Braeckmans, The use of inhibitors to study endocytic pathways of gene carriers: Optimization and pitfalls, Mol. Ther. 18 (2010) 561–569. doi:10.1038/mt.2009.281.

[34] A. Sannidhi, P.W. Todd, T.R. Hanley, Estimation of Intrinsic Magnetic Properties of Single Particles by Particle Tracking Velocimetry, IEEE Magn. Lett. (2019) 1. doi:10.1109/LMAG.2019.2950298.

[35] C. Zhou, E.D. Boland, P.W. Todd, T.R. Hanley, Magnetic particle characterization—magnetophoretic mobility and particle size, Cytom. Part A. 89 (2016) 585–593. doi:10.1002/cyto.a.22866.

[36] Y. Hsiao, S. Peng, K. Lai, C. Liao, Y. Huang, C. Lin, M. Lin, K. Liu, C. Tsai, Y. Ma, Genistein induces apoptosis in vitro and has antitumor activity against human leukemia HL-60 cancer cell xenograft growth in vivo, Environ. Toxicol. 34 (2019) 443–456.

[37] T. Akiyama, H. Ogawara, [30] Use and specificity of genistein as inhibitor of protein-tyrosine kinases, in: Methods Enzymol., Elsevier, 1991: pp. 362–370.

[38] P. Thangavel, A. Puga-Olguín, J.F. Rodríguez-Landa, R.C. Zepeda, Genistein as potential therapeutic candidate for menopausal symptoms and other related diseases, Molecules. 24 (2019) 3892.

[39] C.R. Oliva, W. Zhang, C. Langford, M.J. Suto, C.E. Griguer, Repositioning chlorpromazine for treating chemoresistant glioma through the inhibition of cytochrome c oxidase bearing the COX4-1 regulatory subunit, Oncotarget. 8 (2017) 37568–37583. doi:10.18632/oncotarget.17247.

[40] M.K. Church, K.D. Young, The characteristics of inhibition of histamine release from human lung fragments by sodium cromoglycate, salbutamol and chlorpromazine, Br. J. Pharmacol. 78 (1983) 671–679.

[41] Q. Liu, S. Xia, Z. Sun, Q. Wang, L. Du, L. Lu, S. Jiang, Testing of Middle East respiratory syndrome coronavirus replication inhibitors for the ability to block viral entry, Antimicrob. Agents Chemother. 59 (2015) 742–744.

[42] H. Momattin, A.Y. Al-Ali, J.A. Al-Tawfiq, A Systematic Review of therapeutic agents for the treatment of the Middle East Respiratory Syndrome Coronavirus (MERS-CoV), Travel Med. Infect. Dis. 30 (2019) 9–18. doi:https://doi.org/10.1016/j.tmaid.2019.06.012.

[43] J. Dyall, C.M. Coleman, B.J. Hart, T. Venkataraman, M.R. Holbrook, J. Kindrachuk, R.F. Johnson, G.G. Olinger, P.B. Jahrling, M. Laidlaw, Repurposing of clinically developed drugs for treatment of Middle East respiratory syndrome coronavirus infection, Antimicrob. Agents Chemother. 58 (2014) 4885–4893.

[44] S. Patil, A. Sandberg, E. Heckert, W. Self, S. Seal, Protein adsorption and cellular uptake of cerium oxide nanoparticles as a function of zeta potential., Biomaterials. 28 (2007) 4600–4607. doi:10.1016/j.biomaterials.2007.07.029.

[45] A. Vincent, S. Babu, E. Heckert, J. Dowding, S.M. Hirst, T.M. Inerbaev, W.T. Self, C.M. Reilly, A.E. Masunov, T.S. Rahman, S. Seal, Protonated Nanoparticle Surface Governing Ligand Tethering and Cellular Targeting, ACS Nano. 3 (2009) 1203–1211. doi:10.1021/nn9000148.

[46] K. Murugan, Y.E. Choonara, P. Kumar, D. Bijukumar, L.C. du Toit, V. Pillay, Parameters and characteristics governing cellular internalization and trans-barrier trafficking of nanostructures, Int. J. Nanomedicine. 10 (2015) 2191–2206. doi:10.2147/IJN.S75615.

[47] J.M. Rothberg, K.M. Bailey, J.W. Wojtkowiak, Y. Ben-Nun, M. Bogyo, E. Weber, K. Moin, G. Blum, R.R. Mattingly, R.J. Gillies, Acid-mediated tumor proteolysis: contribution of cysteine cathepsins, Neoplasia. 15 (2013) 1125-IN9.

608963 Particle Characterization in Blood Plasma Using Nanoparticle Tracking Analysis

Abstract:

Particles used for drug delivery must overcome many biological barriers to successfully reach their target. One of those challenges is to evade the immune system and circulate through the body until they find their intended target. Many different engineered particles have been used to improve the circulation time of particles such as by adding poly(ethylene glycol) (PEG) ligands, adjusting particle size and shape, and attaching peptides to the surface. Though some of these tactics have resulted in longer circulation times in the blood, there is still significant off site accumulation of the particles in the liver or spleen due to detection by the immune system. It is difficult to determine why off site accumulation occurs and how changes to the particle structure or surface affect it as there are very few tools available to directly study particles in blood.

Particles for drug delivery are typically characterized using dynamic light scattering (DLS). DLS works well with clean, monodisperse samples. However is does not work well to measure particles in blood as the components of blood also scatter light. Characterizing the size of the particles directly in blood plasma, however, is important to determine the extent of protein corona formation on the particles. With DLS, this is typically done by exposing particles to blood plasma followed by washing or dilution with water until DLS measurements can be done to measure the increase in particle size. By washing or diluting the particles, however, the loosely bound proteins of the soft corona are removed and the measurements do not reflect what the particle would experience *in vivo*. As an alternative, we have developed a method of analyzing particles in blood plasma using nanoparticle tracking analysis (NTA). By using fluorescently labeled particles, particles can be analyzed in complex solutions such as blood plasma. The size of the particle, and thus the size of the protein corona, can be measured in pure blood plasma using this method. In this work, polystyrene particles incubated in blood plasma are measured using NTA in pure blood plasma and in diluted blood plasma and the results are compared to DLS. Surface modification is done to attach PEG ligands of different molecular weights to the particles and the size of the particle is again measured in blood plasma using both DLS and NTA. Measuring the size of the particles by NTA has surprising results with protein coronas on PEGylated particles that are almost 50% of the size of the particle. A dilution study was done comparing the size of the particles measured at different plasma concentrations until the concentration was dilute enough for comparison to DLS measurements. At low plasma concentration, size results are consistent with DLS measurements, however a logarithmic trend is apparent in the particle size with plasma concentration suggesting that by washing a PEGylated particle, most of this soft corona is removed, leaving behind a surface without any protein corona. The use of this characterization method will allow for testing of engineered particles for drug delivery directly in blood plasma to determine the features that improve behavior in the blood.

610084 Characterizing Graphene Oxide Suspensions with Rheological Testing and Modeling, Neutron Scattering, and Electrochemical Performance

Abstract:

Electron transfer in carbon based colloidal particle systems make them a good candidate for a variety of electronics applications from flowable batteries to conductive coatings. In order to develop a toolbox to design application driven conductive suspensions, it is important to be able to understand primary particle morphology, surface chemistry, aggregate morphology and network structure. To this end, our group has synthesized graphene oxide using the modified Hummer's method[1] to use as the basis for our study in colloidal particles in homogenous suspension after ultracentrifugation. The surface chemistry of the graphene oxide in aqueous solution determines the degree of aggregation and the formation of hierarchical structures. The inherent graphene properties and concentration (loading) affect the electronic conductivity, mass transport, and stability of the microstructure in flowable electrochemical applications. Therefore, we have characterized the graphene oxide using electrochemical techniques such as cyclic voltammetry and impedance spectroscopy to evaluate its electronic and ionic conductivity.

Moreover, to understand the evolving microstructure, and its contribution to conductivity we incorporated rheological testing. The evolving microstructure of the graphene contributes to particle (sheet to sheet) interaction. Rheological data shows shear thinning with the graphene suspensions In addition, we show both the linear Small Amplitude Oscillatory Shear and nonlinear Large Amplitude Oscillatory Shear flow experiments as well as results of fitting and predicting the rheological flows with contemporary thixo-elasto-visco-plastic (TEVP) models. We will present our ultra-Small Angle Neutron Scattering results with USANS. From this we hypothesize that our material structure is linked to a breakdown of a hierarchical fractal structure in these suspensions. Preliminary Small Angle Neutron Scattering (SANS) and USANS measurements seem to validate this hypothesis. We show the graphene oxide suspension has a mid q feature that can be associated with the primary particle structure and a large upturn at low-q (inverse aggregate length scale) due to the formation of larger structures in the sample. In order to fully characterize the graphene-oxide we show a combination of SANS and USANS. We plan extract primary particle size, aggregate size and fractal dimension from the SANS/USANS data and correlate these properties directly to the conductivity and linear rheological properties.

Lastly will demonstrate a correlation between our rheological model's prediction of thixotropy, or microstructure with elastic and viscous features of our transient rheological data using Sequence of Physical Phenomena (SPP). We will show how the evolving microstructure contributes to the mechanical properties of the material with Cole-Cole plots. In addition, we will construct the full graphene oxide characterization with the USANS/SANS, rheological modeling, SPP framework analysis, and conductivity of the microstructure itself.

References

[1] Yu et al. Sci Rep (2016)

[2] Horner et al. JOR (2019).

[3] Rogers Rheo. Acta (2017).

[4] Richards et al. JOR (2017).

[5] Hipp et al. JOR (2019).

613840 Operating Characteristics of Fluidized Bed Stripper Internals

Abstract:

Gas-solids fluidized bed strippers have major applications in fluid catalytic cracking (FCC) and fluid coking operations. In the FCC process, the catalyst collected by the reactor cyclones contains significant amounts of hydrocarbon products that have to be stripped out from the catalyst to avoid a loss of valuable products and excessively high temperatures in the regenerator. In a fluid coker stripper, steam is used to remove entrained and adsorbed hydrocarbons from the coke particles descending from the fluid coker reactor to minimize losses of valuable hydrocarbon products. A variety of proprietary and standard internals are in use in commercial and research facilities' fluidized bed strippers. The most common types are disk and donut trays, horizontal sheds, and structured packings. Horizontal grating trays are also utilized but to a much lesser extent. This paper reviews the performance characteristics of fluidized bed stripper internals.

605480 Influences of Biomass Compositions, Particle Sizes, and Fluidization Gases on Fast Pyrolysis

Abstract:

This research investigated the influence of biomass chemical compositions, particle sizes, and fluidization gases on fast pyrolysis in a bubbling fluidized bed. A multistep biomass pyrolysis kinetic was validated using two different experimental data set with detailed yields of bio-char, bio-oil, and bio-gas. The predicted C/H/O compositions in these products also compared well with the available experimental data. The validated kinetics was then used to investigate the influence of feedstock compositions on the pyrolysis yields using systematic MFIX-based simulation campaigns using UQ features in Nodeworks toolkit. The sampling simulations were generated by the Latin hypercube sampling method to leverage space-filling property for the targeted parameter space. A data-fitted surrogate model was then constructed to be used as a fast evaluation tool to quickly estimate the pyrolysis products based on different biomass input compositions without solving stiff and complex chemical reactions with MFIX. The adequacy of the surrogate model was assessed with cross-validation errors to minimize the uncertainty introduced due to surrogate model in lieu of actual MFIX simulations. At last, it was coupled with a coarse-grained discrete element method to model the fast pyrolysis in a bubbling fluidized bed. To accurately capture the residence time of biomass particles, a hybrid drag model which couples the Syamlal-O'Brien drag and a filtered drag was proposed and validated using a cold flow experiment. The predicted biomass particle residence time varied from 3.8 s to 10.8 s when its size increased from 130 micros to 400 micros. The change of particle size also leads to an increase of oil yields and a decrease of bio-char yields. The particle temperatures are uniformly distributed around 738 K in all the tested cases. This CFD model was then utilized to investigate the influence of fluidization gases such as pure Nitrogen and a mixture of Nitrogen and Hydrogen.

609652 Memory Optimization for Particle Access in CFD-DEM Simulations

Abstract:

Computational Fluid Dynamics – Discrete Element Method (CFD-DEM) simulations are employed in modeling large-scale gas-solid systems in energy, pharmaceutical, and petrochemical industries. These simulations play a crucial role in the design of industrial-scale systems that are otherwise difficult to design using laboratory-scale experiments. The computational expense of using a large number of particles in these cases (order of billion or more) inhibit the use of CFD-DEM, which is otherwise a more rigorous and accurate approach compared to continuum approximations of the particle phase. Memory access bottlenecks on current high-performance computing architectures become considerable at this scale. The particle data structures in these simulations have a large memory footprint that does not fit into the processor caches leading to poor computational performance. The focus of this work is on addressing this bottleneck using a space-filling-curve approach for reordering particle data structures in CFD-DEM solvers.

The use of the Morton [1] space-filling curve to improve memory access patterns in the particle data structure is described, and their impact on the performance of representative CFD-DEM simulations is presented. The Morton space-filling curve was applied to co-located particles on uniform and non-uniform k-dimensional tree [2] generated cartesian grids. This optimization technique was applied to reorder the particle data-structure to improve spatial and temporal locality in memory. The performance impact of this technique when applied to two benchmarks cases: the homogeneous cooling system [3] and fluidized bed are presented. This optimization technique leads to a two-fold performance improvement in operations that involve accessing the particle data structure, such as creation of neighbor lists, collisional force calculation and inter-processor data exchange. An example case of a homogeneous cooling system simulation where particles are spatially distributed by thermal motion is described in the attached figure. The particle index, which associates to the position in memory, is not correlated with its spatial location without reordering, leading to cache misses and memory access inefficiencies. On the other hand, spatial reordering of the particle data-structure leads to a twenty percent overall improvement in performance. Some frequently performed operations, such as building neighbor lists, exhibit a significant improvement (about two times) while the expense of reordering particles remains negligible (~ 6% of overall run-time).

References:

[1] D. W. Walker, Morton ordering of 2d arrays for efficient access to hierarchical memory, The International Journal of High Performance Computing Applications 32 (1) (2018) 189–203.

[2] M. Grandin, Data structures and algorithms for high-dimensional structured adaptive mesh refinement, Advances in Engineering Software 82 (2015) 75–86.

[3] W. Fullmer, C. Hrenya, The homogeneous cooling state as a verification test for kinetic-theory-based continuum models of gas-solid flows, Journal of Verification, Validation and Uncertainty Quantification 2 (4) (2017) 044501.

Memory optimization for particle access in CFD-DEM simulations

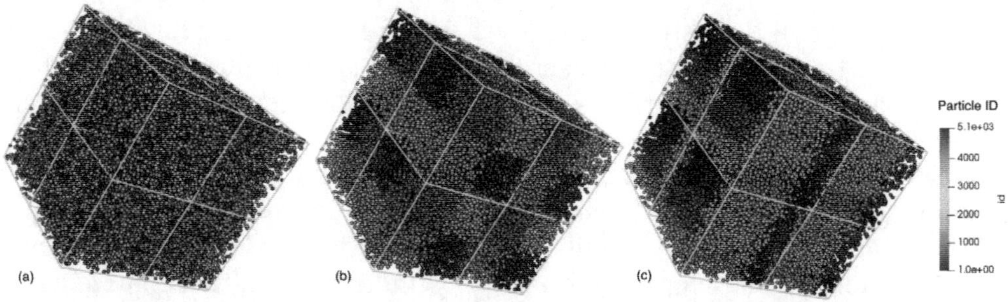

Figure 1.1 Spatial reordering technique to improve spatial and temporal locality in a homogenous cooling system with 40,000 particles simulated using MFiX-Exa with 8 MPI tasks. (a) Default ordering of particles in a grid, (b) kD-tree ordering of particles in the grid (c) Morton ordering of particles in the grid. The particles are stored in memory based on their ID.

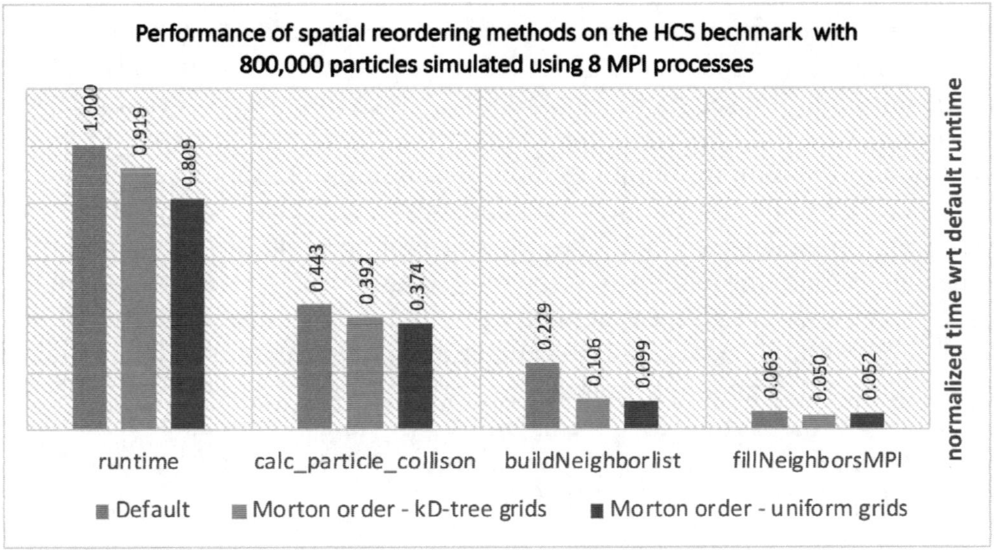

Figure 1.2 Performance impact of spatial reordering techniques on the HCS benchmark simulated using 8 MPI processes.

602178 Particle Model Development and Validation for the Conversion of Boehmite to γ/α-Alumina Driven By Concentrating Light

Abstract:

Solar energy conversion to chemicals and fuels receives progressively more attention. Many of the possible conversion routes incorporate particles or could be transformed into particle-phase reactions. When the particles are diluted into a fluid stream, usually a carrier inert gas or a reacting gas mixture, belong to the large family of aerosol processes. An aerosol process is the central theme of this presentation; particularly, one where solid particles are directly irradiated with concentrated light.

Aerosol phase reactors with a directly irradiated reaction zone can employ direct solar power with potentially higher solar to chemical energy efficiencies than the indirect solar reactors. Many studies investigate particle reactors such as fluidization and moving, but only a few of them focus on aerosol reactors. On the other hand, aerosol phase processes are very similar to solid fuel combustion and gasification, and free-board reactions. Therefore, in this study, we investigated the particle phase reactions of aluminum hydroxides (boehmite or gibbsite) converted to γ- and a-alumina. Two aluminum oxides favored for their unique properties in industrial catalysis and refractory ceramics, respectively.

The aluminum hydroxides as dry particles or water-based solutions have been exposed to various dosages of concentrating light, i.e., variable irradiance levels and exposure duration. The experiments have been conducted at our high flux solar simulator facility, which utilizes six Xe-arc lamps of 6KWe each, able to achieve heat fluxes of more than 5,000 kW/m2. The calcined and irradiated particles have been characterized with the combined use of scanning electronic microscopy (SEM), differential scanning calorimetry (DSC), thermogravimetric analysis (TGA), X-ray diffractometry (XRD), and attenuated total reflectance (ATR). The experimental observations were employed to develop a mathematical model and the results to fit the model parameters. After the successful verification and validation of the model, it was applied to predict the transient evolution of the particles exposed to concentrating light. Overall, the numerical model can predict the particle phase reactions accurately. However, the numerical predictions are very sensitive to the irradiance levels. In other words, the predictions of the particle model can vary significantly depending on the assumed irradiance of the concentrating light, which implies that further work is essential in order to estimate the overall solar-to-chemical energy conversion efficiency.

616737 Advancing the Science of Fluid-Particle Interactions and Transformations in Subsurface Environments and Engineered Processes for a Low Carbon Future

Abstract:

This talk is presented by the recipient of the SABIC Young Professional in Particle Technology Award.

Developing a portfolio of resource aware technologies for greenhouse gas mitigation, utilization, and removal and the utilization of low-value residues to generate value-added materials is essential for realizing our societal goal of achieving a low carbon future. Towards this end, advancing the science of fluid-particle interactions and transformations in subsurface environments and engineered processes and harnessing this understanding to develop novel and scalable technologies is essential. In this context, we will discuss the role of emerging understanding of the organization and transport behavior of nanoconfined fluids as it relates to the storage and utilization of compressed fluids such as CO_2 and CH_4 in geologic environments. The development of novel fluids to enhance permeability in subsurface environments for heat mining and chemical pathways to limit silica scales in geologic reservoirs will be presented. To meet our energy needs in a sustainable manner, we will discuss multi-phase chemical pathways for producing hydrogen with inherent carbon removal and hybrid absorption-crystallization pathways for CO_2 capture and removal, with the potential for utilization. The role of naturally occurring minerals, biomass, low value residues including alkaline residues in advancing these pathways and the potential to produce essential materials such as carbonates for construction applications and battery materials will be presented.

Dr. Greeshma Gadikota is an Assistant Professor and Croll Sesquicentennial Fellow in the School of Civil and Environmental Engineering at Cornell University. Dr. Gadikota directs the Sustainable Energy and Resource Recovery Group. Prior to Cornell, she served on the faculty at the University of Wisconsin – Madison, held postdoctoral research associate appointments at Princeton University and Columbia University, and a research associate appointment at the National Institute of Standards and Technology (NIST). Her PhD in Chemical Engineering and MS degrees in Chemical Engineering and Operations Research are from Columbia University. Her BS in Chemical Engineering is from Michigan State University. She is a recipient of the DOE CAREER Award, AICHE SABIC Award for Young Professionals from the Particle Technology Forum, an invited participant in the NAE Frontiers of Engineering, invited speaker at the NAE German-American Frontiers of Engineering Symposium, and was recognized as a Scialog Fellow in Negative Emissions Science.

601405 Behavior and Properties of Bubble-Free Fluidization

Abstract:

Recently, McLaren et al. (*PNAS*, 2019) reported gravitational instabilities occurring in gas-fluidized beds consisting of light and heavy particles vibrated vertically so as to enable particle motion without gas bubbles rising through the system. This "bubble-free fluidization" state is somewhat different from the bubbling fluidization and homogeneous fluidization states conventionally seen in gas-fluidized beds, since bubbles do not drive particle convection and the bed volume does not expand significantly beyond packed bed conditions. Here, we investigate the convective, diffusive and rheological properties of this bubble-free fluidization state using imaging experiments and computational modeling.

602261 Simulating the Formation of Granular Jets

Abstract:

This study focuses on simulating the formation and growth of granular jets under the impact of a heavy intruder on a bed of small particles. A discrete particle method to track the motion of 150 million individual particles in the bed, coupled with computational fluid dynamics to model the motion of the surrounding air, are used to obtain these predictions. The numerical results agreed with experimental observation that the height of jet diminishes from atmospheric to vacuum conditions. This is the first time the formation and growth of a tall granular jet is predicted numerically. Besides solving an unsolved problem, this study seeks to further explain the mechanisms behind the formation of a granular jet because numerical simulations allow detailed analysis of the flow that may be difficult to acquire with experimentation alone.

Reference: https://doi.org/10.1021/acs.iecr.0c00808

602580 Experimental Investigation of the Existence of the Intermediate Regime in an Aerated-Bed Couette Rheometer

Abstract:

Granular materials represent a large percentage of the materials handled and produced in the chemical and process industries. Such materials are widely used in powder-based unit operations, such as granulation, drying and fluid catalytic processes. Understanding the flow behaviour of those materials is of critical importance as it will help industrial practitioners handle and produce particulates in an efficient and less costly way. Granular materials are widely studied in the quasi-static (slow) and inertial (rapid) regimes. In the slow regime the particle's stress is independent of the applied shear rate. This regime is characterised by the formation, rotation and breakage of particles force chains. The rapid regime is characterised by instantaneous and binary collisions between particles. A third intermediate regime, which develops between the slow and rapid regimes, has recently become the subject of interest of many numerical and experimental studies. In this regime the particle's shear stress becomes shear rate dependent as observed in aerated bed configurations [1] when the gas velocity is lower than the minimum fluidisation velocity, U_{mf}.

The focus of this work is to confirm the existence of the intermediate regime in an aerated virtual Couette rheometer previously developed in our lab [2]. This device is a combination of the FT4 Freeman aeration kit with a new 3D printed cylindrical bladed cell (6 blades). This cell is 70 mm in height and 36 mm in diameter. The experimental procedure is as follows: the cell moves downward through the material placed in the FT4 cup until it reaches a fixed position. The cell is then rotated at different rotational speeds, and the torque, needed to move the aerated powder placed in the cup, is evaluated. The shear stress, *t*, and shear rate, , are computed from the recorded torque and the rotational speed, respectively. This new configuration was calibrated without aeration using a Newtonian fluid with a known viscosity. It was also tested with aerated and non-aerated glass beads powders in the quasi-static regime [2].

In this study, we expand the results previously obtained [2] to values of the fluidisation ratio U/U_{mf} below the minimum fluidisation limit (with U being the aeration velocity). Experiments were performed on an aerated glass bead powder with a particle diameter of 0.08 mm and an aeration ratio ranging from 0 to 0.8; the shear rate was increased from 5 to 42 s^{-1}. The first results showed that, for an aeration ratio lower than 0.6, the shear stress is independent of the shear rate, indicating that the studied powder is in the quasi static regime. For U/U_{mf} higher than 0.6, the shear stress was found to be independent of the shear rate for shear rates values lower than 25 s^{-1}. For higher values of the shear rate, the shear stress was found to increase proportionally with the increasing shear rate, which indicates that the studied sample is no longer in the quasi-static regime. These first observations confirm the existence of the intermediate regime in an aerated bed. Further experiments including different particle diameters will be performed to confirm and expand the first measurements and observations.

References

[1] G. I. Tardos, S. McNamara, and I. Talu, 'Slow and intermediate flow of a frictional bulk powder in the Couette geometry', *Powder Technol.*, vol. 131, no. 1, pp. 23–39, Mar. 2003, doi: 10.1016/S0032-5910(02)00315-7.

[2] L. A. A. Yahia, T. M. Piepke, R. Barrett, A. Ozel, and R. Ocone, 'Development of a Virtual Couette Rheometer for Aerated Granular Material', *AIChE J.*, vol. n/a, no. n/a, p. e16945, doi: 10.1002/aic.16945.

603195 Stochastic Drag Formulations for Particle-Laden Flows

Abstract:

Existing drag force closures for volume-filtered Eulerian-Lagrangian (EL) methods typically capture the mean hydrodynamic force experienced by an assembly of particles. As a result, sources to granular temperature and particle dispersion, arising from neighbor-induced drag force fluctuations, are not captured. This study provides a detailed account of stochastic approaches that may be utilized in EL simulations to account for a distribution of drag forces. The frameworks examined here correspond to Langevin equations for the particle position (PL), particle velocity (VL), and fluctuating drag force (FL). Rigorous derivations of the particle velocity variance and dispersion resulting from each method are presented. The solutions derived herein provide a basis for comparison with particle-resolved direct numerical simulation. The FL method allows for the most complex behavior, enabling control of both the granular temperature and dispersion. A Stokes number is defined that relates the Stokes response time to the integral time scale of the fluctuating force. Formal convergence of the FL and VL scheme is shown for large fluctuating force Stokes numbers -- i.e. rapid uncorrelated drag force fluctuations. In the opposite limit, the fluctuating drag forces are highly inertial and the FL scheme departs significantly from the VL scheme.

607145 Controlled Mixing in a Dynamically Structured, Pulsed Fluidized Bed

Abstract:

Vibration and pulsation of the inlet flow are two ways to enhance the performance of fluidized bed reactors and dryers by homogenizing mixing *[1]*. Furthermore, these perturbations could create a coherent, and sometimes responsive, macroscopic flow structure. However, mixing and solid-gas-contact patterns in such structured systems are still poorly known. In this work we discuss the circulation and mixing behavior in dynamically structured, pulsed beds, and compare them with unstructured pulsed beds and traditionally operated beds, without the use of pulsation. There is scant knowledge on lateral particle mixing in pulsed or vibrated beds. Pulsation and vibration both improve flowability by breaking up long force chains and clusters that restrict flow; therefore, one expects dispersion to improve under an oscillatory flow with the same average superficial velocity. However, a dynamically structured bubbling bed emerges in a different regime. The bed fluctuates around the minimum fluidization conditions and the response to cycles of fluidization and collapse results in a denser system than both the operation at constant flow or a traditional (unstructured) use of pulsation. In an ordered state, bubbles self-organize into a regular triangular lattice *[2]*, but it is unclear how the lattice drives overall solid circulation, mixing and gas-solid contact.

In this contribution we study structured and unstructured bubbling beds of Geldart B particles for various oscillating frequencies experimentally (Figure 1a), and quantify the associated degree of mixing with an effective lateral dispersion coefficient extracted from CFD-DEM simulations, validated in a quasi-2D system (Figure 1b). The mixing and solid circulation in a dynamically structured system exhibits major quantitative and qualitative differences. The organization of bubbles into a pattern contains the solids into a series of mixing compartments that rearrange every pulse. Unlike in a traditional operation, solid dispersion can no longer be described as a diffusive process. It is dominated by advection. A constant exchange rate between mixing compartments establishes a long-range lateral circulation, which depends on the bed dimensions. Solid mixing remains confined to a single type of loop fixed around each compartment and is superimposed to short-range velocity fluctuations caused by the oscillation. The scale of the mixing loops is constant and independent of the bed dimensions (Figure 1c), but it can be externally controlled by the amplitude and frequency of the gas flow oscillations. Here, we report exchange rates between the phases and across compartments, along with the values for an effective lateral dispersion coefficient in dynamically structured beds of Geldart B particles. This information can be used to inform surrogate models and enable the design and scale-up of larger devices that, besides homogenization, present unique advantages, namely scalability and external control. By modulating changes in gas velocity, one can fix the volume of the mixing compartments and, via external control, tailor transport rates and the stresses sustained by the powder. The ability to mix at very low gas velocities in a homogeneous and controllable way opens new ways to tackle heat and mass transfer management, as well as keep elutriation and attrition in check, which are particularly relevant issues for highly exothermic processes, or thermally sensitive and high-value materials.

***Figure 1** – Dynamically structured pulsed fluidized bed. (a) Bubble organization, snapshots at consecutive pulses (b) CFD-DEM simulations. (c) Dispersion of a tracer from the central compartment; effective dispersion coefficient and mixing pattern.*

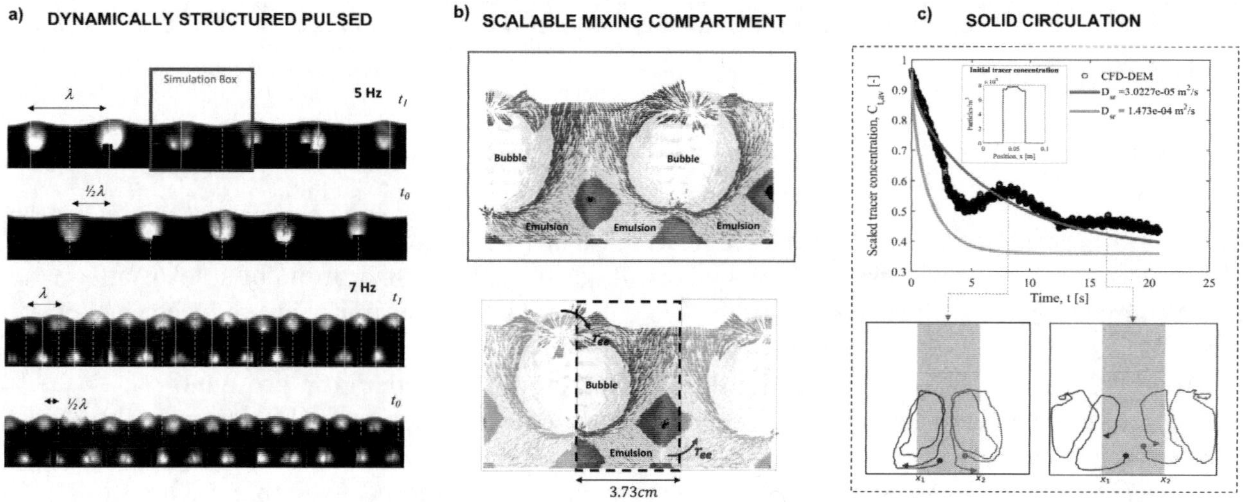

610453 Combined Real-Time Magnetic Resonance Imaging and Numerical Simulation of Gas Bubble Splitting and Coalescence in Fluidized Beds

Abstract:

Gas bubbles in fluidized bed reactors influence heat, mass and momentum transfer between the gas and solid phases and in turn the design of the reactor. In particular, the size of the bubbles is an important parameter, since an increasing bubble size enhances particle mixing, but reduces gas-solid contacting. A precise control of the bubble size distribution is therefore crucial to ensure process continuity and the desired performance of the fluidized reactor. This demands a profound understanding of the mechanisms that lead to bubble splitting and coalescence. However, real-time data of these processes are difficult to acquire in a three dimensional bed due to the opaque nature of granular matter. Magnetic resonance imaging (MRI) has been shown to be a suitable method for the non-invasive measurement of solid fraction and particle velocity.

We have recently implemented MRI acceleration techniques to speed up data acquisition and enable the real-time investigation of granular dynamics *(1)*. In this work, we study the transient splitting and coalescence of gas bubbles in a variety of fluidized bed geometries (i.e. with and without inserts) by means of combined MRI measurements and numerical simulations. Single gas bubbles are injected through a central orifice at the bottom of the bed. These MRI measurements are complemented by numerical simulations using computational fluid dynamics simulation coupled with a discrete element method (CFD-DEM) *(2)*.

Both the numerical simulations and the MRI measurements of the bubble dynamics show very similar dynamics during bubble splitting and coalescence. Additionally, the simulations give insight into the solid motion at the grain level and provide information on the gas flow in and around the bubbles. Combing our experimental and numerical results allowed us to draw conclusions concerning bubble stability and the mechanism that control bubble splitting and coalescence. We are confident that our findings contribute to obtaining a better understanding of the complex interplay between the gas and solid phase in fluidized beds.

1. **Penn, A., *et al.***, "Real-time probing of granular dynamics with magnetic resonance," *Science Advances,* **3** (9), pp. 1-7 (Sep 2017).
2. **Kloss, C., *et al.***, "Models, algorithms and validation for opensource DEM and CFD-DEM," *Progress in Computational Fluid Dynamics,* **12** (2-3), pp. 140-152 (2012).

611233 Effects of Adsorbates on Thermal Transport in Metal-Organic Frameworks

Abstract:

Fast heat transfer during gas adsorption/separations in metal-organic frameworks (MOFs) plays a key role in improving their efficiency for gas storage/capture. Despite its importance, heat transfer in MOFs is not well understood, and whether it is helped or hindered by the presence of adsorbates is an open question. Our computational and experimental research indicate that depending on the nature of guest-framework interaction, adsorbates can increase or decrease thermal conductivity. The presence of physisorbed guests introduces new phonon scattering channels resulting from gas-framework collisions, which serve to reduce thermal conductivity. On the other hand, enhancement in thermal conductivity is observed for e-2–Zn_2(dobpdc) (e-2 = *N*-ethylethylenediamine; dobpdc^{4-} = 4,4′-dioxidobiphenyl-3,3′-dicarboxylate) in the presence of chemisorbed CO_2.

599707 Flare Gas – Carbon Dioxide Assisted Synergistic Biomass Reforming for Hydrogen Rich Syngas Production on Self-Regenerable Carbon Nanofiber (CNF) Supported Fe – Mo_2c Catalyst

Abstract:

Biomass reforming for hydrogen rich syngas production was achieved through synergistic co-processing with flare gas and carbon dioxide over Fe-Mo_2C catalyst supported on carbon nanofiber (CNF). Methane – carbon dioxide activated biomass gasification can produce syngas with H_2:CO of 2 to 4 with H_2 product concentration ranging from 40 to 70 vol.%. Hydrogen rich syngas was generated over 15 cycles on Fe-Mo_2C catalyst at temperatures of 800°C, 850°C, and 900°C with varying Fe loading of 0.5 to 5 wt.% and Mo_2C concentration of 4 wt.%. CNF supported catalyst subjected to multiple cycles showed excellent resistance to deactivation and was discovered to be self-regenerable. The amount of iron loading was instrumental in the nature of support synthesized from pyrolysis at 700°C. Results showed that 0.5Fe catalyst led to synthesis of graphene supported nanoclusters of Fe and β-Mo_2C, whereas 2.5% Fe and 5% Fe loading materialized into CNF / CNT supported nanoparticles of Fe, β-Mo_2C. Conversion and gas yield nosedived to < 20% in 2 of 15 cycles on Fe-Mo_2C-CNF catalyst. These regeneration cycles possibly indicate nucleation or saturation of graphene layer synthesis or CNF growth. Nucleation of regeneration occurred mostly in cycle 2 and continued until saturated regeneration cycle varying in occurrence from subsequent cycles. Self-regeneration was the effect of free-C on the surface and complete conversion of Fe^0 active sites to Fe_3C / $Fe_3C_{1.5}$ active sites. Once elemental Fe was converted to Fe_3C_x, free-C led to 2D graphene layer formation on 2.5% Fe and 5% Fe catalysts. On graphene supported 0.5% Fe catalysts, CNF growth was observed on Fe_3C. Mo_2C active sites were much stable and did not participate in catalyst regeneration process. Self-regeneration mechanism for the CNF supported catalyst was proposed incorporating carbon from lignin oxygenates after being hydrodeoxygenated on β-Mo_2C in the framework of original mechanism from methane reforming perspective.

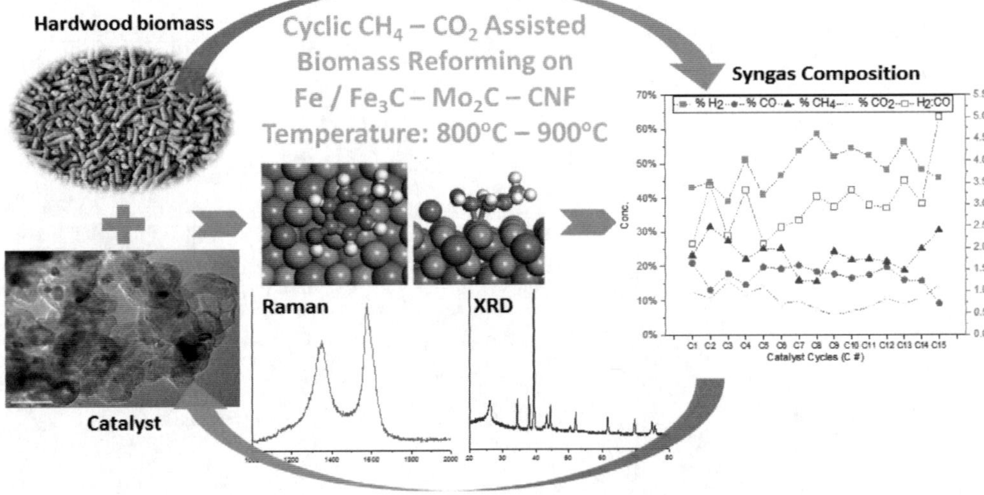

600613 A Sustainable and Scalable Technology to Produce High Performance Mesoporous Silicon for Lithium Ion Batteries

Abstract:

This presentation will provide an overview of our recent efforts on developing a new technology for the sustainable and scalable manufacturing of mesoporous silicon for the application of lithium ion batteries (LIBs). Silicon has the potential as an anode material to increase lithium-ion cell capacity but the associated volume change during lithiation leads to electrode degradation and loss of capacity during cycling. While mesoporous morphologies can potentially address these issues, magnesiothermic reduction (MgTR) of silica to silicon is currently the only method for potential bulk manufacturing. However, due to a distinct lack of mechanistic understanding of this process, predictive and sustainable design of mesoporous silicon is not possible [*J. Mater. Chem. A*, 2018, **6**, 18344].

To better understand the processes, we have systematically studied the process conditions using a model silica system [*J. Mater. Chem. A*, 2020, **8**, 4938]. Further, we performed a complete characterisation of both the reactant and product structures in order to map how reaction conditions affect the properties of the electrode materials and provide a processing-structure-property-performance relationship. Using this knowledge, we demonstrated the design of high performance silicon using any given silica feedstock.

Next, in order to make the process scalable and sustainable, we used three strategies.

(1) We used bioinspired silica (BIS) as a sustainable, economical and scalable feedstock [PCT/GB2016/052705] and the battery performance of porous silicon structures thus produced showed that they were superior compared to other silica sources.

(2) Decreasing the MgTR reaction temperature to make it more sustainable compromises the yield due to an insufficient drive to overcome the activation energy barrier. We addressed this challenge by reducing the particle size of the feedstock to the nanoscale such that their high surface energy could help overcome the energy barrier. We obtained high yields of porous silicon at temperatures previously shown to be too low for the MgTR to even proceed. This phenomenon can be exploited in order to trigger the conversion of larger, commercial sources of silica at low temperatures to porous silicon with high yields.

(3) Scaling up MgTR has proven difficult since, at large scales, the thermal effects of the MgTR are significant and lead to sintering and loss of porosity. By carefully investigating the role of NaCl as a thermal moderator, we identified the critical factors for controlling scale-up. Trials at different scales showed that applying this knowledge of the thermal moderators, we were able to maintain product properties at larger scales.

Overall, we report the first process for the predictive design of high-performance mesoporous silicon for LIBs such that the method is scalable and sustainable. Our results have established pathways for scaling-up MgTR method such that it can now be taken forward to target specific porous silicon properties.

607909 Modular Fluidic Microreactor for Fully Decoupled Precursor Mixing and Reaction Times in Mechanistic Studies of Metal Halide Perovskite Quantum Dot Synthesis

Abstract:

Metal halide perovskite quantum dots are a rapidly growing class of nanomaterials which have sparked significant interest for their broad reaching applications spanning from photocatalysis to optoelectronic devices.[1] Despite the high value of metal halide perovskite QDs, their synthesis optimization and control have been mostly limited to inefficient and inconsistent flask-based processes. Such batch synthesis reactors consume a large volume of often high-cost materials, require extensive manual labor, and possess inconsistent heat and mass transfer properties when moving between experiments and from flask to flask. These issues impair the accuracy and reduce the effectiveness of existing research efforts. In response, recently microscale flow synthesis strategies have been developed towards more efficient and systematic studies of perovskite quantum dots.[2–4] Among these flow synthesis platforms, fluoropolymer tubing-based microreactors are another quickly rising area of research, as they offer low-cost, fully modular, and commercially available microreactor components with quick assembly time as compared to conventional chip-based systems. These fluidic platforms offer micro-liter scale reactions volumes with automated, high-throughput, and non-invasive measurement strategies.[5]

However, as a property inherent to continuous flow processes, reaction time and mixing rates are strongly coupled to the total fluid velocity moving along the flow direction. In this work, we have developed a microscale fluidic microprocessor strategy which can independently control mixing times from 53 ms to 7.3 s with an arbitrarily long residence (*i.e.*, reaction) time. This microscale flow synthesis approach was accomplished by first developing and analyzing a fluoropolymer-based passive micromixer developed using commercially available tubing. We then integrate the passive micromixer module with an automated, material-efficient fluidic sampling module, which allows for two independent flow rates to run simultaneously within the platform. Utilizing the developed modular fluidic platform, "samples" with a specific chemical formulation are pulled from the mixing line and passed to the reaction/analysis line where they continue to react after forming homogenous reactive phase plugs. These plugs move through the flow reactor channel towards a spectral monitoring module for *in-situ* UV-Vis absorption and photoluminescence measurements.

In the next step, we applied the developed modular flow synthesis platform towards the room-temperature synthesis of cesium-lead-bromide perovskite quantum dots [6] and unveiled a mechanistic understanding of the quantum dot growth process through time-resolved spectral sampling. We then conducted a full compositional analysis coupled with variations in the precursor mixing time to attain highly tunable emission colors as well as quantum dot size distributions, quantum yields, reaction yields, and numbers of quantum dots formed. These experiments were coupled with a first principle mesoscopic kinetic Monte Carlo simulation to provide further insights into the mechanism behind the ligand-assisted reprecipitation process at early time scales. The strategies and models developed in this work offer a greater understanding of controlled synthesis of perovskite quantum dots with fast formation kinetics.

References

[1] Q. A. Akkerman, G. Rainò, M. V. Kovalenko, L. Manna, *Nat. Mater.* **2018**, *17*, 394.

[2] K. Abdel-Latif, R. W. Epps, C. B. Kerr, C. M. Papa, F. N. Castellano, M. Abolhasani, *Adv. Funct. Mater.* **2019**, *29*, 1900712.

[3] R. W. Epps, K. C. Felton, C. W. Coley, M. Abolhasani, *Lab Chip* **2017**, *17*, 4040.

[4] L. Bezinge, R. M. Maceiczyk, I. Lignos, M. V. Kovalenko, A. J. deMello, *ACS Appl. Mater. Interfaces* **2018**, *10*, 18869.

[5] S. Marre, K. F. Jensen, *Chem. Soc. Rev.* **2010**, *39*, 1183.

[6] S. Wei, Y. Yang, X. Kang, L. Wang, L. Huang, D. Pan, *Chem. Comm.* **2016**, *52*, 7265.

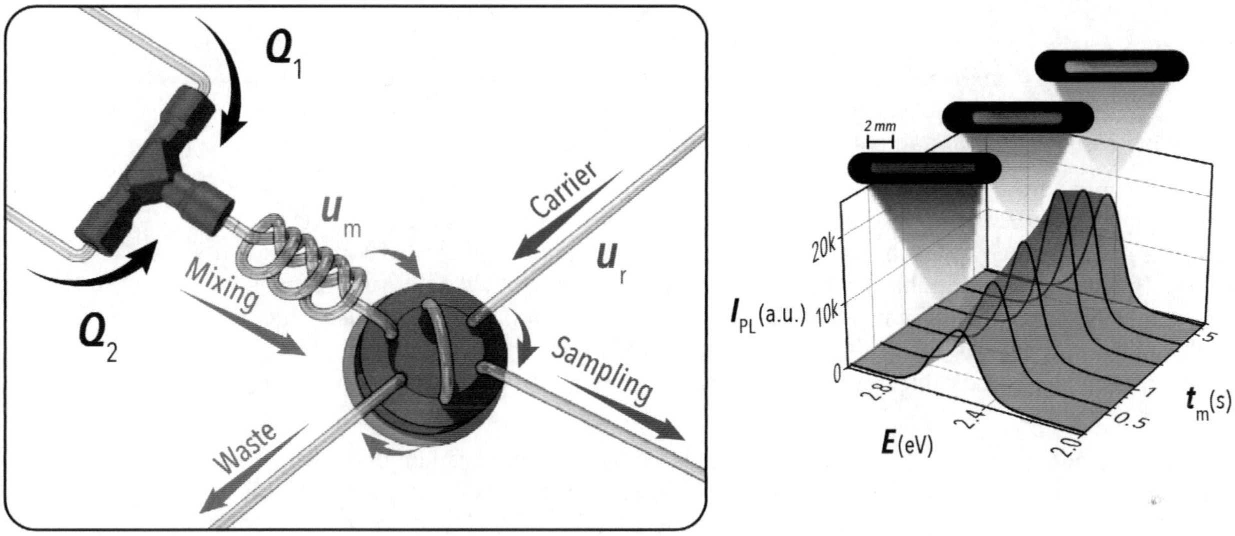

610022 Scalable Biomineralization of AgInZnS Quantum Dots for Photocatalytic Hydrogen Generation

Abstract:

Cadmium-based metal chalcogenide quantum dots are promising semiconductors capable of meeting the demands for photocatalytic H_2 generation, yet their intrinsic toxicity has motivated the search for high-performance, non-toxic alternatives.[1–3] The low toxicity, suitable band gaps, high extinction coefficients, and long photoluminescence lifetimes characteristic of ternary I-III-VI QDs makes them promising candidates especially when Zn is also incorporated to improve quantum yield (QY) and passivate the surface. One particularly promising composition is AgInZnS, owing to the tunability of its band gap (1.8-2.4 eV) to within the range suitable for photocatalytic H_2 generation.[4] However, the low compositional toxicity of AgInZnS is undercut by the high-temperatures and toxic solvents conventionally employed for its synthesis.[4,5]

Biomineralization-based routes to size-controlled QDs offer an alternative, 'green', and potentially scalable synthesis strategy, characterized by low temperatures and aqueous phase processing. Biomineralization of size-tunable CdS QDs, for example, is possible by the aqueous-phase, low-temperature enzymatic (*cystathionine gamma lyase*) turnover of the amino acid L-cysteine to H_2S in buffered solutions of cadmium acetate.[6] In this study, we expand the scalability and versatility of QD biomineralization by 1) establishing strategies for cyclic QD synthesis with facile enzyme recovery and reuse, and 2) elucidating mechanistics insight into single-enzyme synthesis of compositionally complex QDs. Specifically, we establish facile enzyme immobilization strategies on inexpensive, easily recoverable supports that offer sufficient enzyme activity to support continuous cyclic synthesis of QDs with narrow particle size distribution. This facile strategy for immobilized enzyme-mediated QD synthesis opens exciting possibilities for the scalable, 'green', biosynthesis of consistently sized QDs.

We then expand the compositional diversity of the single-enzyme biomineralization approach to AgInZnS QDs. In order to elucidate the AgInZnS formation mechanism, we carry out both one pot and sequential syntheses that identify critical In/Cys ratios affecting photoluminescence, offer insight into the role of Ag content in nucleation and growth, and begin to elucidate In-S complexation and the role of Ag ion exchange in tuning photoluminescence emission wavelength. We will demonstrate synthesis-structure-function relations in the context of photocatalytic hydrogen generation rates. Ultimately, this study advances the scalability and versatility of single-enzyme biomineralization as a means for 'green' synthesis of various non-toxic semiconductors.

References

1. Rao, V. N. *et al.* Sustainable hydrogen production for the greener environment by quantum dots-based efficient photocatalysts: A review. *Journal of Environmental Management* (2019) doi:10.1016/j.jenvman.2019.07.017.
2. Liu, Y. *et al.* One-step aqueous synthesis of highly luminescent hydrophilic AgInZnS quantum dots. *J. Lumin.*(2018) doi:10.1016/j.jlumin.2018.05.040.
3. Zang, Z. *et al.* Tunable photoluminescence of water-soluble AgInZnS–graphene oxide (GO) nanocomposites and their application in-vivo bioimaging. *Sensors Actuators, B Chem.* (2017) doi:10.1016/j.snb.2017.07.144.
4. Tang, X. *et al.* Nanocomposites of AgInZnS and graphene nanoshee photocatalysts for hydrogen evolution. *Nanoscale* (2015) doi:10.1039/c5nr05145b.
5. Lin, P. C., Wang, P. Y., Li, Y. Y., Hua, C. C. & Lee, T. C. Enhanced photocatalytic hydrogen production over In-rich (Ag-In-Zn)S particles. *Int. J. Hydrogen Energy* (2013) doi:10.1016/j.ijhydene.2013.04.125.
6. Spangler, L. C., Cline, J. P., Kiely, C. J. & McIntosh, S. Low temperature aqueous synthesis of size-controlled nanocrystals through size focusing: a quantum dot biomineralization case study. *Nanoscale* **10**, 20785–20795 (2018).

610583 Solution Behavior of Cellulose Nanocrystals in Polydisperse Polymer Solutions

Abstract:

The fabrication of cellulose nanocrystals (CNCs) into high-performance materials has gained significant attention as promising example of a renewable, sustainable, and inexpensive resource for advanced applications. Included in their innate "green" characteristics is a list of desirable chemical and mechanical properties: thermal stability, facile functionalization, and high theoretical tensile strength, that have been shown to translate well into macroscopic properties when used as a nanocomposite. All CNC composite materials require a "in-solution" processing step where solution interactions dictate the aggregation, interfacial, and liquid crystalline behavior. However, elucidating solution behavior is non-trivial and poorly reported for complex systems. We characterize the solution behavior of acetylated CNCs under varying solvent conditions with the addition of a polydisperse technical lignin. Lignin is a branching, heterogenous, biopolymer and provides useful insight into the robustness of CNC composite systems as we evaluate aggregation effects across the whole spectrum of loading conditions. Dilute solutions are characterized using Multi-Angle Light Scattering (MALS) and Dynamic Light Scattering (DLS) and it was observed that the addition of CNCs into lignin solutions improved dispersion and rheological properties for carbon fiber spinning. Included in the analysis are SANS experiments that show distinct conformational behavior for lignin as well as cross polarized light microscopy where we observe disperse solutions. Results show promise for using CNCs as a means for tuning complex phase behavior in multi-component polymeric solutions.

595940 Infrared Temperature Experiments and DEM Simulations of Heat Transfer in an Agitated Dryer

Abstract:

Drying is typically one of the last steps of the manufacturing process for an active pharmaceutical ingredient (API). After crystallization and filtration, the wet bed of API needs to be dried into a powder with a desired moisture content before it can be mixed with other excipients and formulated into a drug product. A common method used by the pharmaceutical industry to carry out API drying is agitated drying, where the API is heated inside a cylindrical vessel while being mixed by an impeller to help improve temperature and moisture uniformity in the particle bed. Agitated drying can be one of the most complicated and delicate phases of the process, and it sometimes remains poorly understood and difficult to optimize. Simultaneous and transient changes in heat transfer, mass transfer, flow, and physicochemical properties of the material during drying make it challenging to thoroughly understand the system. Enhancing fundamental knowledge about the process could be key to enabling the design of efficient and robust operating protocols.

In this work, we isolate the heat transfer element of drying and focus on studying it on its own. The idea is that if we can decouple the different facets of drying and understand them on their own first, we can eventually study their combined interactions more easily. We conduct heat transfer experiments using a laboratory scale agitated dryer and an infrared camera to measure the temperature of the surface of a particle bed over time. The temperature data collected by the thermal camera is used to compute the heat transfer coefficient for the system as well as the temperature standard deviation, giving information about both the mean temperature of the bed and its temperature uniformity. Using this methodology, we investigate how operating conditions, such as the impeller rotation rate, influence heat transfer. The experiments are carried out using glass beads as well as more pharmaceutically relevant materials such as citric acid. Next, we present a comparison between the experimental results and discrete element method (DEM) simulations. The model makes use of a particle-particle and particle-wall conductive heat transfer model to describe the system. Overall, despite the simplicity of the heat transfer model, we find a fairly good agreement between the simulations and the experiments. This suggests that the model can serve as a good starting point towards better understanding how the rate of heat transfer and the temperature uniformity of a particle bed can be optimized in an agitated dryer.

596506 Consequence Prediction for Dust Explosions Involving Interconnected Vessels Using Computational Fluid Dynamics Modeling

Abstract:

Combustible dust explosions continue to present a significant threat toward operating personnel and pneumatic conveyance equipment in a wide variety of processing industries. Following ignition of suspended fuel within a primary enclosure volume, propagation of flame and pressure fronts toward upstream or downstream interconnected enclosures can result in devastating secondary explosions if not impeded through an appropriate isolation mechanism. In such an occurrence, accelerated flame front results in flame jet ignition within the secondary vessel, greatly increasing the overall explosion severity. Unlike an isolated deflagration event with quantifiable reduced pressures (vent sizing according to NFPA 68 guidance), oscillation of pressure between primary and secondary process vessels leads to significant overpressure. Dependent on details of the application such as relative enclosure volumes, relief area, fuel type, suspended concentration, duct size, and duct length, the maximum system pressure in both interconnected vessels can be unpredictable. This study proposes the use of FLame ACceleration Simulator (FLACS) computational fluid dynamics (CFD) modeling to provide reliable consequence predictions for specific case scenarios of dust deflagrations involving interconnected equipment. Supplements toward the minimum calculated required relief area (A_v) were determined through iterative simulation, allowing for reduced pressures (P_{red}) to be maintained below theoretical enclosure design strengths (P_{es}).

596511 Effect of Particle Morphology on Metal Dust Deflagration Severity and Explosion Protection Design

Abstract:

Combustible dust explosions continue to present a significant threat toward industries processing, storing, or pneumatically conveying metal dust hazards. Through recent years, investigations have thoroughly observed the influence of particle size, polydispersity, and chemical composition on dust explosion sensitivity and severity. However, studies characterizing the effect of particle shape (or morphology) on metal dust explosibility are limited and merit further consideration. In this work, three pure aluminum dust samples of varying particle morphology were examined (spherical granular, irregular granular, and flake). To maintain consistency in results obtained, all granular samples were prepared with similar particle size distribution and dispersity. In the case of flake morphology, particle mass was employed as the control parameter rather than particle size due to the obscure shape relative to granular dust samples. Investigations were performed in a Siwek 20 L sphere combustion test vessel and provide a series of measurements of minimum ignition energy (MIE), maximum rate of pressure rise ($\propto K_{St}$), maximum overpressure (P_{max}), and peak combustion temperature as a function of fuel concentration for all morphologies considered. This work aims to apply scanning electron microscopy (SEM) as a means of correlating specific surface area to dust cloud flame propagation dynamics, with the objective of proposing recommendations to standard risk hazard analysis procedures and to existing metal dust explosion mitigation design guidance.

600733 Characterisation of Powder Triboelectric Properties to Prevent Dust Explosions

Abstract:

When two particles are colliding or sliding on each other, electric charges could be exchanged at the surfaces. This contact electrification called triboelectric effect is an old fundamental scientific subject. However, despite many studies dedicated to this subject, the fundamental mechanisms behind the triboelectric effect are not fully understood. In particular, the charging of objects composed by the same material and charging of fine powders need deeper investigations. Even the basic question related to the nature of the transferred charges (electrons, ions or material) is still debated. The difficulties are related to a lack of fundamental knowledge but also to the difficulty to perform measurements when associating the complexities of particulate systems and electrostatics.

In dusty environment, electrostatic discharges might lead to fire and explosions. According to [1] one dust explosion occurs each day in Germany and every tenth explosion is caused by static electricity. According to [2], 70% of 153 industrial accidents that occurred in Japan over the past 50 years were also attributable to static electricity. Therefore, a better knowledge of the physical mechanisms and the existence of improved measurement techniques could have huge positive consequences.

During the last years, we developed a measurement method able to measure the ability of a powder to electrostatically charge during a flow [3]. The instrument called GranuCharge and commercialized by GranuTools is based on this method. This measurement allows to predict the chargeability of grains from a laboratory measurement and therefore to estimate the probability to obtain dust explosion inside a production plant. Moreover, a series of measurements and also numerical simulations have been performed to better understand the fundamental mechanisms leading to triboelectric charge build-up in particulate materials leading potentially to dust explosion.

[1] M. Glor, Ignition hazard due to static electricity in particulate processes, Powder Technology 135, 223 (2003).

[2] A. Ohsawa, Statistical analysis of fires and explosions attributed to static electricity over the last 50 years in Japanese industry, Journal of Physics: Conference Series 301, 012033 (2011).

[3] A. Rescaglio, F. De Smet, L. Aerts, & G. Lumay, Tribo-electrification of pharmaceutical powder blends, Particulate Science and Technology 37, 1024 (2019)

608712 Assessment of Powder Tribo-Charging during Twin-Screw Feeding Operations and Its Sensitivity to Environmental Conditions

Abstract:

INTRODUCTION

The tribo-charging effects during pharmaceutical manufacturing can significantly impact process performance and final product quality. During twin-screw feeding operations, powder tribo-charging can cause particle adhesion to the feeder hopper and at the screw outlet, thus affecting feeding accuracy and consistency[1,2]. Pharmaceutical powders can get charged during feeding processes by the numerous particle collisions and separations within the powder bed and with the feeder surface. The severity of the tribo-electrification depends on several factors, including environmental conditions and intrinsic material attributes of the contacting surfaces[3]. Relative humidity (RH) has been reported to be the environmental factor with a prominent impact on tribo-charging. However, the underlying mechanisms of its effect are not yet fully clear. For diverse material types, it has been reported that there is a specific RH threshold (per material type) after which a reduction in charge magnitude occurs[4]. This study investigates the effect of material-water interactions (induced by changes in RH) and their impact on the powder tribo-charging tendency. The investigated powders were additionally evaluated for their processability during twin-screw feeding with a particular focus on the tribo-charging effects occurring during feeding.

MATERIALS AND METHODS

Materials

Four pharmaceutical powders were investigated in this study: D-mannitol (MAN, Parteck®, Merck), microcrystalline cellulose (MCC, Avicel®, DuPont), micronized paracetamol (PMIC, Mallinkrodt) and magnesium stearate (MgSt, Parteck® LUB, Merck). For the excipients (i.e., MAN and MCC), two different grades were used (i.e., M100 and M200 for MAN; PH101 and PH200 for MCC), whereas for the API (i.e., PMIC) and the lubricant (i.e., MgSt) only one grade was considered.

Methods

All the powders were homogenized and sieved to break large agglomerates prior to the storage at ambient temperature (i.e., 22 ± 2 °C) in desiccators at three levels of RH (i.e., 23, 53 and 75% RH) for 48-72 h. The amount of water present in the samples was determined via loss-on-drying (LOD) and Karl Fischer (KF) titration. Particle size distribution (PSD) was analysed via laser diffraction for fine powders (i.e., MgSt and PMIC) and via dynamic image analysis for the coarse ones (i.e., MAN and MCC grades). Likewise, particle shape was determined via static and dynamic image analysis. The specific surface area (SSA) of the powders was measured via nitrogen physi-sorption and the powder bulk properties were characterized with the FT4 powder rheometer, equipped with different available accessories. For the evaluation of the powder charging tendency at the different RH levels, the GranuCharge™ apparatus (GranuTools, Belgium) was used. The electrostatic charge of the powders was measured after letting flow through the stainless-steel tubes. With regard to the twin-screw feeding experiments, a small-scale feeder (MT-S Hyg, Brabender, Germany) was used to alternately feed the material on a scale (i.e., for mass flow determination) and in a Faraday cup (i.e., for charge measurement). Feeding runs were performed at controlled environmental conditions (i.e., 22 ± 2 °C and 53 ± 4% RH) and the charge acquired after 5 and 12 min from the process start was measured. Statistical analysis was performed to identify relevant correlations between the material charging tendency and both environmental conditions and material attributes.

RESULTS

Powder characterization

Powders were characterized by distinct particle and bulk properties as well as different levels of hygroscopicity. Particularly, MCC and MAN powders consisted of considerably large particles characterized by high aspect ratios and narrow particle size distributions. In contrast, smaller particles with lower aspect ratios and broader distributions were observed for MgSt and PMIC. After storage of the powders at different RH levels, a significant increase in the water content was observed for the hygroscopic MCC grades. For the other non-hygroscopic materials, slight increases were noticed only at high RH (i.e., 75% RH). The increase in the moisture content of the materials at progressively higher RH induced changes in the material bulk properties. For instance, a consistent increase of wall friction angle (WFA) was observed for MAN powders from FT4 measurement, which was found to correlate with a progressively higher charge density measured.

Impact of RH on powder tribo-charging behavior

The measured charge was normalized by the specific surface area of the samples to allow an appropriate comparison among the materials. All the powders were found to charge negatively after the contact with stainless steel, however distinct charge magnitudes were obtained for the different materials at three levels of RH (Figure 1a). Overall, PMIC exhibited the highest propensity to charge, followed by MCC and MAN grades, whereas a minimal charging tendency was observed for MgSt. RH was found to impact the tribo-electrification of the materials to different extents and distinct tribo-charging trends per material type were observed, namely: i) MCC powders showed an increase in charge density from 23 to 53% RH, followed by a pronounced decrease at higher RH (i.e., 75% RH); ii) MAN grades showed a progressive increase in charge density with the increase in RH; iii) PMIC exhibited a decrease in tribo-charging propensity only at high RH; iv) MgSt had a slight decrease in charge density with increasing RH.

Powder tribo-charging during twin-screw feeding operations

All the powders, except MgSt, showed a negative charge polarity also during twin-screw feeding operation (Figure 1b). Moreover, for these materials charge values remained relatively constant between the 2 measurement times (i.e., 5 and 12 min from the process start). For MgSt, an initial negative charge was obtained, which decreased over time and turned positive during the second measurement. This observation was attributable to the particular lubrication mechanism of this material, which allows the material to form a lubrication layer on stainless-steel surfaces and consequently causes an alteration to particle contacts. When comparing the charge values measured after feeding with the one obtained from the GranuCharge™ device at the same RH (i.e., 53% RH), distinct trends were observed. Particularly, the charge density was found to be lower for the excipients (i.e., MAN and MCC grades), whereas PMIC showed to charge to a higher extent. This is in agreement with qualitative observations for this powder with a higher tendency to adhere to surfaces, such as feeder hopper, outlet and screws.

CONCLUSION

This study demonstrates that, depending on the material type, RH impacts powder tribo-charging behavior distinctly. The amount of water sorbed and its distribution depends on intrinsic material properties and thus, affects material bulk properties such as powder tribo-electrification to a different extent. Additionally, the type of unit operation considered can also impact the magnitude and polarity of the acquired charge by changes in the surface contact characteristics and contact mode. Both effects should therefore be considered during the formulation and process development to avoid possible risks in production lines.

REFERENCES

(1) Muehlenfeld, C.; Thommes, M. Small-Scale Twin-Screw Extrusion - Evaluation of Continuous Split Feeding. *J. Pharm. Pharmacol.* 2014, *66*.

(2) Stauffer, F.; Vanhoorne, V.; Pilcer, G.; Chavez, P. F.; Schubert, M. A.; Vervaet, C.; De Beer, T. Managing Active Pharmaceutical Ingredient Raw Material Variability during Twin-Screw Blend Feeding. *Eur. J. Pharm. Biopharm.* 2019, *135*, 49–60.

(3) Naik, S.; Mukherjee, R.; Chaudhuri, B. Triboelectrification: A Review of Experimental and Mechanistic Modeling Approaches with a Special Focus on Pharmaceutical Powders. *Int. J. Pharm.* 2016, *510*, 375–385.

(4) Schella, A.; Herminghaus, S.; Schröter, M. Influence of Humidity on Tribo-Electric Charging and Segregation in Shaken Granular Media. *Soft Matter* 2017, *13*, 394–401.

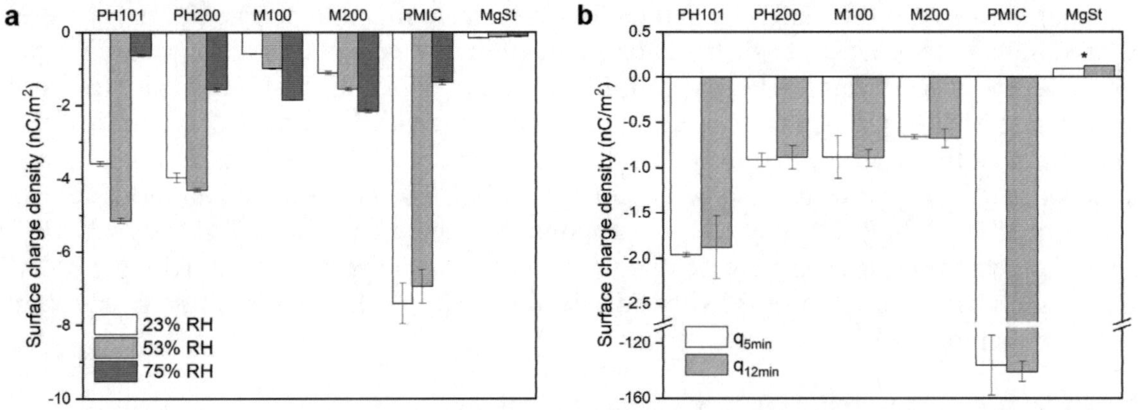

Figure 1 Tribo-charging results: a) at different relative humidity levels; b) after twin-screw feeding. *polarity change of MgSt.

606916 An Approach to Measurement of Volumetric Fill and Flow Rate in a Discharging Hopper Using Helmholtz Resonance

Abstract:

The response of a partially enclosed cavity to acoustic excitation is related to the volume of fluid enclosed, and Helmholtz resonance is increasingly finding application in the measurement of the volume of nonporous objects. Simplistically, an object placed in a cavity changes the free volume of the cavity, and the volume of the object can be found from theory and empirical calibration. Determination of the volume of particulate materials brings challenges which arise from the properties of the individual particles, their size and size distribution. The exploratory work we report here was carried out with a cylindrical discharge vessel 300 mm in diameter and 700 mm high, discharging mustard seeds (approximately spherical, diameter ~2 mm) and agricultural lime (polydisperse). The system was excited by a series of chirps each with a frequency range chosen to span the expected resonant frequency. The response to each chirp was recorded, and resonant frequency determined. Computation time was reduced by using the measured resonant frequency of each excitation chirp as the basis for setting the frequency span of the next chirp in the series; typically, a measurement cycle took less than one second. The material leaving the discharge vessel was also continuously weighed using laboratory scales enabling volumetric discharge rate to be estimated using a (nominal) bulk density, for comparison with the acoustic measurements. The agricultural lime behaved similarly to a solid object, and degree of fill could be found with an accuracy of approximately 2%. The response of the mustard seeds was different, exhibiting a negative correlation with degree of fill. This is thought to result from the open structure of the particle bed, and the interaction of this with the free volume in the discharge vessel.

600750 Powder Temperature and Moisture Behavior Effects on Bag Filling Rates from an Industrial Installation; How to Quantify This Effect

Abstract:

Automatic bag filling is a well known and often used technology to bag solids for further transport.

Bag filling devices often use vibrations and vacuum suction to get rid of entrapped air in between the solid material. This is done to efficiently fill the bags, but more importantly to reduce the amount of entrapped air, which may lead to instable piling of the bags, causing collapse of staged bags. To understand the bag filling process, investigations have been done on the use of vibration and vacuum suction to get a more dense product thus preventing the instable bag piling. In this work, the effect of temperature of the product coming from up-stream processing has been investigated. It will be shown what the effect of temperature is on the rate of bag filling and why this may lead to slower filling of bags in automated bag filling lines. A typical step change in the processing speed is observed.

598340 Rational Design of Modified-Release Dosage Forms

Abstract:

Oral modified release (MR) dosage forms are developed by altering the kinetics and/or site of drug release/absorption in order to achieve specific clinical objectives, such as improved patient compliance, optimized efficacy and reduced adverse events. This presentation introduces basic design principles and different MR delivery technologies. It also discusses the importance of understanding material, formulation, process and their interplays in rational product and process development along with case studies.

602132 Strategies to Develop IV Formulations of Poor Water-Soluble Compounds and Their in Vitro Evaluation

Abstract:

Although a drug might be developed and formulated for oral delivery, the same compound is commonly administered as a solution via an IV injection/infusion in order to determine absolute bioavailability to support early development. A common challenge to develop IV formulation is the low solubility of drug candidates, risking phlebitis due to precipitation upon injection. Other parameters that need attention include the pH variation (too high or too low) that can cause irritation, and osmolality differences that may cause significant fluid shifts between the intracellular and extracellular spaces. Thus, the development of a stable and safe IV formulation requires an adequate selection of solubilizers and evaluation of the safety of the solution prior to administration.

In this study, a weak acid drug with poor water-solubility required a minimum of 40-fold increase in aqueous solubility for IV delivery. To achieve increased aqueous solubility, multiple strategies were assessed including screening of solvents and co-solvents, in situ salt, and pH control. Phlebitis risk was evaluated by in vitro static dilution using surrogate plasma and observation of flocculation in rat plasma by optical microscopy. In conclusion, a pH-buffered solution containing in situ sodium salt allowed for up to 133-fold increase in aqueous solubility of the compound showing no risk of precipitation while complying with pH and tonicity requirements for IV administration.

602618 Integrated Product & Process Development

Abstract:

During drug product development, science-based design of formulation and manufacturing process is essential for achieving and assuring consistent quality products. This presentation discusses rational product development based on the integrated product and process knowledge. It addresses the importance of fundamental understanding of physicochemical, mechanical and biopharmaceutical characteristics of API, formulation, process and their interactions during product design, development, scale-up and commercial manufacturing with case studies in ensuring consistent quality, robust drug delivery performance and process.

602619 Predictive and Model Based Product & Process Design

Abstract:

Use of predictive technologies and in silico modeling can significantly enhance speed, efficiency and effectiveness of drug product and process design. This presentation will discuss several case studies, in which multidisciplinary sciences and engineering principles, coupled with high power computing, are used to guide formulation development, manufacturability assessment, and drug substance solid form risk assessment.

605621 Rational Design of Amorphous Solid Dispersion

Abstract:

Amorphous solid dispersion (ASD) is a highly effective drug delivery technology for enabling oral absorption of insoluble APIs. This presentation discusses basic principles of formulation design & characterization, and importance of understanding material, formulation, dissolution (LLPS) and stability in rational product development using case studies

605997 Improvement of Drug Substance Bulk Powder Properties through Process Crystallization

Abstract:

The existing recrystallization process for a drug substance produced crystalline needle-shaped aggregates with low bulk and tap density and very poor flow properties. The particle properties led to poor performance of the drug substance during formulation. The goal of this work was to develop a process to yield drug substance with improved bulk powder properties for future formulation development and manufacturing work.

The approach taken for this work involved the use of computational and experimental solvent screens to rapidly narrow down the choice of solvent systems and working conditions. The computational work also had the advantage of significantly shortening development timelines.

Ultimately a crystallization process was developed which produced drug substance with enhanced particle morphology and significantly increased bulk density.

611726 Singly-Twinned Growth of Si Crystals upon Chemical Modification

Abstract:

During crystallization a continuum of patterns could emerge due to the interplay of growth kinetics, material or solution chemistry, and crystallographic defects. The coherent twin boundary is widely known to catalyze growth in pristine crystals including polycrystalline Si. Much remains unknown about the impact of changing the chemical environment of the crystallization process through the deliberate addition of trace metallic species – termed chemical modification. Pristine Si has been reported to grow through the classical model of two parallel twin planes acting in concert to enable steady-state propagation of the solid-liquid interfaces. Here, we achieve a new vision on the growth process *via in situ* synchrotron X-ray microtomography and further corroborated by *ex situ* crystallographic investigation. We find that steady-state growth is impossible in chemically-modified alloys that consist of trace (0.1 wt.%) Sr. This is because the Sr modifier poisons the concave re-entrant grooves, thereby de-activating the advantage of the twin-plane re-entrant edge mechanism and leading to a singly-twinned interface. This study may serve as a proxy to chemically-modified crystallization pathways of eutectic Si in Al-Si alloys and, more broadly, as a framework for the crystallization-mediated synthesis of materials.

597305 Elucidating the Influence of Hydroxycitrate on Brushite Crystallization Via Bulk and Interfacial Studies

Abstract:

Kidney stone diseases (nephrolithiasis) affect 12% of the world population and can occur on all ages, sexes and races. Of all the cases, calcium stones are the most prevalent and 20% of these are calcium phosphates (CaP). Brushites, or dicalcium phosphate dihydrate, are one of the types of CaP stones and known to induce calcium oxalate mineralization, and to be converted to less soluble form hydroxyapatite. Kidney stone incidence has increased about three times for the past decades. Efforts in identifying alternative inhibitors of stone growth remain one primary strategy in preventing the increasing occurrence of kidney stones. Once identified and obtained clinical clearance, this may provide patients more choices for therapeutics against pathological calcification. Fundamentally, carboxylates are known to inhibit crystal growth of these minerals and one example is hydroxycitrate (HCA) with have effectively curbed calcium oxalate mineralization. To tackle this goal, we employed a combination of bulk and atomic force microscopy studies to elucidate the impact of HCA on the crystal size, morphology, number density of CaP. Analysis shows a 50 % reduction in the number of crystals formed and an increase in the aspect ratio at a very small inhibitor concentration (Ca/inhibitor=25000). Amplitude and Frequency Modulated AFM studies show that surface roughness increased through time in the presence of the HCA. Changes in the periodic features at near atomic resolution was also observed in the presence of HCA implying possible modifier-crystal surface interactions. Results obtained in this study may serve as a step in understanding the inhibitors' effect on CaP crystals and in designing preventative drugs for kidney stone disease.

602826 Islatravir Penultimate Step Crystallization: Purification of an API in the Presence of High Protein Loading from an Enzymatic Cascade

Abstract:

Islatravir is a novel nucleoside active pharmaceutical ingredient (API) currently in clinical trials for both HIV treatment and prevention. As reported recently in *Science (*Huffman et. al. 2019*)*, the API synthesis under development involves a 9-enzyme 3-step in-vitro cascade without intermediate isolations until the final API molecule is assembled. This process represents an incredibly efficient synthesis of the API molecule, but introduces significant challenges to the downstream purification of the product. Enzymes, host-cell proteins, bacterial endotoxins, residual nucleic acids, and organic impurities/related substances all must be removed to generate a drug substance of adequate purity for human dosing.

The primary challenge in the purification development has been on enzyme/protein removal. The presence of high levels of protein tends to make filtrations very slow, so extractive work-ups are often preferable, but the relatively low solubility of the API in most organic solvents was a formidable obstacle to this approach. Several extractive work-ups were developed but each posed stability concerns.

Additionally, enzyme immobilization was pursued for the enzymes with the highest loadings in the reaction, which improved downstream processing but added a significant cost and supply-chain complexity.

In the final reaction, the API crystallizes from the reaction mixture, and even wth several enzymes removed via immobilization, filtration was still very challenging. Filtration was optimized by adding filter aid (diatomaceous earth), reducing shear in the reaction, adding seed crystals to the reaction, and agitating the batch intermittently during filtration. Optimizing the filtration step to isolate the crude API made it possible to produce quality API without the need to develop an extractive work-up or use immobilized enzymes.

601168 Development of a Crystallization Process for the Final Step of the Biocatalytic Synthesis of Islatravir

Abstract:

Merck recently reported a biocatalytic cascade for the manufacture of Islatravir, a potent HIV reverse transcriptase translocation inhibitor [1]. The API step of the synthesis requires removing protein and other impurities accumulated throughout the synthesis. Also, the low dose of the drug requires API particle size control in the final step to mitigate content uniformity concerns in the tablet. This industrial case study will provide an overview of the development of a pure step that meets purity and particle size targets. The presentation will highlight the development of a novel crystallization process to meet the desired particle size distribution while maintaining a rapid filtration rate of the product crystals. The scale-up crystallization performance and impurity rejection of the process will also be compared to a more 'traditional' seeded crystallization using jet milled seed. The development of the process was motivated by the desire to eliminate the cost and industrial hygiene liabilities associated with dry milling API as seed.

[1] Huffman et al. "Design of an in vitro biocatalytic cascade for the manufacture of islatravir." Science, 2019, 366 (6470), 1255-1259.

598747 Impact of the Process Conditions of an Evaporation to Dryness Operation Performed in a Conical Screw Vacuum Dryer on the Isolated Product Psd: Lab- and Pilot-Scale Studies

Abstract:

Conical Screw Vacuum Driers (CSVD) are frequently used in the pharmaceutical industry for drying heat-sensitive powders to a very low volatile matter content in a low shear mixing environment. This type of dryer can also be used to perform evaporation to dryness operations. Evaporation to dryness operations, routinely executed in laboratory preparations but less frequently at industrial scale, aim solely to recover solids from solutions by complete evaporation of the solvent. A manufacturing process industrialized by UCB-Pharma includes an evaporation to dryness operation executed in a CSVD. It follows a multi-column chromatographic separation, yielding the purity level required and it comes before a re-slurry of the powder in a suitable solvent. After solid-liquid separation and drying, the powder isolated from the re-slurry operation is milled to meet the API granulometry specifications of a mono-modal distribution, characterized by a d(v, 0.5) and a d(v, 0.9) within defined target ranges.

Following a process optimization exercise to increase productivity of the chromatographic unit, the solvents used were changed. This also resulted in a significant change in the typical Particle Size Distribution (PSD) observed after the evaporation to dryness operation. Risks associated to not knowing to what extent the final dry milling operation was able to mitigate those unpredicted changes in the control of PSD, triggered this comprehensive investigations on the CSVD process at different laboratory and manufactory scales. Given the change of performance driven by the solvent changes, an additional aim of this work was also to identify process conditions for the CSVD that might have been able to yield directly material within specifications.

A Design of Experiment (DoE) approach was used to determine the impact of the process conditions and the scale-up of the evaporation to dryness operation investigated here on the PSD of the isolated powder. A first DoE plan was carried out using a lab-scale CSVD (300 g scale) under well controlled process conditions. A second DoE was also reproduced at pilot plant-scale (30 kg scale) using appropriate scale-up criteria for those process parameters identified as critical, like the evaporation rate. The heat transfer characterisation of the equipment at laboratory and pilot plant scale allowed for establishing suitable procedures to control critical scale-up process parameters like jacket temperature and operating pressure.

The complexity of the mass and heat transfer phenomena occurring during the evaporation to dryness were elucidated and rationalised in a multiphase process steps while executing the experimental plan (Figure 1). Each process step was clearly defined by specific rheological behaviours. Evidence of patterns and reproducible mechanistic behaviours were found to be linked to specific operating conditions independently from the scale used for the operation. However, macroscale phenomena linked to the batch size of the operation together with exacerbation of the mixing conditions at scale led to mixing zone segregation. These differences ultimately significantly impacted the reproducibility of the performance of the operation under the same operating conditions at scale.

The complexity of the phenomena and phases involved was rationalised in a model behaviour and optimal process operating conditions and procedures to maximise process robustness. The operation was successfully validated at final production scale (300kg) with material produced according specifications over a number of validation batches.

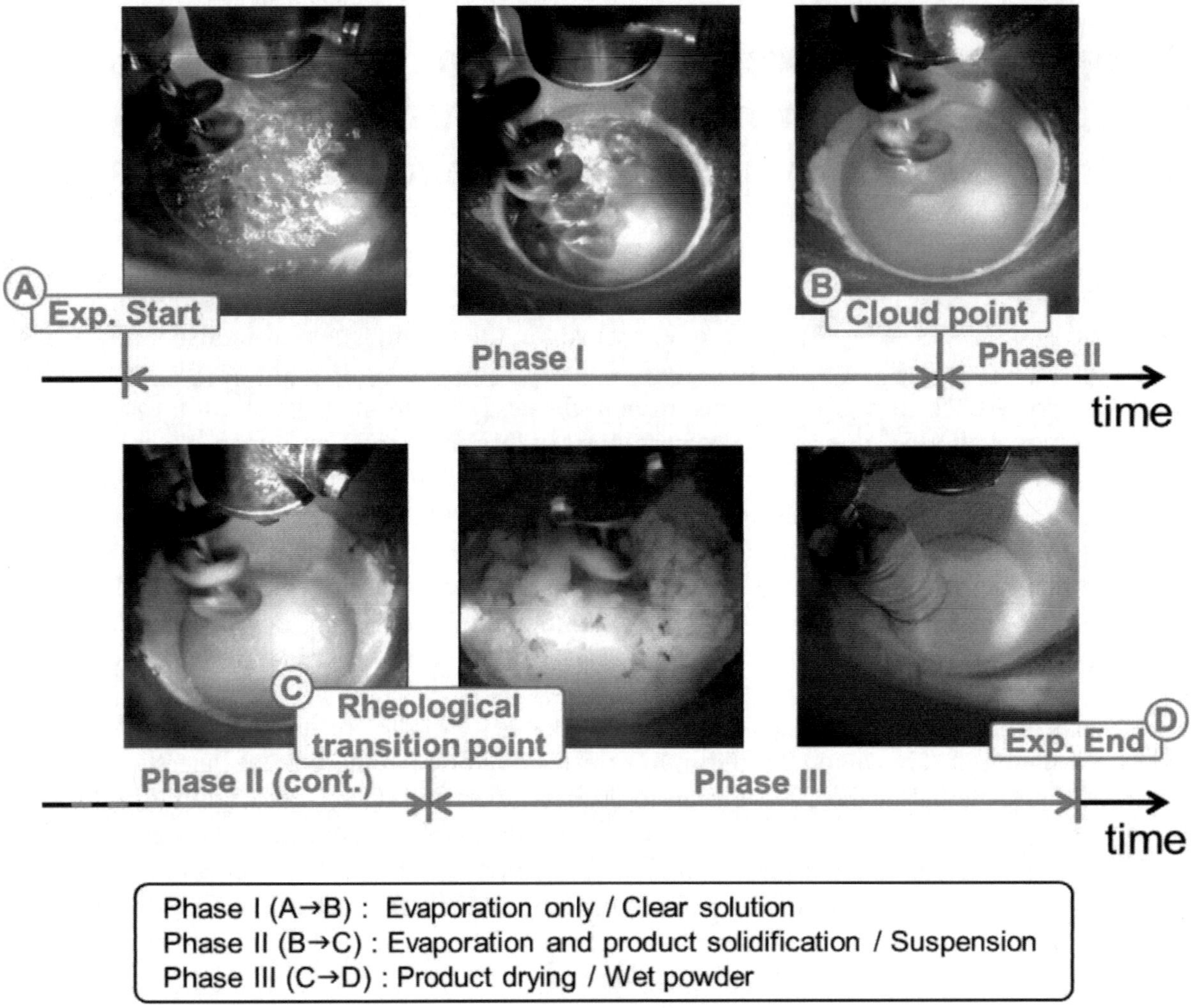

Figure 1 : Observed process phases during the evaporation to dryness experiments

610946 A Generalized Approach for Rapid Aqueous MOF Synthesis By Controlling Solution pH

Abstract:

Metal organic frameworks (MOFs) are a promising class of crystalline materials with potential to significantly advance the fields of catalysis, separations, biotechnology, and sensing, amongst others. Comprised of metal ion nodes and organic linkers, lab-based syntheses have excelled at producing novel MOFs with tuned pore size and chemical activity. However, developing a crystallization technique that meets high throughput scaling requirements and is environmentally benign remains a challenge. Currently, applying classical crystallization techniques to MOF formation are limited due to the multi-component nature. In this work, we approach understanding MOF formation through a reaction-based perspective to better understand the non-classical crystallization pathway. We demonstrate that tuning the reactant speciation, namely labile metal ion nodes and deprotonated organic acid linkers, via pH has a significant effect on MOF yield. Using this approach, we develop a generalizable technique to rapidly produce a range of MOFs, including UiO-66-NH_2, UiO-66, HKUST-1, and ZIF-L, under ambient, aqueous conditions. We further this understanding by tuning the labile metal node and deprotonated linker concentrations to produce space-time yields (STY) exceeding 2250 kg m^{-3} day^{-1}, which is among the highest reported STYs for HKUST-1 and ZIF-L and an order of magnitude higher for aqueous zirconium-based MOF synthesis. Finally, we demonstrate UiO-66-NH_2 completes crystallization in 5 minutes at room temperature with 70% of crystallization of UiO-66-NH_2 occurs within 30 seconds, which is among the fastest reported reaction times in literature. With this work, we demonstrate a generalizable pathway for scaling MOF crystallization to high STY under mild and benign synthesis conditions and lay ground work for understanding the complex multi-step crystallization kinetics of MOFs.

601122 Quantification of Interparticle Forces and Its Variation with Temperature in a Gas-Solid Fluidized Bed

Abstract:

Gas-solid fluidized beds are broadly employed in chemical and metallurgical industries for gas phase catalytic and gas-solid reactions, as well as solids processing. While most industrial fluidized beds are presently operated at temperatures well above ambient, in the near future they will be more commonly utilized under extreme operating conditions to process complex raw materials. The performance of these units is considerably influenced by the heat and mass transfer rates, which are impacted by the bed hydrodynamics. Therefore, a comprehensive understanding of the bed hydrodynamics under elevated operating temperatures is required for a successful design, operation, and scale-up of high temperature fluidized beds. The design and performance evaluation of high temperature fluidized beds are typically accomplished by changing the gas properties in hydrodynamic correlations, which are based purely on hydrodynamic principals. These correlations are generally capable of accounting for changes in the gas properties and their corresponding effects on the bed hydrodynamics, while neglecting the changes in the level of interparticle forces (IPFs) owing to the changes in the operating conditions. As the level of IPFs may appreciably change with the operating temperature, an in-depth understanding of these forces and their influences on the fluidization characteristics are critical. The complexity of IPFs and unavailability of a reliable measurement method that quantifies the magnitude of IPFs at high temperature are the primary reasons why these forces and their effects on the bed hydrodynamics have generally been neglected. The main objective of this study is, therefore, to develop simple and reliable methods to quantify the magnitude of IPFs and its variation with temperature in a gas-solid fluidized bed without disturbing the particulate system.

The first part of this study was dedicated to quantifying the variation of the level of IPFs by temperature in a gas-solid fluidized bed. To this end, the variation of minimum fluidization velocity (U_{mf}) with temperature for 225-micron silica sand and 83-micron FCC particles, as well as 67-micron glass beads was studied. It was shown that the difference between the magnitude of IPFs at a given temperature and that at ambient conditions is proportional to the deviation of the experimental U_{mf} from the prediction of an appropriate correlation. The correlation must be capable of adequately accounting for the effect of gas properties on U_{mf}, while matching the experimental data at ambient conditions. The difference between the experimental and predicted U_{mf} was then translated into the force that the fluidizing gas is required to exert on the particles to overcome thermally induced IPFs.

The second approach was established on the overshoot in the bed pressure drop profile at U_{mf} due to the difference between the minimum fluidization voidage (ε_{mf}) and the loose bulk voidage (ε_0). The ε_{mf} and ε_0 were measured in beds of 83-micron FCC particles and 67-micron glass beads at temperatures ranging from ambient up to 600°C. It was shown that the ratio of $\varepsilon_{mf}/\varepsilon_0$, which was called the "dynamic Hausner ratio", increases as the level of IPFs grows. A model was, hence, proposed to quantify the magnitude of IPFs as a function of the dynamic Hausner ratio.

The magnitudes of IPFs quantified by the above approaches were in the same order as the particle's weight for 67-micron glass beads and 83-micron FCC particles, which is analogous to the findings reported in the literature for similar powders. In addition, the results of both approaches were in accordance with the temperature variation of the normalized peak area, i.e., the area under the overshoot between the pressure drop profiles in the increasing and decreasing velocity passes at a given temperature divided by that at ambient conditions, as a qualitative indication of the magnitude of IPFs.

601088 Professor Daizo Kunii' Encounter with Chemical Engineering

Abstract:

Daizo entered Tokyo Imperial University (now University of Tokyo) , dreaming to succeed as a designer for high performance aircrafts. He had to give up his dream, however, because aircraft manufacturing was prohibited in Japan following the end of WWII. The department of aeronautics was renamed as the department of applied mathematics when he graduated.

He aspired to find a job as a researcher in engineering field and his chance meeting with Professor Sakae Yagi brought about his later accomplishments in the field of "chemical engineering," which was quite new at that time of Japan. He joined Yagi's laboratory at University of Tokyo and was given a chance to work on design calculation of industrial furnaces. He had coauthored two books on this topic with Yagi when he became the age of 30. Through these experiences, he acquired sense and knowledge on thermodynamics and geometrical modeling of practical reactors, as well as organizing complex systems.

He started works on heat transfer in fixed bed catalytic reactors, and derived Yagi and Kunii's theoretical equation of the effective thermal conductivities in particle beds, which is one of the smart works in this period. At the end of 1950s He spent two years at Northwestern University with Professor J. M. Smith working on heat transfer in porous rocks and after returning to Japan, his topics of research have been extended to fluidization, moving beds, simultaneous heat and mass transfer, solid reactions, thermal cracking, combustion in complicated systems, etc.

All of these topics have evolved out of Professor Daizo Kunii's willingness to challenge high peaks in the fog so that he could contribute to the well-being of the public. He was really a man of Samurai.

613084 Torrefaction of Biomass Residues in Pulsating Fluidized Beds for Pellet Production

Abstract:

Torrefaction is a mild pyrolysis, which maximizes solid yield, producing a high percentage of 'biochar' (70-80%wt) with improved grindability, prolonged durability, lower hydrophilicity and slower biological and thermal degradation compared to raw biomass. Torrefaction has been explored for pre-treatment of biomass residues to make torrefied pellets, which can be applied as a high quality feedstock in gasification for high quality syngas production and as a substitute for coal in thermal power plants and metallurgical processes. This presentaiton provides an overview of the advances made at UBC on developing a novel pulsating fluidized bed reactor for biomass torrefaction. The reactor is intended to address two major challenges in biomass torrefaction: stable fluidization of irregular biomass particles and narrow particle residence time distribution for uniform solids product. To achieve stable fluidization of sawdust particles with a narrow residence time distribution, a gas-pulsating horizontal fluidized bed reactor has been developed. The fluidization hydrodynamics, particle residence time distribution, biomass drying and torrefaction performance were performed systematically in both a cold model unit and a hot model unit. The suitable operating envelop of the reactor including fluidization velocity, pulsation frequency, reactor temperature and solids residence time were explored. The results demonstrated the superior performance of the novel reactor for torrefaction of biomass particles.

601075 Daizo Kunii, the Chemical Engineer

Abstract:

I graduated from the University of Tokyo (UT) wherein I learned a lot from Prof. Daizo Kunii at the Kunii Laboratory. Every day at lunchtime, he used to fondly share his experiences with us. After graduating from the Aerospace Engineering Department, Young Daizo Kunii had the opportunity to work in the Chemical Engineering Department of UT, which was an entirely different field. He said he made every effort for his boss to recognize him and finally succeeded when he came up with a new scientific theory on radiation heat transfer with his strong background in mathematics. Simultaneously, he also summarized a design method for a furnace.

Nowadays, the importance of inter-, multi-, and trans- disciplinary research is recognized on every issue, but Prof. Kunii demonstrated it more than half a century ago. He had the opportunity to study the development of a new process for coal gasification and waste treatment. Moreover, he was attracted by fluidized bed technology. He recognized the importance of an engineering viewpoint, solution engineering, and needs-oriented research during this period. He always told us that organizing different researchers is similar to conducting an orchestra; the chemical engineer being the conductor. Interestingly, one of the research topics of Kunii laboratory is the methodology of invention wherein a new idea is created from other ideas with processes similar to mathematics (multiplication, transfer matrix, reverse matrix, etc.).

Prof. Daizo Kunii was a chemical engineer. I owe a lot to him for his teachings about the engineering sense and the method of invention, namely: mixing the unmixable and organizing. Hence, after graduating from UT, I focused on the fusion (multiplication) of different researches to create a new research field, similar to what Prof. Kunii did during his younger years; this was achieved by organizing different fields of researchers.

Particularly, materials can be categorized into three phases: gas, liquid, and solid. However, some physical chemistry textbooks said that there is one more phase, which is the supercritical phase. Thus, I tried to use the supercritical state for several reactions (organic, inorganic, coal, petroleum, biomass, wastes, etc.) and in material synthesis.

Supercritical water provides an excellent hydrothermal synthesis reaction atmosphere to control phase behavior, reaction kinetics, and equilibrium with minimal changes in temperature and pressure. Interestingly, in the supercritical state, water and organic matter form a homogeneous phase, thus, resulting in organic-inorganic hybridization. This new process of mixing the unmixable can lead to the development of new materials, which cannot be synthesized by any other methods, such as nanomaterials with extremely high catalytic activity, hybrid materials of ceramics and polymers, and fluidic ceramics. Our proposed supercritical hydrothermal synthesis process was commercialized not only in Japan but also in other countries. With this, I have a feeling that I can finally realize what I learned from Prof. Daizo Kunii, that is, chemical engineering.

Multiphase CFD on Emerging Computing Technologies

Madhava Syamlal
Senior Fellow, Computational Engineering
National Energy Technology Laboratory

Multiphase CFD on Emerging Computing Technologies

Madhava Syamlal
Senior Fellow, Computational Engineering
National Energy Technology Laboratory

NATIONAL ENERGY TECHNOLOGY LABORATORY

Quiz!

1. Moore's law is the observation that the number of transistors in a microchip doubles about every two years. But what's *Dennard Scaling*?

2. Nothing in life is free! Then, what do computational scientists mean by "free flops"?

3. The theoretical peak speed (*Rpeak*) of today's supercomputers is measured in PFlop/s. What *fraction of Rpeak* could CFD codes likely achieve?
   a) 1 %
   b) 10 %
   c) 20 %
   d) 40 %
   e) 80 %

4. CFD codes primarily use double-precision floating-point numbers. What's *transprecision* computing?

U.S. DEPARTMENT OF ENERGY 4

MFIX: NETL's Open Source CFD Software

NATIONAL ENERGY TECHNOLOGY LABORATORY

MFiX Multiphase Computational Fluid Dynamics (CFD) software for Energy Technology Development

3 Decades
of development history

6,000+
registered users

200+
downloads per month

400+
citations per year

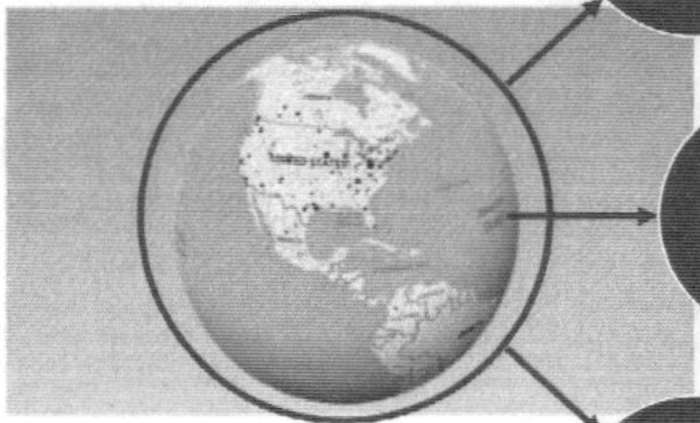

1 — **MFiX** Exa — Exascale Computing

2 — Wafer Scale Engine

3 — **MFiX-AI** — AI/ML for speed up

U.S. DEPARTMENT OF ENERGY

5

Simulation for Gasifier Design: Industrial Scale Up and Next-Generation Systems

NATIONAL ENERGY TECHNOLOGY LABORATORY

- MFIX-DEM (coarse grained) used for designing a 15 MWth gasifier for service in Alaska
- A comprehensive simulation program has provided key insight in coal gasifier performance for a **5x scale up**
  - Performance over a range of loads
  - O_2 enrichment and biomass co-feed performance
  - Oxygen-blown gasification with steam enrichment for hydrogen production
- This helps de-risk the design of the $46 million facility

PI: Dr. William Rogers (NETL)
Co-PI: Dr. Mehrdad Shahnam (NETL)

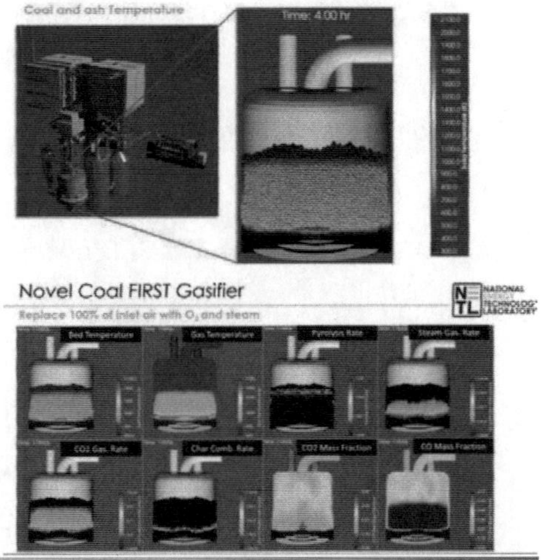

UAF Worley energy | chemicals | resources HMI U.S. DEPARTMENT OF ENERGY Office of Fossil Energy

U.S. DEPARTMENT OF ENERGY

6

MFIX-TFM used for troubleshooting and redesigning a nuclear waste treatment reactor

NATIONAL ENERGY TECHNOLOGY LABORATORY

- Simulations used to resolve technical issues related to start-up operations of Idaho Waste Treatment Unit fluidized bed reactor.
- In less than 8-months, with 3-FTEs, MFIX, and Joule supercomputer:
  - Screened several reactor designs under various operating conditions
  - Eliminated the need to test each design under a range of operating conditions at full-scale potentially ***saving $10's million***.
  - The new design was successfully tested at full-scale using a simulant during the summer of 2018 preventing a long-term delay to the start-up operations **saving *$100's million*** *in additional costs*.

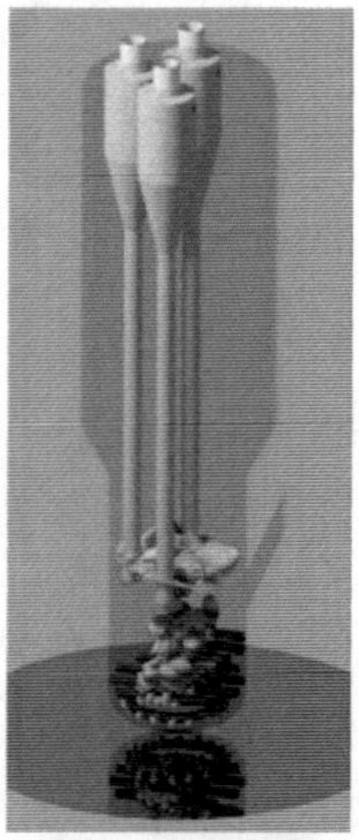

PI: Dr. Chris Guenther (NETL)

INL Idaho National Laboratory

U.S. DEPARTMENT OF ENERGY OFFICE OF ENVIRONMENTAL MANAGEMENT

Computing Technologies Then and Now

NATIONAL ENERGY TECHNOLOGY LABORATORY

Prof. Dimitri Gidaspow: Pioneering contributions to computational gas-particle flow and fluidization

	Gidaspow Lab (circa 1984)	NETL – MFS team (2020)
On prem Computer	Prime 550	Joule 2.0 69th of Top500*
Speed (GFlop/s)	<0.001	3,600,000
Memory (GB)	0.001	253,000
Off prem Computer	Cray-1 at Westinghouse	Summit at ORNL 2nd of Top500*
Speed (GFlop/s)	0.16	200,000,000
Memory (GB)	0.008	2,802,000

From Giga to Exa, via Tera & Peta*

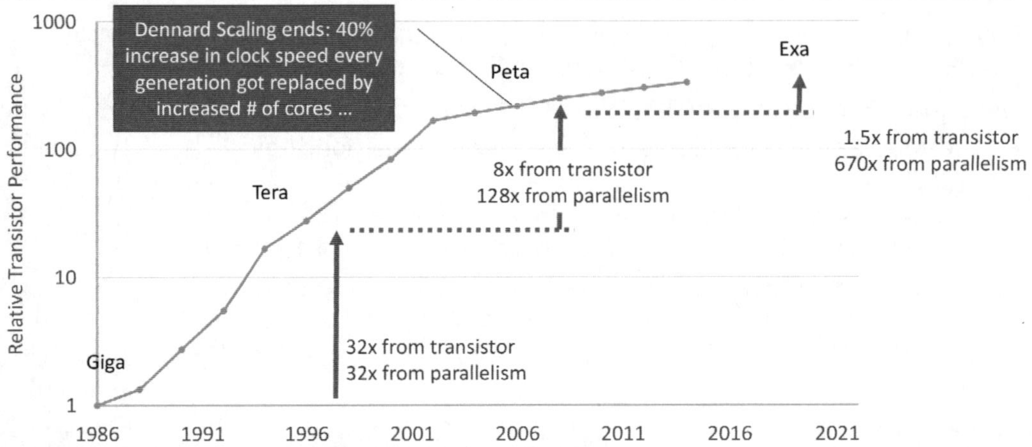

*Source: S. Borkar, Intel

CFD codes must accommodate non-uniform memory architecture & use increased parallelism

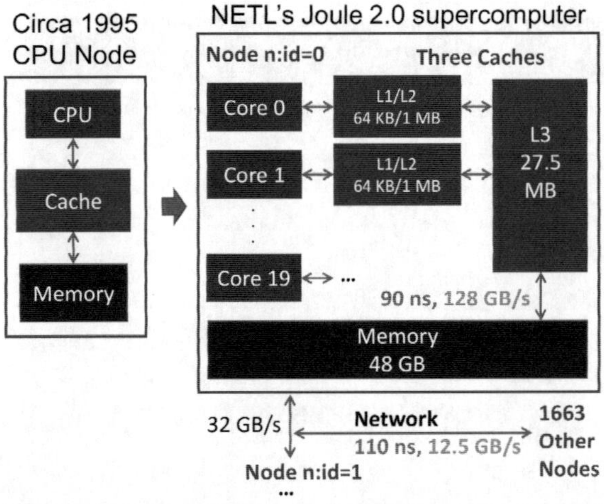

- When data is not in the core (registers), it needs to be fetched from cache, memory, or network
- Memory fetches have
  - start up time or latency (ns)
  - a 'flow rate' or bandwidth (GB/s)
- Trouble is that over the last 25 years memory access speed has not kept up with processor speed*:
  - Flop *vs* network latency ~ **30x**/decade
  - Flop *vs* memory latency >**100x**/decade
  - Flop *vs* network bandwidth **9x**/decade
  - Flop *vs* memory bandwidth **4.5x**/decade

*McCalpin JD. 2016. Memory Bandwidth and System Balance in HPC Systems. Invited talk at Supercomputing 2016, Salt Lake City, Utah.

Speed of CFD calculations is limited by bandwidth and latency

NATIONAL ENERGY TECHNOLOGY LABORATORY

- Most computational tasks of CFD are limited by memory and network bandwidth
  - DAXPY: y[i] = a*x[i] + y[i]   i = 0 ... N-1
    - 2 Flop = 1 multiply + 1 add
    - Fetch 3*8 B (keep 'a' in a register)
    - Operational Intensity (OI): 2/24 Flop/B
  - SpMV*: For 7 non-zero diagonals, OI = 7/52 Flop/B
- Dot products are limited by memory and network latency: sum += x[i] * x[i]   for i = 0 ... N-1
  - Sum across all cores and broadcast results back to all cores → 2 $\log_2(N_{core})$
  - Speed/step ~ 1 Flop/110 ns = 0.01 GFlop/s

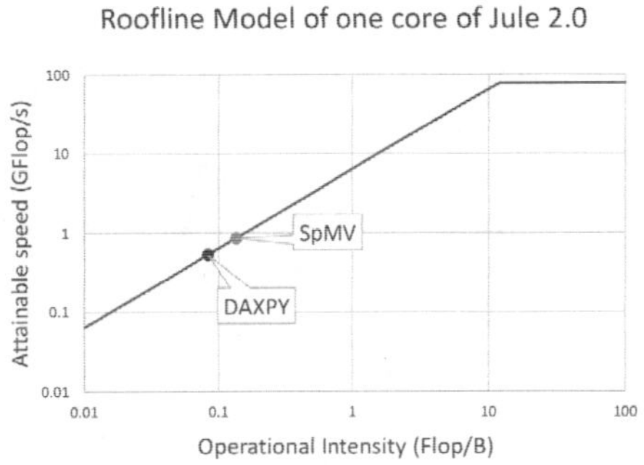

*Gropp WD, Kaushik DK, Keyes DE, Smith BF. Toward realistic performance bounds for implicit CFD codes. In D. Keyes D, Ecer A, Periaux J, Satofuka N. Fox P. editors. Proceedings of Parallel CFD'99, pages 233—240. Elsevier. 1999.

U.S. DEPARTMENT OF ENERGY 11

Parallel computations must balance core and network performance

NATIONAL ENERGY TECHNOLOGY LABORATORY

- Every time a core fetches data by *cache lines;* i.e., multiple bytes, say 64 B, at a time
- The core looks for data first in its cache
  - If data is found, it is a *cache hit* ☺
  - Otherwise it is a *cache miss* ☹; then the core looks for data in a higher-level cache or memory
- When more cores are used, the local problem size may fit within the cache
  - But the communications over the network increases

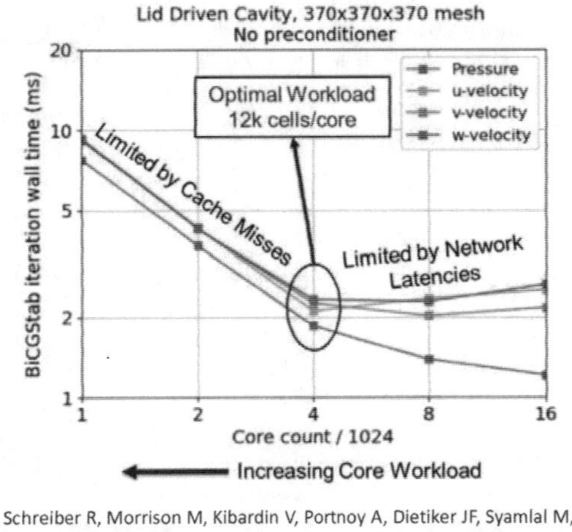

Rocki K, Van Essendelft D, Sharapov I, Schreiber R, Morrison M, Kibardin V, Portnoy A, Dietiker JF, Syamlal M, James M. Fast Stencil-Code Computation on a Wafer-Scale Processor. arXiv:2010.03660v1 [cs.DC] 7 Oct 2020

U.S. DEPARTMENT OF ENERGY 12

Other emerging challenges in high performance computing

NATIONAL ENERGY TECHNOLOGY LABORATORY

- Energy-aware computing
- Heterogeneous compute nodes including GPUs
- New approaches for analyzing large volumes of data; e.g., machine learning, *in-situ* processing
- Application-based defensive I/O (checkpointing)

Functionality	Energy (pJ)	Performance (ps)
Core/FP Unit	50	125
Register bank	10	250
Cache/SRAM	100	2000
Memory/DRAM	1000	100,000

Source: T. Hoefler. Energy-aware Software Development for Massive-Scale Systems. Keynote at EnA-HPC, 2011, Hamburg, Germany

1. Exascale Programming Challenges, DOE-ASCR report 2011; 2. Shalf J, Dosanjh S, Morrison J. Exascale Computing Technology Challenges. 9th International Conference on High Performance Computing for Computational Science, VECPAR 2010, Berkeley, CA, USA, in June 2010

1 MFIX-Exa will exploit Exascale computing to enable industry relevant CFD-DEM simulations

NATIONAL ENERGY TECHNOLOGY LABORATORY

MFiX + AMReX = MFiX Exa

NETL's opensource software suite, widely used by academia and industry

Framework developed by ECP's Block-Structured Adaptive Mesh Refinement Co-Design Center

A planned 1000X improvement over the CFD-DEM state-of-the-art in 2017

Node: 1 HPC and AI Optimized AMD EPYC CPU
4 Purpose Built AMD Radeon Instinct GPU

NATIONAL ENERGY TECHNOLOGY LABORATORY — M. Syamlal (PI), J. Musser (Co-PI)

BERKELEY LAB — A. Almgren (Co-PI), J. Bell (Co-PI)

Boulder — C. Hrenya (Co-PI), T. Hauser (Co-PI)

ECP EXASCALE COMPUTING PROJECT — U.S. DEPARTMENT OF ENERGY Office of Science

DOE launched the Exascale Computing Project (ECP) in Fiscal Year 2017

NATIONAL ENERGY TECHNOLOGY LABORATORY

- ECP is producing applications, a software stack, and exascale hardware technology to form a capable exascale computing ecosystem.
- 800 researchers (22 laboratory and agency partners; 39 universities) engaged in:
  - 66 software projects
  - 24 science application projects
  - 5 co-design centers
- **10^{18} Flop/s is not enough; Applications must run 50x faster compared to 20 PFlop/s machines.**

Exascale Systems 2021-2024

FRONTIER **ORNL** AMD/Cray

ANL Intel/Cray

LLNL TBD

Alexander F et al. 2020. Exascale applications: skin in the game. Phil. Trans. R. Soc. A 378: 20190056. http://dx.doi.org/10.1098/rsta.2019.0056.

ECP EXASCALE COMPUTING PROJECT U.S. DEPARTMENT OF ENERGY Office of Science NNSA National Nuclear Security Administration

https://www.exascaleproject.org/

MFiX-Exa challenge problem: CFD-DEM simulation of a 50 kW Chemical Looping Reactor

NATIONAL ENERGY TECHNOLOGY LABORATORY

Oxygen carrier	200 µm diameter X 5 billion particles Composition: Fe_2O_3, FeO, and an inert substrate 12 unknowns per particle
Fluid	Grid: 1 mm cell size X 330 million (active) cells Composition: CH_4, H_2, H_2O, CO, CO_2, O_2, and N_2 12 unknowns per fluid cell
Chemical Reactions	Air reactor: $2FeO + \frac{1}{2} O_2 \rightarrow Fe_2O_3$ Fuel Reactor: $4\ Fe_2O_3 + CH_4 \rightarrow 8\ FeO + CO_2 + 2\ H_2O$
Operating Conditions	Temperature: 1100 K Pressure: 1 atm

1. Musser J, Fullmer W, Syamlal M, Almgren A. MFIX-Exa Challenge Problem Specifications: 50 kW Chemical Looping Reactor Simulation. ECP internal report, February 15, 2019.
2. Bayham S, Straub D, Weber J. Operation of the NETL Chemical Looping Reactor with Natural Gas and a Novel Copper-Iron Material," NETL-PUB-20912, 2017.

Adjust the computational effort locally to maintain a uniform level of accuracy throughout the reactor

NATIONAL ENERGY TECHNOLOGY LABORATORY

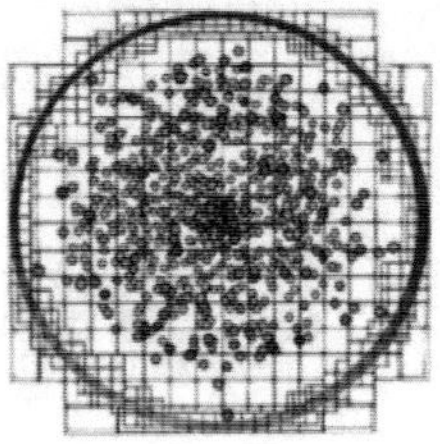

Near-wall mesh refinement, also with corner pruning, for internal pipe flow

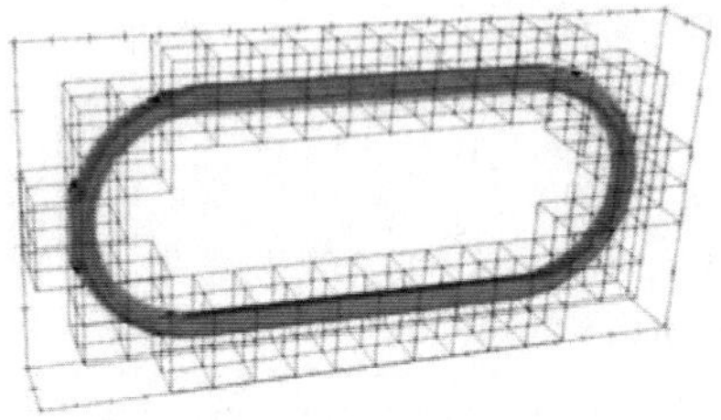

Mesh pruning for a simple loop system

For the challenge problem, this will reduce the mesh size by ~10x

Manage work dynamically to effectively use computational resources

NATIONAL ENERGY TECHNOLOGY LABORATORY

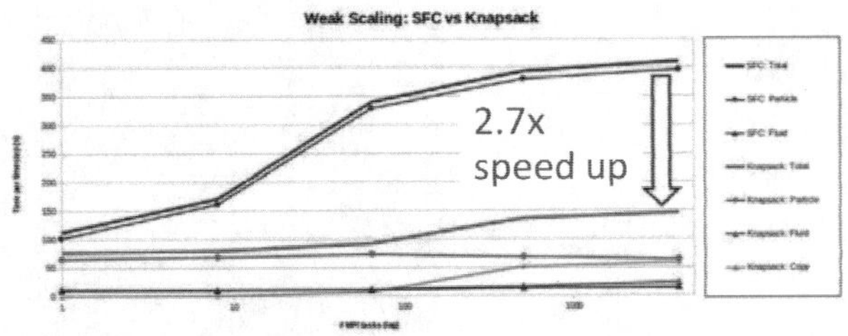

Use dual grids because fluid and solids computations scale differently

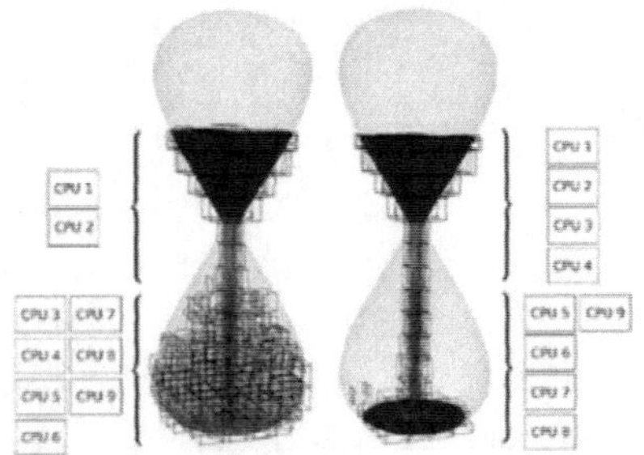

Dynamic load balancing for solids relying on runtime cost measurements

Rosso M et al. Assess, profile and document the performance of MFIX-Exa using basic EB geometries. WBS 2.2.2.04, Milestone ECP-ADSE21-15. December 2018.

MFIX-Exa demonstrates efficient scaling on 15% of Summit supercomputer

NATIONAL ENERGY TECHNOLOGY LABORATORY

- Conducted scaling study of dense (30% solids by volume) gas-particle flows
- Largest simulation contained over one billion computational cells and more than 160 million DEM particles on 4096 GPUs
- Weak scaling showed only a 6.6x increase in time to solution with a 512x increase in problem size.
- Excellent scaling performance demonstrates significant progress towards Exascale readiness.

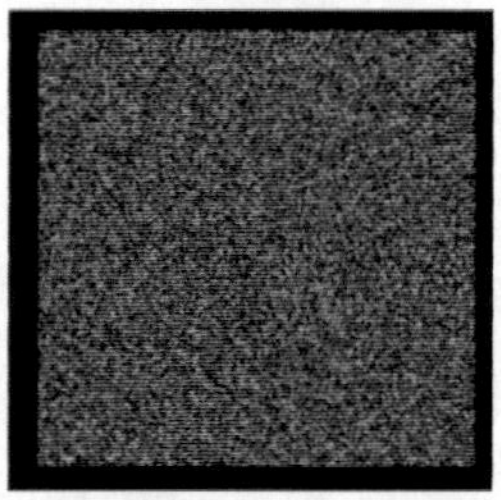

MFIX-Exa simulation of homogeneous cooling system used in scaling study

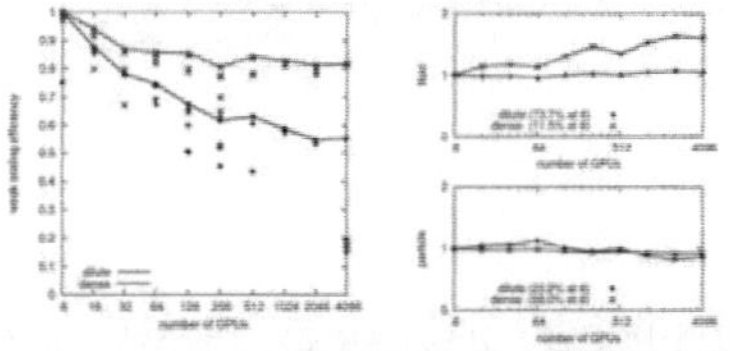

MFIX-Exa Team:
(LBNL) Ann Almgren (co-PI), Andrew Myers, Oscar Antepara, and Michele Rosso
(NETL) Jordan Musser (co-PI), William Fullmer, Roberto Porcu, and Deepak Rangarajan

U.S. DEPARTMENT OF ENERGY 19

2 Using the new wafer-scale processors that offer significantly higher bandwidth

NATIONAL ENERGY TECHNOLOGY LABORATORY

	Joule 2.0	CS-1
Size ratio	52	1
Cores	66,560	400,000
Memory Bandwidth (TB/s)	0.256	9,600
Interconnect Bandwidth (Tb/s)	0.1	100,000
Memory (TB)	160	0.018
Power (kW)	425	20

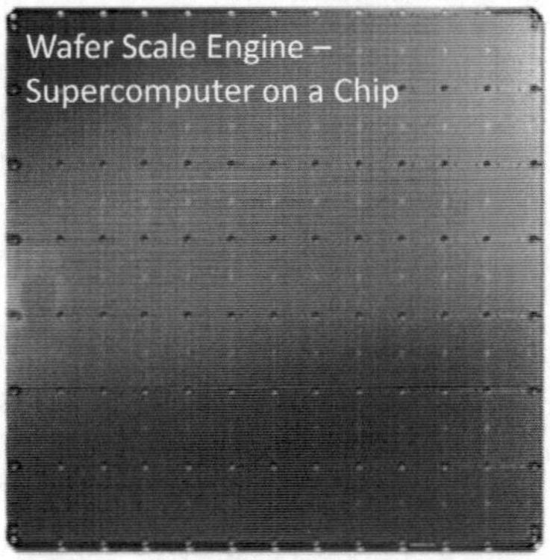

cerebras

Rocki K, Van Essendelft D, Sharapov I, Schreiber R, Morrison M, Kibardin V, Portnoy A, Dietiker JF, Syamlal M, James M. Fast Stencil-Code Computation on a Wafer-Scale Processor. arXiv:2010.03660v1 [cs.DC] 7 Oct 2020

U.S. DEPARTMENT OF ENERGY 20

0.86 PFlop/s achieved on a CS-1 for the BiCGStab linear equation solver

NATIONAL ENERGY TECHNOLOGY LABORATORY

370x370x370 ≈ 51 M cells	Joule 2.0	CS-1
Wall time/iteration (µs)	2100	6
Achieved speed (TFlop/s)	1.1 (DP)	371 (MP)
% of theoretical max speed	0.35	30
600x595x1536 ≈ 548 M cells		
Wall time/iteration (µs)		28
Achieved speed (TFlop/s)		860 (MP)
% of theoretical max speed		30

DP – Double Precision, 64-bit floating point (fp64)
MP – Mixed Precision, fp16 and fp32

Rocki K, Van Essendelft D, Sharapov I, Schreiber R, Morrison M, Kibardin V, Portnoy A, Dietiker JF, Syamlal M, James M. Fast Stencil-Code Computation on a Wafer-Scale Processor. arXiv:2010.03660v1 [cs.DC] 7 Oct 2020

Also see paper 630a of this conference

Mixed precision and Transprecision Computing

NATIONAL ENERGY TECHNOLOGY LABORATORY

- A CFD solver on CS-1 would need to use mixed precision computing
  - BiCGStab solver used fp32 for 4 dot products and fp16 for the remaining 40 + and * operations per iteration per mesh point
  - Linear equation algorithms for mixed precision computing have been developed[1]
  - NASA Fun3d CFD code includes fp32 and fp16 calculations
- Features of *Transprecision* computing[2]:
  - Use adaptive precision during computation
  - Offload large part of computation to low power HW units; e.g., PULP accelerator (https://pulp-platform.org/)
  - Recover accuracy by end of calculation (up to required level)

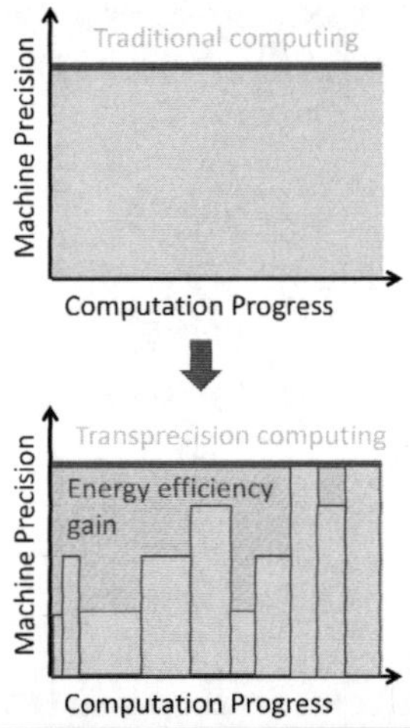

1. Baboulin M et al. Accelerating scientific computations with mixed precision algorithms. Computer Physics Communications 180 (2009) 2526–2533; Carson E. Higham NJ. Accelerating The Solution of Linear Systems by Iterative Refinement in Three Precisions. SIAM J. Sci. Comput. 40, A817--A847 (2018).
2. Diamantopoulos D. Transprecision Computing. OpenPower Summit Europe. Amsterdam. October 2018.

3 MFIX-AI leverages emerging AI/ML computing technologies

NATIONAL ENERGY TECHNOLOGY LABORATORY

MFiX + = MFiX-AI

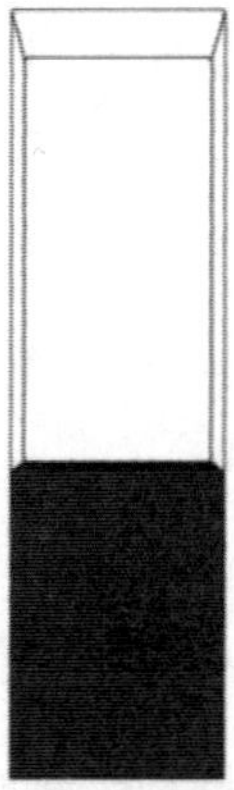

- Create a single, unified framework for doing both CFD and AI/ML on emerging hardware for getting speed boosts from
  - Hardware
  - Software and Algorithms
- TensorFlow, the most widely used AI/ML framework, was chosen as the platform
- Wrote a first-generation MP-PIC granular motion solver that is almost 10x faster than MFiX classic

MFIX-AI Team:
(NETL) Dirk Van Essendelft (PI), Terry Jordan, Mino Woo, Tarak Nandi

3.8 Million Cells, 16.8 Million Parcels

U.S. DEPARTMENT OF ENERGY 23

Concluding Remarks

NATIONAL ENERGY TECHNOLOGY LABORATORY

- Evolution of computing technologies and some challenges
- Adoption of emerging technologies for advancing multiphase CFD through MFIX open-source software
  1. Exascale computing
  2. Wafer scale engines
  3. AI/ML for speed up

Contact: Madhava.Syamlal@netl.doe.gov

U.S. DEPARTMENT OF ENERGY 24

THANK YOU!

VISIT US AT: www.NETL.DOE.gov

@NETL_DOE

@NETL_DOE

@NationalEnergyTechnologyLaboratory

CONTACT:
Madhava Syamlal
Madhava.Syamlal@netl.doe.gov

597957 Chemical Looping – Oxidative Dehydrogenation of Ethane Using Surface Modified Mixed Oxide Particles: Kinetic Studies and Process Demonstration

Abstract:

Ethylene is commercially produced from ethane via high-temperature pyrolysis in the presence of diluting steam a.k.a. steam cracking. The process requires high temperature (>1100 K) and a large steam generation load, making it energy intensive with an energy consumption of 17-21 GJ/tonne ethylene. When compared to steam cracking, Chemical Looping – Oxidative Dehydrogenation (CL-ODH) has the potential to substantially reduce the energy demand and to increase the single-pass ethane conversion. The ability of a redox catalyst for selective hydrogen combustion (SHC), i.e. selectively oxidizing hydrogen co-product from ethane dehydrogenation, represents an effective strategy for CL-ODH because it can shift the reaction equilibrium and facilitate autothermal operation.

In this work, redox kinetics of unpromoted and Na_2WO_4-promoted $CaMnO_3$ based redox catalysts were investigated in the context of CL-ODH. In the case of unpromoted $CaMn_{0.9}Fe_{0.1}O_3$, the reduction kinetics follow reaction-order models. The activation energy for C_2H_4 reduction is approximately three times greater than that for H_2 reduction. The greater dependence on active lattice oxygen concentration and the larger energy barrier under C_2H_4 reduction result in an order-of-magnitude decrease in the reduction rate. Despite of its favorable SHC properties, $CaMn_{0.9}Fe_{0.1}O_3$ redox catalyst leads to significant CO_2 formation in CL-ODH of ethane. To further improve the redox catalyst's selectivity towards hydrogen combustion, a Na_2WO_4 promoter was impregnated on $CaMn_{0.9}Fe_{0.1}O_3$. The reduction of Na_2WO_4-promoted $CaMn_{0.9}Fe_{0.1}O_3$ follows an Avrami-Erofe'ev nucleation and nuclei growth model. The addition of Na_2WO_4 more significantly suppressed C_2H_4 combustion relative to H_2 oxidation. As a result, the reduction rate of Na_2WO_4-promoted $CaMn_{0.9}Fe_{0.1}O_3$ under H_2 was three orders of magnitude greater than that under C_2H_4, demonstrating its excellent SHC properties.

In addition to $CaMn_{0.9}Fe_{0.1}O_3$ based redox catalysts, a number of other mixed oxide redox catalyst particles with different alkali metal salt promoters were investigated, showing promising performance at different operating temperature ranges. Large packed bed studies were performed for over 1000 cycles to demonstrate the long term stability and performance of a few promising "low-temperature" redox catalysts. Autothermal operation and favorable heat of reactions in both reduction and oxidation steps were demonstrated. In addition, long-term fluidized bed studies were also performed on a "high-temperature" redox catalyst, showing near 70% single-pass ethylene yield and excellent long-term stability.

Both the kinetic studies and long-term experiments validate the feasibility and attractiveness of the CL-ODH approach.

598694 Linear Vibration and Mixing Behavior in a Continuous Spatial Particle ALD Reactor

Abstract:

Continuous spatial Particle ALD reactors enable powders to be continuously coated with sub nanometer films by sending the powder through alternating regions of precursor gas using a process called linear vibration. Linear vibration systems use a self-reversing actuator to vibrate the bed, producing steady and uniform powder flow. Some materials do not flow as easily using linear vibration, particularly as the powder diameter decreases. Fine powders are frequently used in Particle ALD, so understanding and predicting their flow behavior is essential for process control. Another important aspect of process control is guaranteeing equal exposure time of each particle to the precursor gas. For thick powder beds, effective powder flow and top-bottom mixing must occur simultaneously to guarantee equal exposure time with high powder throughput. In this work, the relationship between linear vibration and mixing in a continuous spatial Particle ALD reactor is investigated. An MFIX-DEM model was developed and the results were compared to in-situ images of dyed glass microsphere mixing during continuous bed vibration.

607725 Environmentally Sustainable Pathways for Extracting Valuable Elements from Electronic Wastes

Abstract:

Electronic wastes including printed circuit boards (PCBs) are some of the fast-growing wastes threatening our environment. Meanwhile, industrial mining continues to provide materials for fast-growing electronic industry. Thus, there is a strong need for sustainable resources for electronic materials. Interestingly, the high metal content of PCBs makes it an excellent source for urban mining and the recovery of metals from PCBs will provide a sustainable pathway to reduce the conventional mining and the landfilling of electronic wastes. Within the PCB board, the metal connectors contain considerable amount of coated gold, thus making them the most valuable resource to recover. In terms of total quantity, copper is a valuable resource that can be recovered from PCBs.

Unfortunately, the current resource recovery methods for PCBs, including intensive physical separation, pyrometallurgy, and hydrometallurgy with strong acids, are energy intensive, and environmentally hazardous. Most importantly, the complex compositions of PCBs connectors include multiple layers of metals, fiber glass board, solder masks, surface finishing, etc., which increases the difficulty to recover valuable resource without generating environmental and human health concerns.

In this study, a greener method to extract the metals (e.g. Au, Pd, Cu) from waste PCBs connectors is investigated, which aims to lower the environmental and human health hazards generated during the extraction process. This study employs supercritical CO2 to facilitate physical separation and/or hydrometallurgy extraction. Our study shows that the utilization of CO2 can reduce the use of strong acids during metal leaching, and inherently recover precious and base metals at different stages and in different forms, thus enhancing the overall sustainability of the PCB treatment processes. The feasibility and mechanisms of using CO2 (either in its liquid or supercritical forms) to enhance the metal recovery is analyzed and the role of CO2 in the physical and chemical alterations of polymer-metal matrix in PCBs is investigated.

611021 Numerical Study of Particle Dynamics during Flame Spray Synthesis of Nanoparticles

Abstract:

A numerical study of a process, such as flame spray pyrolysis (FSP), in which many different phenomena take place simultaneously, is important in order to gain insight in the characteristics of the final product. The properties of the synthesized particles depend on the composition, structure and size of agglomerates/aggregates and constituent primary particles [1]. Hence, the description of their evolution within the highly reactive and turbulent flame environment is of major importance for nanoparticle production by FSP. In this regard, population balance models (PBM) have been considered as an effective tool for the simulation of nanoparticle dynamics [2].

In the present study, modelling of particle dynamics during FSP is performed assuming that there is no surface growth. Monomers are formed by instantaneous chemical reactions and grow to clusters and particles due to collisions. As FSP is a complex process, mainly because of formation and growth of particles by agglomeration/aggregation and sintering, a bivariate population balance equation is solved by the quadrature method of moments (QMOM) that better describes the particle population [3]. This approach takes into consideration polydispersity of the produced particles. Particle number concentration, sintering rate and residence time of nanoparticles are calculated and the obtained results are compared with results from the monodisperse model of Kruis *et al.* [4], which describes the size distribution by its mean value. The obtained size values are compared with experimental data by nitrogen adsorption measurements. Coupling the control of growth processes with accurate characterization tools and modelling approaches is prerequisite for process scale-up and consequently development of high-performance materials.

References

1. Tsikourkitoudi V., Karlsson J., Merkl P., Loh E., Henriques-Normark B., Sotiriou G.A., Flame-made calcium phosphate nanoparticles with high drug loading for delivery of biologics. *Molecules* 2020;25:
2. Tsantilis T., Kammler H.K., Pratsinis S.E., Population balance modeling of flame synthesis of titania nanoparticles. *Chemical Engineering Series* 2002;7:2139-2156.
3. McGraw R., Description of Aerosol Dynamics by the Quadrature Method of Moments. *Aerosol Science and Technology* 1997;27:255-265.
4. Kruis F.E., Kusters K.A., Pratsinis S.E., Scarlett B., A Simple Model for the Evolution of the Characteristics of Aggregate Particles Undergoing Coagulation and Sintering. *Aerosol Science and Technology* 1993;19:514-526.

605692 Novel Nanoscale Hybrid Electrolytes with Unique Tunability for Sustainable Energy Storage

Abstract:

The recent growth in the worldwide installation of renewable energy has allowed the first steps towards decarbonization of power sector. However, further penetration of renewable energy into the grid is challenged due to the limitations of storage of intermittent renewable electricity. Thus, vast research efforts are on-going to develop electrochemical energy storage systems such as batteries and dense energy carriers. Electrolyte design and selection plays a key role in the development of these electrochemical energy systems, mediating transport and solubility of the reactive species to ultimately deliver high energy, current and power density systems.

A novel class of nanoparticle organic hybrid materials (NOHMs) has been developed that can serve as breakthrough electrolytes with concentrations of electrochemically active species that are not achievable in other conventional solvents. NOHMs based mixtures have shown higher conductivity compared to pure polymer and due to the tunability of the constituent polymers, NOHMs can be tailored as versatile electrolyte media for redox flow batteries and other electrochemical systems, such as electrochemical conversion of CO_2 to dense energy carriers, NOHMs-based electrolytes have shown high CO_2 solubility, as well as the ability to selectively complex redox active species such as copper and zinc, increasing their effective concentration in solution and enabling improvements in energy densities.

NOHMs-based electrolytes have been found to have highly tunable physicochemical and electrochemical properties. The electrochemical behavior of NOHMs-complexed copper (II) species has been explored, indicating changes of the overall conversion mechanism through stabilization of the copper (I) state. This electrochemical behavior can be tailored by varying the solution pH, by which the degree of copper (II) chelation can be controlled. Tailoring transport properties is also of particular importance in NOHMs-based electrolytes mixtures, as they are challenged by inherently high viscosities, impacting charge transport critical in electrochemical performance. These physicochemical properties can be tuned by controlling the intermolecular interactions between the polymeric canopy and the surrounding fluid. In particular, NOHMs have been found to be highly responsive to ionic stimuli, with the addition of even low salt concentrations inducing large reductions in the viscosity of NOHMs-based electrolyte mixtures. Alterations in the degree of polymer swelling and the conformational structure of the NOHMs polymer canopy with ionicity have been probed via Dynamic Light Scattering and Diffusion NMR to explain measured bulk physicochemical properties.

Overall, NOHMs have been found to be electrolyte materials enabling high concentrations of electrochemically active species and with highly tunable transport properties, making them versatile electrolytes for a variety of electrochemical energy storage systems.

601257 Computational Materials Screening and Experimental Validation of Hydrogen Environmental Barrier Coatings Via Atomic Layer Deposition

Abstract:

Hydrogen environmental barrier coatings can be used to reduce hydrogen embrittlement of structural materials by reducing diffusion into susceptible substrates. This protection is especially important in hydrogen pipelines and fields such as nuclear and fuel cell applications where hydrogen is used as a fuel source. The higher the operational temperature, though, the fewer options for material choice. A nanoscale film of the current state-of-the-art material for nuclear thermal propulsion (NTP), tungsten, was previously shown to increase the temperature at which hydrogen reacts with the substrate. However, the increase was still far from the operational temperature of these nuclear applications. In this work, density functional theory (DFT) calculations for energy barriers of atomic hydrogen diffusion were used to compare likelihood of various refractories in preventing hydrogen diffusion. The larger the energy barrier, the less likely hydrogen is to diffuse through the material. The bonding behavior of hydrogen as it diffuses into the surface of these materials and the rearrangement of atoms to accommodate it were also analyzed. To experimentally confirm these predictions, the material predicted to perform best, BN, was deposited via atomic layer deposition (ALD). The films were then characterized and the samples tested in a high temperature hydrogen environment using differential thermal analysis. Results showed that BN performed as predicted, preventing reaction of hydrogen with the coated sample above 1400°C.

609529 Understanding and Predicting the Efficacy of Cold Flow Improvers

Abstract:

Unwanted crystallization of wax components in fuels and crude oils has a major impact on product performance, composition and overall usability. Currently, this issue is addressed through the development of chemistries known as cold flow improvers (CFIs) that depress the cloud point of products. However, different crude feedstocks, with different wax component compositions, mean a range of CFIs have to be utilized/tested to ensure efficacy. This is further exacerbated in diesel fuels with the ever-growing use of biofuels, which, alter fuel composition and create a higher proportion of wax components within the fuel[1]. Presently, cloud-point depressant tests are performed to establish which CFI has the greatest impact on wax crystal suppression, but it is not possible to predict when a wax nucleation event may occur within a product that is in a state of thermodynamic and kinetic metastability. This work focuses on a method that enables the prediction of wax crystallization in the presence of CFIs, through accelerated nucleation testing using the isothermal by design (IbD) methodology [2]. This is coupled with wax solubility and nucleation kinetic analysis, whereby the mechanism of nucleation is divulged.

A polythermal screen of 5 CFIs was performed for the model system of eicosane in toluene solution, with the KBHR approach[3,4] undertaken to assess nucleation kinetics. Eicosane in toluene was found to nucleate through instantaneous nucleation (IN), whereas all CFIs were found to alter the nucleation pathway to progressive nucleation (PN). This is important to product flow assurance, as IN would cause catastrophic wax nucleation throughout the product, whereas the capability to alter this to PN would mean crystals would not appear all at once and flow would still be possible. One CFI was found to have the largest efficacy, increasing the metastable zone width (MSZW) to nucleation by 3 times, shown in figure 1, and was used for further study.

For the IbD methodology a second miscible solvent was required, in order to perform antisolvent crystallization experiments. For this purpose, acetone was selected, owing to the low solubility of waxes in this solvent. The solubility in the absence and presence of the CFI over a large compositional range of toluene:acetone mixtures in 10vol% increments was assessed, with a sharp decrease found with increasing acetone content. IbD results were obtained following a workflow previously developed by the author [2]. Induction times (the time to crystallization after a level of supersaturation is reached) were reduced by 4 orders of magnitude over the range of supersaturations studied, demonstrating the ability to accelerate nucleation. Furthermore, results were found to lie on one curve, meaning extrapolation of results obtained for short induction times could be used to determine those at longer induction times. These findings are extremely important for predictive testing as it enables experimental work to be conducted in a short time frame, but results obtained can be used to determine effects over longer timescales that would be present in industrial product applications and are not accessible in normal laboratory experimental work.

References

1) ATC - Technical Committee of Petroleum Additive Manufacturers in Europe, 2013, 68.

2) P. L. Kaskiewicz, G. Xu, X. Lai, N. J. Warren, K. J. Roberts, C. Morton, P. Dowding and N.

George, Organic Process Research and Development, 2019, 23, 1948–1959.

3) D. Kashchiev, A. Borissova, R. B. Hammond and K. J. Roberts, Journal of Crystal Growth, 2010, 312, 698–704.

4) D. Kashchiev, A. Borissova, R. B. Hammond and K. J. Roberts, Journal of Physical Chemistry B, 2010, 114, 5441–5446.

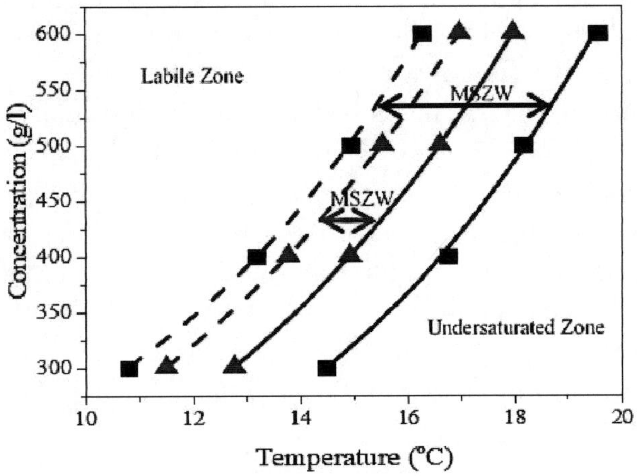

Figure 1. MSZW comparison for eicosane in toluene (blue) and eicosane:CFI in toluene (black). Solid lines show solubility lines. Dashed lines show supersolubility lines.

609592 Isothermal By Design: Accelerated Nucleation of Aqueous and Non-Aqueous Solution Systems

Abstract:

1. **Introduction**

Nucleation is a phenomenon that can occur in a matter of seconds or a period of years depending upon the ease with which a system will crystallise. Conventional crystallisation methods do not allow for slow nucleating systems to be studied in a relatively short and useful timeframe, instead systems have to be left supersaturated for an inordinate time period or the nucleation events are not observed. Also, a vast number of pharmaceutical crystallisations are performed under high supersaturation conditions using antisolvent (drown out) crystallisation methodologies, but currently it is not easy to determine key kinetic crystallisation parameters at such high supersaturations prior to industrial scale crystallisations. Both of these issues are addressed and overcome in this work.

Isothermal by Design (IbD) is a methodology used for studying nucleation kinetics over a large range of supersaturations, utilising an antisolvent crystallisation procedure. It has been shown to achieve high levels of supersaturation that are unachievable through conventional crystallisation methodologies, whilst enabling key nucleation parameters to be calculated[1]. A key aspect of IbD is its ability to take into account the enthalpy of mixing associated with adding an antisolvent to a solution through simple calorimetry calibrations, ensuring isothermal conditions are maintained.

This work focuses on results obtained utilising IbD for an aqueous pharmaceutically relevant system of p-aminobenzoic acid (pABA) in ethanol:water mixtures and a non-aqueous fuel and hydrocarbons relevant system, eicosane:additive (C20:A) in toluene and acetone mixtures. Nucleation kinetics for both systems were examined using classical nucleation theory (CNT).

2. **Materials & Methods**

2.1. Materials

pABA was supplied by Alfa Aesar, ethanol absolute and acetone were supplied by VWR, toluene was supplied by Honeywell, C20 was subpplied by Sigma-Aldrich, the additive was supplied by Infineum Uk Ltd. and deionized water was sourced on site at the University of Leeds.

2.2. Experimental Procedure

All experiments were performed on the Technobis Crystal16 system[2] at a 1 ml solution scale.

Calorimetry calibrations were performed in order to understand and take account of the enthalpy of mixing associated with antisolvent addition processes. Solutions of pABA in ethanol:water 70:30 wt% saturated at 293 K were held between 293 - 301 K in 2 K increments, containing various initial solution volumes from 0.2 - 0.8 ml. Water (antisolvent), held at different temperatures, was added to the solutions to make the final solution volume up to 1 ml. The exothermic spike in solution temperature was recorded, using a digital thermometer inserted into the solutions, and recorded. Final solution temperature was plotted as a function of antisolvent temperature for each volume studied and a linear regression was fitted through the points. Extrapolation of the linear regression to the initial solution hold temperature determined at what temperature the antisolvent must be to maintain solution temperature and ensure isothermal conditions were achieved throughout the experiments. For C20:A (99.9wt%:0.1wt%) in toluene solutions, saturated at 278 K, the same procedure was completed using acetone as an antisolvent, with solutions held at 283 K.

Induction time experiments were performed to enable key nucleation kinetic parameters to be calculated through CNT analysis. These experiments were performed in the same manner as the calorimetry calibrations, but the determined 'isothermal' antisolvent temperature was used for each temperature and initial volume of solution, no digital thermometer was present to ensure nucleation was not affected by the large temperature probe heterogeneous surface and induction times were determined through transmission data.

Data analysis was performed as outlined in a previous research article[1].

3. **Results & Discussion**

3.1. Calorimetry Calibrations

The antisolvent temperature was found to have a pronounced effect on the solution temperature of both the aqueous and non-aqueous systems. For both systems studied, at the solution temperatures used, antisolvent addition caused an exothermic effect, in which, the solution temperature was raised, even with a lower temperature antisolvent. Therefore, in order to ensure an isothermal system was used, antisolvent temperature had to be offset (lowered) to negate the exothermic enthalpy of mixing.

3.2. Induction Time

Induction times were reduced from around 10,000 s - 1 s, for the pABA system studied, over a range of supersaturations from 1.1 - 3.7. A similar range of induction times were probed for the C20:A system, but over a much smaller range of supersaturations, due to the metastable zone width (MSZW) of the C20:A system being much smaller than for the pABA system, 3 - 4 K in comparison to 4 - 8 K, respectively. The higher supersaturations achieved for the pABA system using IbD were much higher than those that could be achieved through conventional isothermal crystallisation methods, where a solution is cooled to a set supersaturation and the induction time to nucleation is measured. This is due to the initial cooling period required to reach a certain level of supersaturation, which limits achievable supersaturations. In comparison, for the pABA system, IbD achieved levels of supersaturation up to 5, but above 3.7 the system nucleated too fast to be detected by the instrumentation used.

The ability to accelerate the induction time by orders of magnitude with relative ease, for both aqueous and non-aqueous crystallising systems, demonstrates IbD's efficacy with regards to accelerated nucleation testing.

3.3. Nucleation Kinetic Parameters

Values of effective interfacial tension (), critical nucleus radius () and number of molecules within the critical cluster () were calculated from the induction time vs. supersaturation data and CNT. For all solutions studied, and were found to decrease, in a non-linear fashion, as a function of supersaturation. Vales of showed strong compositional dependence, with large ranges of values of the mixed solvent systems studied as well as between the aqueous and non-aqueous systems.

The ability to determine nucleation kinetic information at high supersaturations for both aqueous and non-aqueous crystallisation systems is very important in an industrial setting, enabling accurate information about the system to be known and the operation conditions to be fine-tuned to improve crystallisation processes. Furthermore, this is extremely important for accelerated testing, where accurate kinetic data must be obtained at high supersaturations, so that accurate extrapolations of solution behaviour can be made to 'real life' solution conditions.

3.4. Nucleation Mechanism

The large range of supersaturations studied through IbD for pABA solutions enabled a nucleation mechanism change from heterogeneous nucleation (HEN) to homogeneous nucleation (HON) to be observed at a transition supersaturation of around 1.5. This was indicated through a sharp change in over values collected for a given solution temperature[3,4]. No such mechanism change was observed over the range of supersaturations studied for the non-aqueous C20:A system.

The knowledge of nucleation mechanism changes within a crystallising system is very important in an industrial setting as it enables changes in operating conditions to account for potential beneficial crystallisation control.

4. Conclusions

IbD was used to study two contrasting crystallisation systems, an aqueous system of pABA in ethanol:water mixtures with water antisolvent and a non-aqueous system of C20:A in toluene with acetone antisolvent. Antisolvent calorimetry calibrations showed that the antisolvent temperatures used had to be reduced in comparison to the solution temperature, to take into account the exothermic enthalpy of mixing associated with antisolvent addition, and ensure isothermal conditions were maintained for induction time analysis. Induction times were reduced by orders of magnitude for both systems studied, with the higher supersaturations and lower induction times being accessible compared to conventional isothermal crystallisation methodologies. This is particularly important with regards to accelerated nucleation testing capabilities and probing nucleation kinetics at higher supersaturations. Key nucleation kinetic parameters were determined through induction time data and CNT, allowing a large range of solution conditions to be probed and understood, as well as demonstrating IbD's ability to determine key nucleation information over a vast range of supersaturations. A nucleation mechanism change was also observed for the aqueous pABA system studied.

References

1) Kaskiewicz, P. L. et al. Isothermal by Design: An Accelerated Approach to the Prediction of the Crystallizability of Slowly Nucleating Systems. Organic Process Research and Development 23, 1948–1959 (2019).

2) Technobis. Crystal16. Available at: https://www.crystallizationsystems.com/crystal16.

3) Söhnel, O. & Mullin, J. W. A method for the determination of precipitation induction periods. Journal of Crystal Growth 44, 377–382 (1978).

4) Kim, K. J. & Kim, J. K. Nucleation and supersaturation in drowning-out crystallization using a T-mixer. Chemical Engineering and Technology 29, 951–956 (2006).

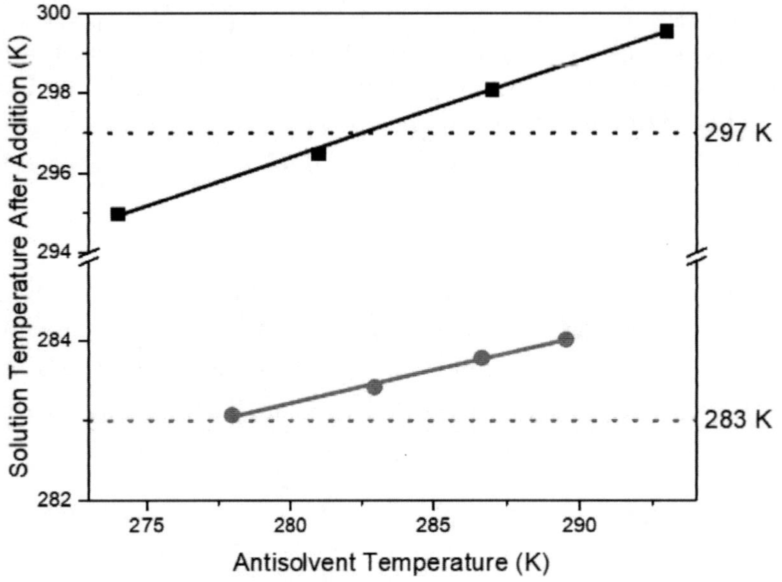

Figure 1. *Calorimetry calibrations - effect of antisolvent temperature on solution temperature after antisolvent addition. Water added to pABA in ethanol:water 70:30 wt% solution, initial solution volume of 0.5 ml (■). Acetone added to C20:A in toluene solutions, initial solution volume of 0.6 ml (•). Horizontal dotted lines show temperature solutions were held at (black dotted line for pABA solutions, red dotted line for C20:A solutions. Break applied to graph between 285 - 294 K.*

604658 High-Throughput Continuous Flow Synthesis of Peg-Functionalized Iron Oxide Nanoparticles for Biomedical Applications

Abstract:

Magnetic nanoparticles, in particular iron oxide, have been extensively investigated as they are a promising biotechnological platform, with applications such as MRI contrasting agents, drug delivery, cell labelling and sorting and magnetic hyperthermia. However, their large scale manufacturing to translate the promising research results into the real world still remains a challenge. Large scale synthesis of iron oxide nanoparticles of controlled size has been demonstrated but typically relies on high temperature thermal decomposition in the presence of hydrophobic surfactants. A second and lengthy ligand exchange step is then necessary, in order to stabilize the particles in water/biological environments.

In this work, for the first time, iron oxide nanoparticles were synthesised and functionalised in a single setup using a platform of microreactors connected in series. The nanoparticles obtained have a highly stable polyethylene glycol (PEG) shell anchored with a bisphosphonate/catechol moiety, ensuring water dispersity and stability in biological media. Carboxylic acid end chain groups allow for further derivatization of the magnetic nanoplatform (*e.g.* with fluorophores). The synthesis rig can reproducibly generate large amounts of these high quality stable iron oxide nanoparticles, enabling their deployment in real world applications.

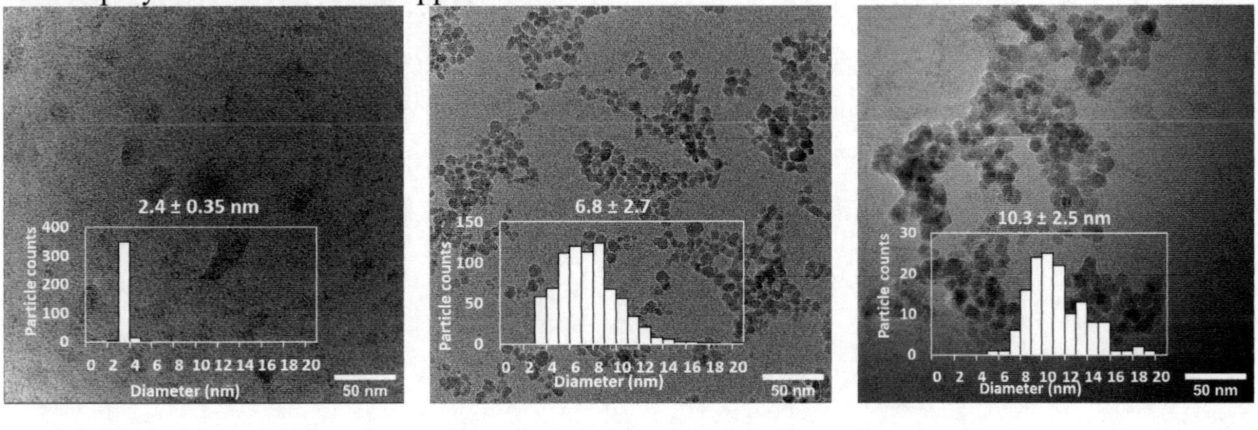

608335 Design of Nanoscale Hybrid Electrolytes Enabling Combined CO2 Capture and Conversion

Abstract:

Due to the rapid cost reduction of renewable energies, it is now possible to convert CO2 to chemicals and fuels using electrochemical pathways. In addition to catalyst development and reactor design, it is also important to consider the design of electrolytes that can effectively deliver CO2 to electrodes with desired product selectivity and reaction rates. Novel reactive materials have been shown to improve CO2 solubility, tune the selectivity of the CO2 reduction reaction and enhance reaction rates. In this study, Nanoparticle Organic Hybrid Materials (NOHMs)-based fluids have been designed as hybrid electrolyte systems that are capable of combined CO2 capture and conversion, without the need for a separate solvent regeneration step. Developing NOHMs-based electrolytes requires a fundamental investigation into the structuring and organization of these complex nanoscale fluids. A study of the mixing behaviors of NOHMs with a variety of secondary fluids has revealed that physical properties such as hydrodynamic size, viscosity and diffusion coefficient are highly dependent on the hydrogen bonding ability and size of the secondary fluid molecule. Additionally, it was discovered that the viscosity and the diffusion coefficient of the untethered polymer were found to scale according to theoretical predictions in the semi-dilute unentangled regime, while this was not observed for NOHMs-based fluids. Lastly, the CO2 solubility of NOHMs-based electrolytes has been measured and a change in the selectivity of the CO2 reduction reaction has been identified in the presence of these nanoscale electrolytes. The results of this work highlight the characterization and performance of a novel reactive medium designed for combined CO2 capture and conversion applications.

604811 Microencapsulated UV Filters@ZIF-8 Sunscreens for Broad Spectrum UV Protection

Abstract:

Sunscreens play a vital role in protecting human skin from photodamage upon UV exposure, but their effectiveness is limited by degradation and phototoxicity. Photoactivation of UV filters in sunscreen can generate reactive oxygen species, with accompanying cytotoxicity. Herein, we report stable, safe, fluorescent, and long-lasting sunscreens produced by microencapsulation of organic UV filters octylmethoxycinnamate (OMC) and butyl-methoxydibenzoylmethane (BMDM) alone and in combination into mesoporous zeolitic imidazole frameworks (ZIF-8). The visibly transparent OMC@ZIF-8, BMDM@ZIF-8, and OMC+BMDM@ZIF-8 sunscreens displayed complete broad-spectrum photoprotection against UV light even after 24 hours. Photostability testing of the designed sunscreens demonstrated the photo-protective nature of mesoporous ZIF-8 for stabilizing these UV filters. The sun protection factor (SPF) values for 10% (w/w) OMC@ZIF-8, BMDM@ZIF-8 and OMC+BMDM@ZIF-8 sunscreens decreased from 29.8 to 26.0, 32.5 to 26.4, and 62.1 to 53.0, respectively upon 24 hours of UV exposure. No release of the UV filters from the encapsulated formulations was observed. The ZIF-8 particles are too large to be transported across the stratum corneum. Thus, our findings suggest that zeolitic framework-encapsulated sunscreens could provide prolonged UV-protection efficacy and ultra-high SPF with reduced potential for degradation, phototoxicity, and transport across the skin.

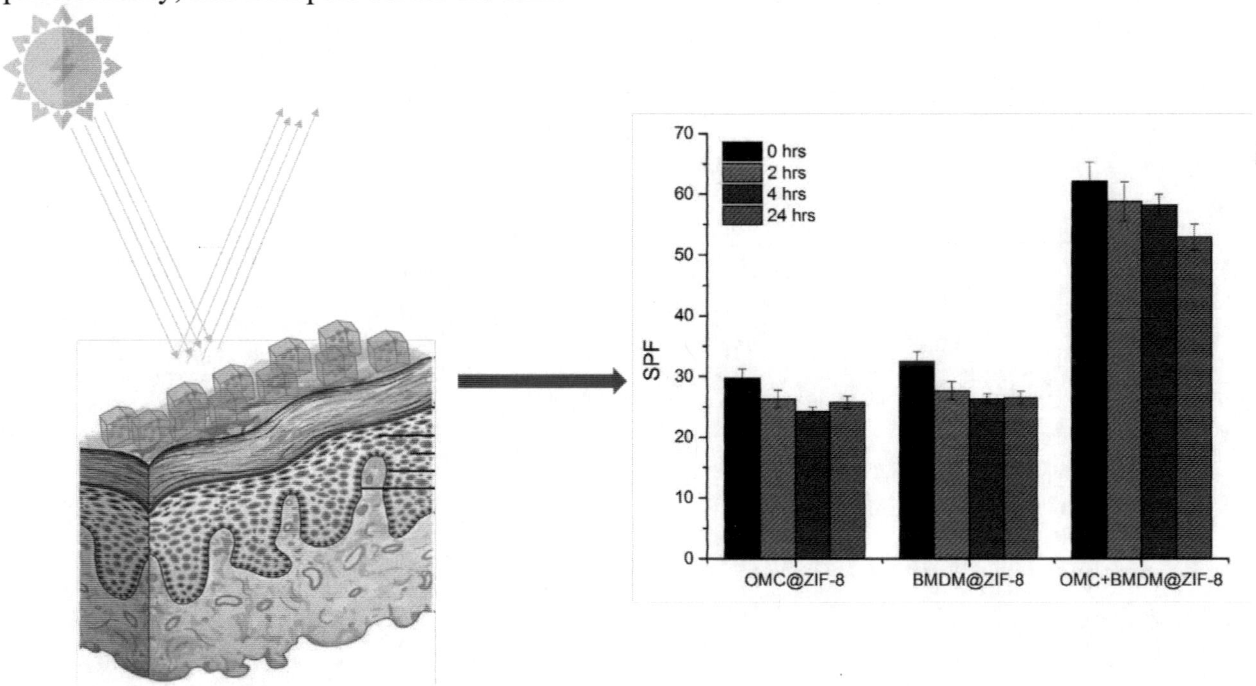

604719 Deformation and Wear in Nanocomposites

Abstract:

Fundamental atomistic and molecular processes have been applied to tribological contacts. Wear mechanism can occur from abrasion, adhesion and by fatigue.. Nanocomposites depict elastic or elastic plastic behavior and particle to particle interactions depending on the extent of deformation and dispersion of particles in the matrix. However, these interactions are balanced in each nanocomposite systems by the surface treatments (chain grafting, filler modification) and the molecular weight of the matrix. Nano-indentation of polymer films undergoes strain softening and strain hardening depending on the stiffness of the surface at different depths. Our results on polycarbonate and on polyester containing nanoparticles of silica showed that there is an increase in modulus and hardness up to a certain concentration in threshold of concentration. As the level of nanoparticles increases beyond a threshold level, aggregates form which results in weakening of the structure. Polymer nanoparticles interface is found to be weak as filler is non-interacting promoting interfacial slip at the filler- matrix junctions. At critical stress level for silica –polymer interfacial slip is reached and filler-matrix sliding is activated resulting in a decrease in the composite's stiffness. Wear and friction are computed considering the effect of the interface between the hard filler and relatively soft polymers.

616955 Implementation and verification of numerical solution of the bivariate population balance equation in a finite-element open-source framework

Abstract:

Polydisperse multiphase flows appear in a large number of processes in the chemical industry. The population balance equation (PBE) is frequently used for the modelling of the evolution of such dispersed-phase systems. Monovariate PBE models, characterized by one specific property (or one internal coordinate), can be effectively managed by a variety of numerical techniques, while the solution of bivariate (or, in general, multivariate) PBEs is still considered a challenging task. While in the past much research has been conducted on the coupling of the solution of the PBE to computational fluid dynamics (CFD) packages, numerical methods have typically been computational expensive. An implementation of an accurate and computationally affordable numerical method for solving multivariate PBEs is still required. Notably, to date, an implementation of DQMOM for solving the bivariate PBE is not available in any open-source CFD package as a standard feature. The present work focuses on the implementation of the direct quadrature method of moments (DQMOM), a favourable numerical method for the PBE-CFD coupling and computationally economical for solving the bivariate PBE in a finite element (FE) open-source code – Fluidity. This FE framework is highly-parallelised with adaptive-mesh refinement on fully-unstructured meshes. In order to investigate the accuracy of the bivariate PBE solution in the proposed framework, transient homogeneous bivariate aggregation and bivariate breakage simulations were performed and verified against analytical solutions, showing excellent agreement. The results show that the proposed finite element framework produces precise numerical solutions for the bivariate PBEs and is a robust tool that can be exploited for the modelling of complex polydisperse multiphase flows.

595142 Deciphering the Effect of Morphology of Fillers in Thermal Conductivity of Nanofluid Using Diffusion-Based 3-D Modeling

Abstract:

In this work, a Finite Element Analysis (FEA) model has been developed using COMSOL Multiphysics® to estimate the effective thermal conductivity of nanofluids. Representative Volume Element (RVE) containing copper (Cu) nanofillers in ethylene glycol (EG) was created using MATLAB. Different morphologies such as spheres, cylinders, and triangular plates have been used as fillers in this study.

First, the required number of non-overlapping filler elements are placed inside the RVE randomly, and the domain is discretized. Then the steady-state heat conduction equation was solved for the composite using a cubic geometry, as shown in Figure-1. Temperature boundary condition was used in one pair of opposite surfaces, and the lateral surfaces were insulated. The effective thermal conductivity was obtained by using the steady state heat flux through the composite.

This model predicts around 26% increase in thermal conductivity in compared to the base fluid when rod-like filler was used at 0.15 vol% loading. In contrast, only 0.5% improvement is obtained for the particle for the same loading. The enhancements obtained for the spherical copper nanoparticles are in close agreement with the Hashin Shtrikman (HS) lower limit or the Maxwell model. Nanoplates show enhancements in between those of particles and wires. These simulation results suggest that elongated morphologies provide better enhancement than spherical geometries.

Since the morphology effect is very strong, significant effect of particle aggregation is also expected. For a given volume fraction, aggregated particles may form elongated aggregates which can increase the effective thermal conductivity significantly. This point has been confirmed by the simulation, and it is also found that the degree of enhancement is dependent on the nature of aggregate: a more elongated structure will produce more enhancement. For identical volume fraction (0.6 vol%), merely changing the state of aggregation changes the enhancement from 1.8% to 3.8%. All these shreds of evidence clearly indicate that the 'anomaly' usually reported for nanofluid is a morphology effect created by unstable colloids. Whenever a well-dispersed colloid is formed, such anomaly vanished producing the enhancement predicted by Maxwell's theory. Possibly this feature has created much of the confusion on the subject.

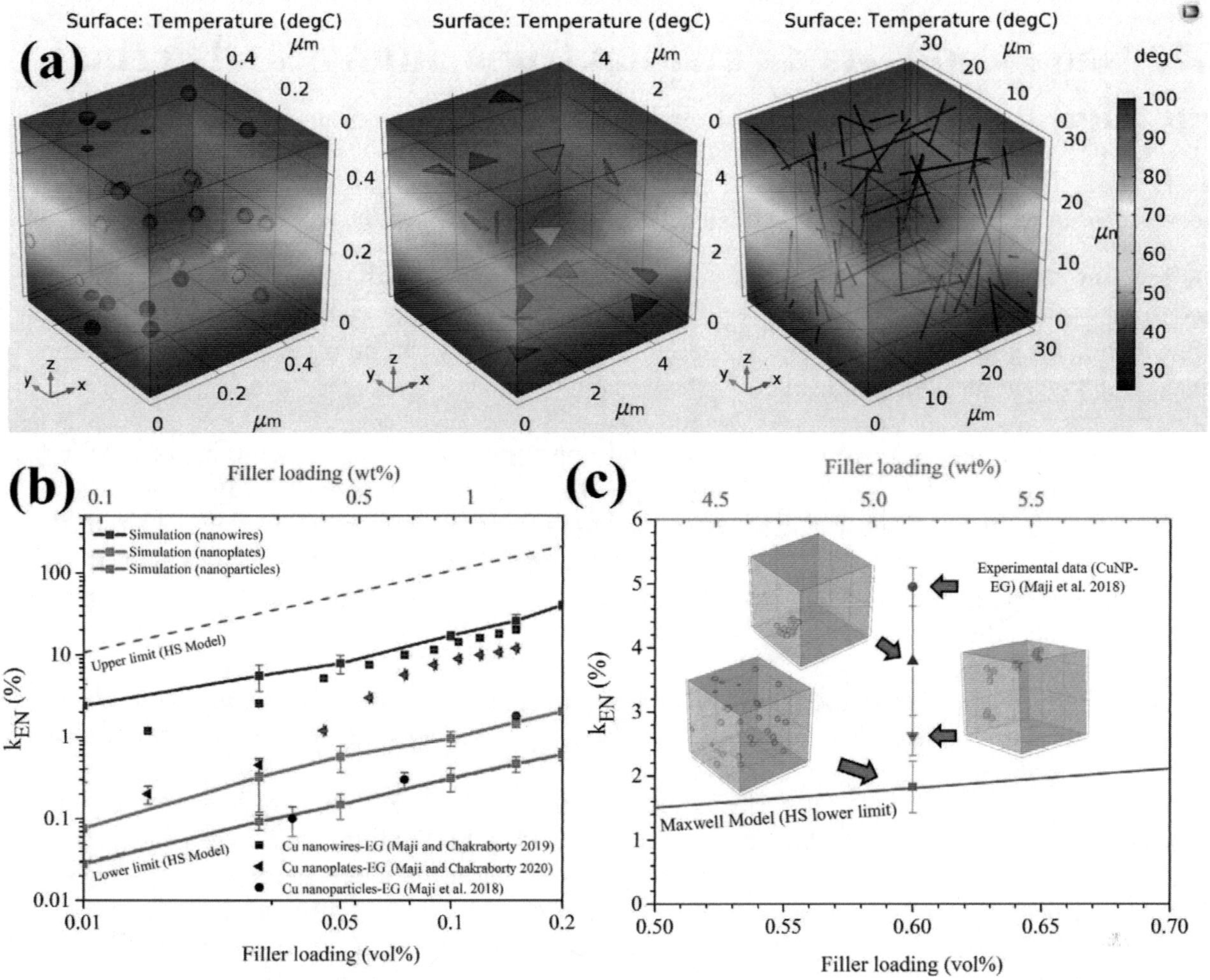

Figure 1. (a) Temperature distribution in RVE with copper (Cu) fillers of different morphology, (b) Thermal conductivity enhancement in Cu-EG nanofluid, (c) Effect of aggregation in thermal conductivity

599027 Nano Fluids As a Potential Opportunity for Thermal Application

Abstract:

Modern developments in nanotechnology have instigated new budding heat transfer fluids. The curiosity in nano fluids as novel coolants is proved through a numerous research conducted in the field of heat transfer and coolants. The development of such fluids can be by suspending the nano-particles on the known regular base fluids.

These Nano-fluids have been proven to be a potential opportunity to be used as the heat transfer fluids especially as the coolants due to its heat capacity. Due to the smaller in size, leading to the larger surface areas, these fluids can showcase superior thermal properties compared to the conservative heat transfer fluids or coolants. These fluids have improved thermo-physical properties and heat transfer performance as well. Utilization of these fluids, not only appreciably develop heat transfer capabilities but also boost the stability of suspension.

The additional benefit of utilizing the nano-fluids is their utilization for the 'miniature technologies' as they work suitable with the tiny channels which can be clogged due to the larger particle size if used.

Nano fluids illustrate enhanced stability, rheological properties and considerably soaring thermal conductivities. The improved thermal behavior of nano fluids could offer a foundation for an massive innovation for heat transfer intensification, which is of chief significance to numerous industrial sectors including transportation, power generation, micro manufacturing, thermal therapy for cancer treatment, chemical and metallurgical sectors, as well as heating, cooling, ventilation and air-conditioning. The present paper aims to offer an insight in the field of nano fluids through detail evaluation of the physio-chemical, thermo-physical and heat transfer characteristics of nano fluids for practical heat transfer applications.

599112 Coarse-Grained DEM Simulation of a High Shear Mixer

Abstract:

A discrete element method (DEM) is widely used for numerical simulations of various powder handling processes. However, the DEM is still computational costly. In order to overcome this issue, a coarse-grained (CG) model has been proposed. In the CG model, a group of original particles is replaced to a single CG particle which is larger than the original particle. Thereby, a large-scale DEM simulation can be performed using much fewer particles than that in the original system. We have developed a new CG model for a powder shear flow, which is named as the CGSF: Coarse-grained method for Granular Shear Flow. We have demonstrated that the CGSF is applicable to the coarse-grained DEM simulation of a rotating drum. However, its applicability to other powder handling processes was still unknown. We here present for the first time a coarse-grained DEM simulation of a vertical high shear mixer by using the CGSF. It was demonstrated that by using the CGSF the powder shear flow in the high shear mixer can be simulated accurately from the view point of the particle velocity, kinetic energy, and powder mixing.

601848 Numerical Simulation of Packing Structure and Compression Process of Cohesive Bimodal Particles.

Abstract:

All-solid-state lithium ion batteries (ASSLIBs) are composed of electrode active materials and nonflammable inorganic solid electrolytes instead of flammable liquid electrolytes. ASSLIBs are expected as next-generation batteries for electric vehicles due to their high energy densities and safety. Bulk-type solid state batteries with advantages of high capacity are prepared by compressing electrode and solid electrolyte powders. Hence, voids between the particles reduce the energy density. To solve this problem, the densification of ASSLIBs is necessary. The straightforward approach to the further densification includes an addition of bimodal particles with coarse and fine particles. An increase in packed active materials by adding fine particles will improve the energy density. The densification also leads to a reduction of internal resistance due to an increase in conductive paths. Here, it is crucial to predict the particle size ratio and the mixing ratio of fine particles for optimal packing structure by the aspects of a void fraction and a coordination number. Several studies have revealed that adhesion forces dominated in the case of fine particles, such as battery materials, strongly affects the packing structure of cohesive particles. However, the optimal packing structure of adhesive bimodal particles has not still been analyzed theoretically.

In this study, the effects of the particle size ratio, the volume fraction of fine particles, and the cohesion force on the packing and compression process were investigated by using the discrete element method (DEM). EDEM2019.1 (DEM Solutions Co. Ltd.) was used for DEM calculation. The Hertz-Mindlin with JKR model (hereinafter called JKR model) was applied as a contact model to describe the adhesive forces between particles. Adhesive forces were as a function of surface energy in the JKR model.The simulation parameters were as follows: particle size ratio α = 1-4, volume fraction of fine particles S_{vf} = 0-0.5, and surface energy γ = 0-0.2 J/m^2. Particle properties were assumed to be $LiNi_{1/3}Co_{1/3}Mn_{1/3}O_2$ (NCM) of positive active materials; however, only young's modulus was halved to reduce calculation costs. The particles were randomly generated above the cylindrical container to form a stable packing under gravity. To simulate the compression and decompression processes, the upper punch was vertically moved. A pressure at 400 MPa was applied to the powder bed in the compression process. The lower punch remained stationary throughout simulation. All compacts were conducted at a constant loading and unloading rate of 25 mm/s. The total simulation time was 0.09 s to obtain the stable packing structure after compression.

In the non-cohesive condition, void fractions of particle size ratio 1 and 4 were 0.410 and 0.313, respectively, demonstrating that the densification was achieved by the addition of fine particles. In addition, under the all volume fractions of fine particles, void fractions in cohesive conditions were higher than those in non-cohesive conditions. Moreover, in the cohesive condition, the void fraction and the contacts between fine particles increased with the cohesive forces. On the other hand, the densification effect due to the addition of fine particles decreased as the cohesive force increased. After the compression at 400 MPa, the cohesive forces had little effect on the void fraction at either particle size ratio.

608402 PLA-paclitaxel Nanoparticles in Lung Cancer Nano-Delivery Systems

Abstract:

Paclitaxel (PTX) or Taxol is widely used in chemotherapeutic treatment of breast, lung, Kaposi sarcoma and ovarian cancers. Besides effectiveness in therapy, paclitaxel causes various severe side effects by preventing healthy cell division. The hydrophobicity of paclitaxel is related to the performance of the drug. The formulation of paclitaxel in a nano-delivery systems is showing a promising future for cancer treatment. Polylactic acid (PLA) is a degradable polymer, non-toxic, degrades into generally regarded as safe monomers, and has shown compatibility with paclitaxel. Through localizing the release of PTX, this work aims to improve the efficacy of PTX against lung cancer. Emulsion techniques illustrate one approach to nanoparticle formation PTX at 0% (blank), 1% and 10% by mass was added to a polymer-solvent mixture and formulated into nanoparticles by a W/O/W emulsion with poly-vinyl alcohol (PVA) added as a surfactant Particle size and morphology of the particles were evaluated using scanning electron microscopy (SEM). The hydrodynamic diameter and surface charge of the particles were evaluated through the use of dynamic light scattering (DLS). The release kinetics of the nanoparticles were evaluated in buffer solution over time. Encapsulation efficiency (EE%) was calculated as the cumulative amount of drug released divided by the total drug amount added to the formulation. The efficacy of PTX loaded PLA-nanoparticles was tested against the A549 cell line, pulmonary adenocarcinoma epithelial cells. This study shows the potential of nano-carries as a drug delivery method for chemotherapeutics against lung cancer.

610042 Sonication Effectively Reduces Particle Diameter in Hemoglobin Nanoparticles Made By the Co-Precipitation Method

Abstract:

Purpose/Introduction: A variety of techniques have emerged for creating nano- and micro-particles. Protein based particles are especially appealing for biomedical applications due to their inherent biocompatibility. The method of co-precipitation, wherein a carbonate matrix is filled with a protein before being dissolved, creates a porous particle with substantial surface area, with high potential for drug or oxygen delivery applications. However, published methods using this technique produce particles with diameters in the submicron or micron range, generally regarded as incompatible with intravenous use due to rapid macrophage-mediated clearance. This study describes the use of ultrasound (US) sonication to achieve smaller hemoglobin nanoparticles (Hb-NPs) by the co-precipitation method which would be suitable for preclinical investigation as an oxygen therapeutic or drug delivery platform.

Experimental Design/Method: Hemoglobin was extracted and purified from expired human red blood cells and dissolved in a solution of manganese chloride. An equimolar amount of sodium carbonate was added to this solution, resulting in rapid precipitation of particles and particle aggregates. The entrapped hemoglobin was chemically cross-linked before dissolving the manganese carbonate particles with EDTA. Particles were washed into buffered saline and then sonicated with a 500 W probe sonicator for a variable duration and intensity to reduce overall size and break apart aggregates.

Major Findings/Results: We found that sonication effectively reduces particle size by an order of magnitude, reducing diameters from several micrometers to below 500 nm. A larger sonication probe tip (1/2") was more effective at creating a monodisperse particle distribution than a smaller (1/16") tip, even when the ratio of probe surface area to sample volume was fixed. Increased US intensity reduced diameter to a certain threshold size of about 250 nm, beyond which no significant difference was seen. Duration of sonication appeared to be a less important parameter than anticipated, with no difference seen with treatment greater than 5 minutes. Hb-NPs demonstrated a capacity to bind and release oxygen through stopped-flow deoxygenation kinetics as well as oxygen binding equilibria studies.

Summary of your interpretation of the data or conclusion: The reduction in particle diameter shown in these results opens the door to a wider variety of applications for nanoparticles made with the co-precipitation technique. These Hb-NPs exemplify one potential application, using the large surface area inherent to the method to facilitate oxygen transport by the hemoglobin protein.

608715 Investigation of the Effects of Surface Coatings on the Properties and Biodistribution of 64-Cu Labeled Nanoparticles for Use in Mouse Pancreatic Cancer Model

Abstract:

Positron Emission Tomography (PET) is a biomedical imaging modality that can construct highly detailed images of target regions by detecting gamma rays produced from the collision of positrons and electrons within a patient's body. Nanoparticles (NPs) acting as contrast agents have improved the sensitivity of PET-targeted imaging, as localized clusters of NPs increase the signal intensity in a particular region. A prominent concern that comes with the use of PET-active NPs relates to their biodistribution, which is affected by particle size and surface coatings. Previously, Lu et al. has optimized use of the hydrophobic molecule 1,4,8,11,15,18,22,25-Octabutoxy-phthalocyanine ("762Pc") for chelation of 64-Cu into 762-core NPs to create PET-active NPs with a poly-ethylene glycol (PEG) surface coating. However, little work has been done on the effect of other surface coatings on copper loading for NP radiolabeling. Here, we: (1) formed and characterized the stability of NPs with different surface coatings (2) optimized their copper chelation kinetics (3) evaluated biodistribution in an animal model. Toward this end, we utilized the self-assembly Flash NanoPrecipitation (FNP) process to formulate water-dispersible, poly-ethylene glycol (PEG), poly-acrylic acid (PAA) and poly-dimethylethylmethacrylate (DMEMA)-coated NPs. These coatings were selected based on varying charges imposed on the nanoparticle surface. By tuning the core concentration during FNP, 762Pc encapsulated particles were formed with sizes that can be tuned from 90 to 160 nm. 64-Cu radiolabeling trials on these NPs revealed that PEG and PAA-coated NPs took up > 99% of the initial 64-Cu activity after a 3- hour incubation in a pH 5.5 buffer at 37°C - reaching over 500uCi of activity. Subsequent animal studies will be conducted in a mice model to evaluate how surface coating and size affects NP biodistribution using PET. We hope that these results demonstrate the potential of different surface coatings to be used in PET-imaging and diagnostic applications.

610218 Can You Improve Dissolution Rate of Micronized Poorly Water-Soluble Drugs after Dry Coating with Hydrophobic Silica?

Abstract:

Dry coating of silica has been shown to be very helpful with enhancing flow and bulk density of all types of powders[1]. However, its effect on the dissolution of poorly water-soluble drugs has not been well-understood, most reports have dealt with a coating of hydrophilic silicas[2,3]. A major challenge with fine powders, in particular those below 30μm, is significant agglomeration due to high cohesion that delays their dissolution rate[2]. For such fine powders, dry coating with glidants, typically silica nanoparticles, is very effective in reducing cohesion and hence the extent of agglomeration[3,4]. It is possible that if the agglomerate sizes are reduced greatly, even the use of hydrophobic silica could promote their dissolution rate after coating due to the increased surface area that gets exposed to the dissolution medium. However, the previous reports have not quantified agglomerate size reduction after dry coating, along with its competing relationship with the surface hydrophobicity on the drug dissolution rate[3,4]. In this work, the investigation is carried out to fill these gaps and develop a more quantitative understanding of the relationship between cohesion reduction and agglomerate size reduction. The latter objective will be based on proposing a predictive model that is based on the dimensionless parameter that captures the effect of interparticle cohesion, in the form of the granular Bond number. Griseofulvin (d_{50} of 10 μm) and ibuprofen (d_{50} of 20 μm) were selected as model drugs, which were dry coated with either hydrophilic (A200) or hydrophobic (R972P) fumed silica via LabRAM at varying theoretical surface area coverage (SAC) of silica. Agglomerated and primary particle sizes were evaluated by two different particle sizers, a gravity-driven dispersion method, and a compressed air dispersion method, respectively. Following USP <711> protocol, USP IV dissolution testing was conducted in de-ionized water. Surface hydrophilicity was measured via the modified Washburn method[5,6]. Inverse gas chromatography was employed to measure the surface energy of the dry coated powders. The results show that the dissolution rate significantly influenced by the changed surface hydrophilicity and the reduced agglomerate size. The multi-asperity model was employed to calculate the bond number for this work[7]. The applicability of the previously presented mathematical correlation between a bond number to an agglomerate ratio[8] was tested by comparing the predicted agglomerate size to the measured agglomerate size keeping in mind that it does not account for the polydispersity of the powders and process shear effect from the dry coating. The outcomes from this study help answer the original question about if and when even hydrophobic silica coating can help improve the dissolution rate of poorly water-soluble drug through the understanding of the impacts of the surface hydrophobicity and the agglomerate size.

References

1. Yang, J.; Sliva, A.; Banerjee, A.; Dave, R. N.; Pfeffer, R., Dry particle coating for improving the flowability of cohesive powders. *Powder Technology* **2005,** *158* (1-3), 21-33.
2. de Villiers, M. M., Influence of agglomeration of cohesive particles on the dissolution behaviour of furosemide powder. *International Journal of Pharmaceutics* **1996,** *136* (1-2), 175-179.
3. Han, X.; Ghoroi, C.; To, D.; Chen, Y.; Davé, R., Simultaneous micronization and surface

modification for improvement of flow and dissolution of drug particles. *International Journal of Pharmaceutics* **2011,** *415* (1-2), 185-195.
4. Han, X.; Ghoroi, C.; Davé, R., Dry coating of micronized API powders for improved dissolution of directly compacted tablets with high drug loading. *International Journal of Pharmaceutics* **2013,** *442* (1-2), 74-85.
5. Washburn, E. W., The dynamics of capillary flow. *Physical Review* **1921,** *17* (3), 273-283.
6. Thakker, M.; Karde, V.; Shah, D. O.; Shukla, P.; Ghoroi, C., Wettability measurement apparatus for porous material using the modified Washburn method. *Measurement Science and Technology* **2013,** *24* (12).
7. Chen, Y.; Yang, J.; Dave, R. N.; Pfeffer, R., Fluidization of coated group C powders. *AIChE Journal* **2008,** *54* (1), 104-121.
8. Castellanos, A., The relationship between attractive interparticle forces and bulk behaviour in dry and uncharged fine powders. *Advances in Physics* **2005,** *54* (4), 263-376.

602762 Crosslinked Peg-Nanoparticles Prepared through Inverse Flash Nanoprecipitation and Thiol-Michael Reaction and Its Potential to Delivery Hydrophilic Biologics

Abstract:

A hydrophilic nanoparticle composed of multi-armed PEG-acrylate and a multi-armed PEG-thiol was fabricated through multi-inlet vortex mixer (MIVM), using Pluronic copolymer as stabilizer and DMF/diethyl ether as solvent/anti-solvent system. Subsequent thiol-michael addition created a crosslinked PEG-nanoparticle (PEG-NP). Using different hydrophilic molecules as drug models, influences of formulation composition on PEG-NPs properties and performance were studied (e.g., ratio of solvent/anti-solvent on NP size, NP stability; percentage of co-solvent on NP stability and crosslinking; selection of PEG reactants on retention and release profile of model drugs; PEG network on bioavailability of drug, etc.) Due to the step-growth character of thiol-michael addition, thiol- or acrylate-bearing PEG-NP could be prepared by off-stoichiometric feeding of the multi-armed PEG reactants, which can facilitate further modification of the crosslinked PEG-NPs, such as hydrophobic surfacing and conjugation of targeting ligand. Based on the biocompatibility of PEG segments, the NP stability from the PEG-network, the scalable and mild fabricating condition, we anticipate that the crosslinked PEG-NP could be a promising candidate for delivery of hydrophilic biologics, such as protein and gene.

603606 Drug Delivery System for Platinum Nanoparticles, Anticancer Agent for Triple Negative Breast Cancer with Low Toxicity

Abstract:

Breast cancer is the second leading cause of cancer death for women. It is classified based on the overexpression of receptors on the cancer cell membrane. Targeted therapies to these receptors have decreased the mortality of certain breast cancers. Current therapies target the overexpression of progesterone (PR), estrogen (ER) or human epidermal growth factor 2 (HER2). In addition, chemotherapy, surgery, or radiation are often combined to increase the effectivity of the treatment. However, 10-19% of breast cancers are triple negative, defined by the lack of these three receptors. Moreover, triple negative breast cancer (TNBC) tends to metastasize faster than other kinds of breast cancer, resulting in a worse prognosis and often chemotherapy as the only treatment for metastasis. Many cancer patients need to abort chemotherapy due to complications arising from its side effects, and this is a primary contributor to mortality. Therefore, new treatment options need to be developed for TNBC with low toxicity to healthy cells.

Nanotechnology is a multidisciplinary field that has many different applications including cancer treatment and detection. Noble metal nanoparticles such as gold, silver or palladium, have shown major potential in the field of medicine and pharmaceutics. In particular, platinum nanoparticles (Pt NPs) have been reported with promising anticancer results. The activity of nanoparticles is based on small size and high surface area; these properties allow them to penetrate biofilms as well as influence intracellular mechanisms. Recent studies indicate that Pt NPs can be used as a therapy for cancer showing a limited toxicity to healthy cells. Hence, in this work we developed a new treatment option for TNBC based on Pt NPs. Moreover, to enhance the circulation at the tumor site, a delivery system was proposed. Poly(lactide-co-glycolide) (PLGA) particles were synthesized as a delivery system for Pt NPs. PLGA is a copolymer of poly(lactic acid) (PLA) and poly(glycolic acid) (PGA), those two monomers are metabolized by the body, making PLGA toxicity minimal. The encapsulation of Pt NPs within PLGA nanoparticles was performed by three different methodologies: nanoemulsion, electrohydrodynamic co-jetting and nanoprecipitation.

Pt NPs anticancer activity was tested performing cell viability experiments with TNBC and comparing with a common chemotherapeutic drug (cisplatin). As the major drawback of chemotherapy is the toxicity, cell viability experiments were also performed on fibroblasts. The results showed no apparent toxicity of Pt NPs for healthy cells in contrast to cisplatin. Encapsulation of Pt NPs within PLGA was performed by three different methodologies. The highest encapsulation rate was achieved by nanoemlusion with a 60% loading percentage. In summary this work provides a new treatment for TNBC based on Pt NPs with no apparent toxicity for healthy cells, overcoming one of the major drawbacks of current treatments.

610089 Enabling Nano-Insecticides: Biodegradable Nanoparticle Biodistribution in *Aedes Aegypti* Mosquitoes

Abstract:

Mosquito-borne illnesses cause millions of cases and hundreds of thousands of deaths every year, and insecticide resistance poses a threat to attempts to control their populations. Successful insecticides must migrate through the cuticle, distribute among physiologically relevant tissues, and reach sufficient accumulation at the target site to achieve a toxic effect. Complex and distinct physicochemical requirements imposed by each biological barrier contribute a significant, cumulative hurdle for successful delivery of new insecticides. Micro- and nanoparticle drug delivery carriers are designed to navigate such biological barriers to improve therapeutic and prophylactic payload delivery to target sites. Their pharmacological potentiation capabilities have been relatively underexplored for insecticides, but could expand the palette of insect control strategies and improve the activity of existing compounds.

Microparticles and nanoparticles were synthesized from two formulations of rhodamine B end group-functionalized polyanhydride copolymers and used to treat *Aedes aegypti* mosquitoes by three field-relevant exposure routes: surface, topical, and *per os*. Microparticles were observed to associate with mosquitoes to a much lesser extent than nanoparticles, regardless of chemistry, after treated surface exposure. Nanoparticle chemistry-associated differences were observed for both external labeling and labeling of the pharmacologically relevant malpighian tubules, midgut, and ovarian internal tissues. Both nanoformulations significantly labeled at least one internal tissue in all three exposure routes, while soluble rhodamine B was unable to label these tissues after topical exposure. This provides powerful evidence that these nanocarriers can guide payloads past the cuticle into target tissues that are unable to accomplish such transport on their own. Across all exposure routes, the more amphiphilic nanoparticle chemistry achieved more intense labeling of internal tissues than the more hydrophobic chemistry. Neither particle chemistry under any exposure route caused increased mortality for the first five days of any treatment.

Collectively, these observations indicate polyanhydride nanoparticles capabilities to navigate arthropod vector biological barriers, motivating further exploration of nanocarriers for active insecticidal compound delivery. Particle chemistry and exposure route can be tailored to target insecticidal molecules to specific, pharmacologically relevant tissues within insects, or enable the broad internal distribution of such compounds within *Ae. aegypti* mosquitoes. Such capabilities could contribute to the next generation of insecticidal interventions by improving potency and potentiating novel insecticidal mechanisms.

606600 PLGA & PLA Microcarrier Systems for Treatment of Parkinson's Disease

Abstract:

In recent times, polymeric microcarriers have gained favor to act as loading vessels for the delivery of various kinds of drugs to the required target site in a controlled manner. The principle of action is leaching of the drug from the polymer or degradation of the polymer matrix, allowing the therapeutic agent to be released from the microcarrier system. A significant advantage of this method is when the correct type of polymer and carrier fabrication system is chosen, it allows the optimization of the drug release profiles. It is intended to utilize polymeric microcarriers for the administration of the drug, Levodopa (L-dopa) via the respiratory system to treat Parkinson's Disease (PD). Preliminary studies have shown that L-dopa destabilizes in aqueous media and hence can predictably be degraded in the pulmonary environment. The aim of this study is to combine the polymeric microcarrier system formulated with chemical stabilizers to prevent L-dopa degradation. It can increase the duration of well-controlled phases of PD symptoms and reduce the number of dosages required. Two major classes of biocompatible polymers, Poly (D,L)-Lactide-co-Glycolide acid (PLGA) with molecular weights ranging from 25,000 kDa to 107,000 kDa) and Poly (L)-Lactide Acid (PLA) with molecular weights ranging from 5000 kDa to 20,000 kDa are considered. The microcarriers are characterized by Scanning Electron Microscope (SEM), Anderson Cascade Impactor (ACI), X-ray Diffraction Spectra (XRD), Thermogravimetric Analysis (TGA), Differential Scanning Calorimetry (DSC), etc. Pharmacodynamics study is carried out in simulated lung fluids and the therapeutic stability of L-dopa in microcarriers are assessed in simulated mucous model.

601876 Metal-Free Carbon Electrocatalysts for Efficient Energy Conversion and Storage

Abstract:

Green and renewable energy technologies, such as fuel cells, batteries, and water-splitting systems, hold great promise to solve current energy and environmental challenges. However, noble metal catalysts (*e.g.*, Pt, Pd, RuO_2, IrO_2) are generally needed to promote the *hydrogen evolution reaction* (HER) for hydrogen fuel generation from photo-electrochemical water-splitting, *oxygen reduction reaction* (ORR) in fuel cells for energy conversion, and *oxygen evolution reaction* (OER) in metal-air batteries for energy storage. The high cost of precious metal-based catalysts and their limited reserve have precluded these renewable energy technologies from large-scale applications.

Along with the recent intensive research efforts in non-noble metal based ORR catalysts, we have previously demonstrated that vertically-aligned nitrogen-doped carbon nanotubes could actively catalyze ORR *via* a four-electron process free from the CO poisoning effects with a 3-time higher electrocatalytic activity and better long-term durability than that of commercial Pt/C catalysts. The improved catalytic performance was attributed to the doping-induced charge transfer from carbon atoms adjacent to the nitrogen atoms to change the chemisorption mode of O_2 and to readily attract electrons from the anode for facilitating the ORR.

Subsequently, it was demonstrated that various graphitic carbon materials, doped with heteroatoms of different electronegativities from that of carbon atom, physically adsorbed with certain polyelectrolytes, and even without any apparent dopant or physically adsorbed polyelectrolyte, could also exhibit good ORR performance. More recent studies have further demonstrated that certain heteroatom-doped carbon nanomaterials could act as metal-free bifunctional catalysts for ORR/OER in metal-air batteries for energy storage, and even ORR/OER/HER trifunctional catalysts for self-powered water-splitting to generate hydrogen fuel and oxygen gas from water.

In this talk, I will summarize some of our work on the metal-free catalysts based on carbon nanomaterials for various energy-related reactions, along with an overview on the recent advances and perspectives in this exciting field.

602186 Electrospun Particle/Polymer Fiber Mats As Fuel Cell and Battery Electrodes

Abstract:

Electrospinning is gaining popularity as a convenient, robust, and scalable technique for fabricating non-woven mats of sub-micron diameter polymer fibers. Although not as well studied, the technique can also be used to prepare particle/polymer fiber networks with high intra- and inter-fiber porosity. Such fibrous mats can be used as porous electrodes in fuel cells and batteries, where a high electrode/electrolyte interfacial area is needed. In the present talk, recent experimental work on new nanofiber mat compositions and architectures for hydrogen/air fuel cell cathodes and lithium-ion battery electrodes will be presented. The fuel cell work is focused on high particle content fibers containing Pt/C or Pt-alloy catalyst powder and a polymer binder based on Nafion® perfluorosulfonic acid. The fuel cell power output using such a cathode is very high for a low Pt loading (0.10 mg/cm^2), near 1.0 W/cm^2 for H_2/air feeds. Additionally, the fiber mat cathodes exhibit excellent durability, as determined by power losses after accelerated voltage cycling tests. For Li-ion battery applications, single and dual fiber anode mats have been prepared and evaluated, where the fibers contain a polymer binder (e.g., poly(acrylic acid), polyamidimide, or polyvinylidene fluoride), Si nanoparticles as the electrochemically active material, and carbon powder as an electron conductor. Here, the gravimetric and areal capacities of the fiber anode are considerably higher than those for a conventional slot-die coated slurry anode. Inter-fiber voids allow for volume changes in Si during lithiation/delithiation, thus minimizing capacity fade during charge/discharge cycling (up to 1,000 cycles). Experimental details for both fuel cell and battery electrodes will be presented in this talk, including electrospinning procedures, fiber mat characterization data, and device performance.

611471 Bottom-up All Aqueous Assembly Approach for the Synthesis of Lightweight 3D Carbon Nanocomposite Aerogels for Electrochemical Energy Storage and Conversion Applications

Abstract:

Novel Bottom-Up All Aqueous Assembly Approach for the Synthesis of Lightweight 3D Carbon Nanocomposite Aerogels for Electrochemical Energy Storage and Conversion Applications

Brigit A. Duffy[1], Gabrielle M. Milanesa[1], An B. Vu,[1] Jordan M. Davis[1], Duncan R. Day[1], Pamela L. Sheehan[2], Preston C. Haney[2], Harry L. Moore Jr[2], F. John Burpo[1], Enoch. A. Nagelli[1*]

[1]Department of Chemistry & Life Science, Chemical Engineering Program, United States Military Academy, West Point, New York 10996

[2]U.S. Army Combat Capabilities Development Command, Army Futures Command, CCDC Armaments Center, Picatinny Arsenal, New Jersey 07806

Carbon nanomaterials constitute an ideal miniaturized composite platform for integration with other solid-state electrode materials. Self-assembly has been recognized as an effective strategy for the bottom–up synthesis of 3D macrostructures using graphene and CNTs as building blocks. [1, 2] We demonstrate a novel, all-aqueous, and scalable 3D platform materials design process for the synthesis of carbon nanocomposite aerogels to serve as porous ultralightweight electrodes for the next generation energy storage and conversion applications. Specifically, the negatively charged surface functional groups on graphene oxide and oxidized CNTs will be the sites for electrostatic coordination of positively charged noble metal cations and polyelectrolytes from aqueous solutions. This enables the incorporation of carbon nanomaterials with any polyelectrolyte and noble metal nanostructures with precise connection of each of the individual components while maintaining their unique properties.[3-5] Through solution-based self-assembly followed by critical point drying and chemical reduction, we can develop hierarchical porous 3D nanocomposite aerogels comprised of carbon nanomaterials with any precious or noble metal nanostructures (Pt, Pd, Au, Ru, Ag, and Ru) and polyelectrolyte. Electrochemical techniques such as cyclic voltammetry, linear sweep voltammetry, galvanostatic charge-discharge, and impedance spectroscopy are used to determine the electrical and ionic conductivity, electrocatalytic activity, and supercapacitor performance of the aerogels as lightweight high-power and energy density electrodes.

KEYWORDS: Three-dimensional Nanomaterials, Nanocomposites, Aerogels, Electrocatalysts, Batteries, Fuel Cells

CONTACT: Enoch A. Nagelli, Email: enoch.nagelli@westpoint.edu

References

[1] L. Dai, “Intelligent Macromolecules for Smart Devices: From Materials Synthesis to Device Applications”, Springer: Berlin, 2004.

[2] L. Dai, (Ed.) “Carbon Nanotechnology: Recent Developments in Chemistry, Physics, Materials Science and Device Applications”, Elsevier: Amsterdam, 2006.

[3] M. Yanga, Y. Houd, N. A. Kotov, “Graphene-based multilayers: Critical evaluation of materials assembly techniques” Nano Today 2012, 7, 430.

[4] E. Nagelli, R. Naik, Y. Xue, Y. Gao, M. Zhang, and L. Dai “Sensor arrays from multicomponent micropatterned nanoparticles and graphene” Nanotechnology 2013, 24, 444010.

[5] E. Nagelli, L. Huang, A. Q.-Z. Dai, F. Du, L. Dai, “3D Vertically-Aligned CNT/Graphene Hybrids from Layer-by-Layer Transfer for Supercapacitors” Part. Part. Syst. Charct. 2017, 34, 1700131. Article in Special Issue of Graphene Oxide Liquid Crystals.

[6] F. J. Burpo, E. A. Nagelli, L. A. Morris, J. P. McClure, M. Y. Ryu, J. L. Palmer. “Direct Solution-Based Reduction Synthesis of Au, Pd, and Pt Aerogels.” J. Materials Research. 2017, 32 (22), 4153-4165. (Invited Paper) Featured in MRS Bulletin: “Metallic aerogels synthesized in one-step process,” December 18, 2017.

610947 Tailoring Internal Pores for Mitigation of Volume Expansion in Silicon Graphene Hybrid Anodes for Lithium Ion Batteries

Abstract:

The current state of art Lithium ion batteries (LIB) are unable to satisfy the demands for electric vehicle (EV) applications due to their low energy density. To improve their energy density, silicon as anode material has been very promising due its extraordinary theoretical capacity. However, silicon undergoes huge volume expansion(~ 400%) during lithium-ion storage. This leads to the pulverization of the electrode structure over several cycles of lithiation and delithiation and consequently severe capacity decay. Thus, mitigation of this volume expansion has been a stumbling block for its commercialization. One approach to address this issue is to use silicon carbon-based composites and engineering void spaces for the effective accommodation of silicon volume expansion. The internal pores give breathing room for the silicon to expand, which maintains the structural integrity. Many scientists have researcher methods to mitigate the silicon volume expansion by engineering voids; however, there is a lack of direct correlation in terms of the size of pores created and the mitigation of volume expansion achieved. Additionally, the various approaches studied in the literature to create pore spaces invariably involve the use of expensive fabrication methods.

In our study we are using a two-step approach to form tailored internally porous silicon graphene hybrid anodes. In the first step, we form directly deposited silicon graphene hybrid anodes via air-controlled electrospray, and in the second step we use a mild heat treatment to thermally remove poly acrylic acid, which serves as our sacrificial template. This approach provides structural integrity and volume expansion accommodation. In order to prove our hypothesis of mitigation of silicon volume expansion via the creation of internal pores, we also study in-situ volume expansion via a dilatometer and compare the resulting mitigation of volume expansion for Si/Gr with pores and Si/Gr without pores. Our conclusions show the interplay of creation of pores and the mitigation of volume expansion over a number of charge/discharge cycles.

610999 Scalable Surface Application of Stabilized Lithium Metal Particles (SLMP) for Enhancing Silicon-Based Anode in Full-Cell Lithium-Ion Batteries

Abstract:

Silicon has been considered as one of the most promising replacements for graphite anodes in next-generation lithium-ion batteries due to its superior energy density. However, one prominent issue with silicon-based anode materials is the substantial consumption of lithium ions due to the large volume expansion of silicon upon lithiation and the continuous formation of the solid-electrolyte interphase. In a full-cell setting where silicon is paired with a commercial cathode, this irreversible consumption of lithium ions is especially detrimental to the limited lithium-ion reserve of the system, leading to a low first-cycle Coulombic efficiency and a shortened cycle life. Several pre-lithiation strategies have been explored to mitigate this problem, and the use of stabilized lithium metal particles (SLMP) is being demonstrated as a practical pre-lithiation method compatible with scaled-up industrial applications. Particularly, a SLMP slurry was designed with hexanes as a compatible solvent and a nonionic surfactant additive at 1%wt serving as the stabilizer for maintaining a uniform SLMP suspension. This slurry was then spray-coated onto the prefabricated silicon electrode, forming a uniformly distributed and well adhered SLMP layer on the surface. By optimizing the amount of SLMP being applied, we are able to demonstrate improvements in the electrochemical performance of silicon-based full-cell batteries with both increased first-cycle Coulombic efficiency and capacity retention. We will also attempt to provide a description of SLMP's effects at the electrode surface throughout the electrochemical cycling of the batteries.

610144 Corn Stover-Derived Biocarbon from Hydrothermal Liquefaction for Supercapacitor Applications

Abstract:

Biocarbon for supercapacitor anodes has become a frontrunner for renewable energy storage materials. Its unique physical characteristics allow for enhanced rate capabilities, increased ion transport, all leading to sustainable scale-up. Current work focuses on the use carbon-rich residue from the corn stover refining. Value added chemicals were first extracted by enzymatic hydrolysis, and the resultant lignin-rich unhydrolyzed solids (UHS) derived from preprocessing of corn stover were subjected to hydrothermal liquefaction (HTL). Bio-char samples derived from investigating the HTL reaction kinetics at various temperatures, initial purge, reaction time, and biomass loading, were subjected to thermal treatment at 400 °C under nitrogen for 2 h and followed by the activation at 850 °C under Ar for 3 h to obtain the biocarbon material. Electrochemical results show that select samples reach capacitances above 200 F g^{-1} at 5 mV s^{-1}. Physical and electrochemical characterizations such as BET, XRD, SEM, and ATR-FTIR, Raman, cyclic voltammetry, and chronopotentiometry were carried out to fundamentally probe the biocarbon. This work provides an insight into how pretreatment conditions affects the performance of biocarbon for supercapacitor applications.

600493 Nanoalloy Catalysts for Boosting H_2 Production from Hcooh Decomposition: A First-Principles Study

Abstract:

The oxidation of small organic molecules has been widely studied owing to fundamental interest and relevance to fuel technology. Among these molecules, formic acid (HCOOH) has received much attention as a source of hydrogen because of low-toxicity, transportability, and easy-storability. HCOOH may decompose via dehydrogenation or dehydration, depending on the reaction conditions. It was found that the activity of the noble metal catalysts for the H_2 production via HCOOH decomposition was of the following order Pd > Rh > Pt > Au. Since the carbon monoxide (CO) generated from HCOOH dehydration can poison the Pt-based electrode in a fuel cell, a highly selective catalyst toward HCOOH dehydrogenation is required and such desired selectivity can be obtained by modifying the chemical and physical properties of surface by adopting the bimetallic Pd-M alloy catalysts, where M is transition metal. In this talk, I will deliver the underlying mechanism for the enhancement of H_2 productivity and selectivity in metal alloy catalysts. The followings are the main presenting topics. (1) We found that the H_2 production rate strongly depends on the Pd shell thickness in Ag-Pd core shell catalyst, in particular, the Pd thinnest shell (Pd monolayer) showed the highest activity. (2) In the bimetallic Pd/M catalyst, we discovered that HCOOH decomposition strongly relies on the variation of surface charge polarization (ligand effect) and lattice distance (strain effect), which are caused by the heteronuclear interactions between surface Pd and core M atoms. Especially, the contraction of bond distance between the surface Pd atoms and the increase of charge density in surface Pd atoms compared to the pure Pd (for example, Pd/Cu, Pd/Rh) are responsible for the enhancement of the selectivity to H_2 formation via HCOOH decomposition. (3) We employed Pd_3M-Pd core–shell models (M = Sc, Ti, V, and Cr) and discovered the Pd/Pd_3Sc and Pd/Pd_3V surfaces exhibit particularly enhanced selective H_2 production capability. Significant charge transfer from the subsurface Sc atoms to the surface Pd atoms and subsequent extremely low level of d band occupancy (<0.1) around the Sc atoms are identified as a key factor in deriving the modification of the Pd surface electronic structure. In contrast, the promotion by V is mainly attributed to the increase of the activation energy barrier of the CO producing reaction path as a result of the severe upshift in the transition state energy along with the relatively negligible modification of the initial state energy for the rate-determining step. Our theoretical calculation offers the fundamental mechanism of HCOOH decomposition on metal alloy catalysts and also provides physical and chemical intuition for the next generation bimetallic catalyst for hydrogen and fuel cell applications.

600729 Rational Design of CeO_2-Based Catalysts for Enhanced Water-Gas Shift Reaction

Abstract:

Water-gas shift reaction (WGSR) is important for chemical processes that use hydrogen resources. It has been widely used in fertilizer and petroleum refining industry. Nowadays, the WGS reaction is also aimed at preventing carbon monoxide (CO) poisoning which deactivates fuel cells. In the hydrogen flow supplied to the fuel cell, CO is contained, which is adsorbed on the anode to cause inactivation. Therefore, before the hydrogen is supplied to the anode, the WGS reaction process is performed to oxidize the contained CO to prevent poisoning. Supported metal catalysts are mainly used for the WGS reaction, and the support reducibility of these catalysts is widely known as an important factor that affects the activity. WGS reaction is catalyzed by redox and associative mechanisms, so the reducibility of catalyst plays an important role in the activity. Controlling the support reducibility can form a large amount of oxygen vacancies (O_V) on the catalyst surface. These O_V increase the lattice oxygen mobility, and help oxygen more easily participate in the oxidation reaction. Therefore, many studies to improve the support reducibility are actively underway. In this study, we prepared supported metal catalysts that controlled the reducibility of CeO_2 by metal-doping and morphology modification. First, we identified whether the catalyst structure was well formed from each approach. Subsequently, the synthesized catalyst was confirmed to have higher activity than un-modified catalysts and a reducibility test was conducted to prove that it was related to reducibility. Furthermore, we designed the catalyst using two strategies simultaneously to improve the reducibility described above.

600527 DFT Calculations and Machine Learning Approach to Predict Catalytic Properties of Nanoscale Electrocatalysts in Solution for Clean Fuel Generation

Abstract:

Innovative design of functional nanomaterials has been prolonged challenge to scientific community and industrial sector, especially regarding with structural stability on the thermodynamic and electrochemical aspects. It is due to the key functionality of the materials in wide range of applications, such as Li-air batteries, Photovoltaic devices, and fuel cells. Advanced experimental tools and computational theories attempted to deploy various candidates, but their performances in operations did not meet the practical level requirements.

For example, one of the major huddles toward to the solution is how to accurately and quickly explore the gigantic material space with varying the structure and composition of catalysts to identify promising catalysts and propose design principles for the performance optimization. Over the last several decades, first-principles calculations have been considerably developed for especially the objective through an efficient analysis of large-scale database. It is noteworthy that the approach enabled an identification of a key descriptor on the catalytic activity and the high-throughput screening of the best candidate beyond the conventional Pt. Unfortunately, most of the candidates were not applied to the real devices, due to the serious structural degradation issue via chemical or/and electrochemical reaction paths. It means that just electronic structure-level analysis for the catalytic activity may not be enough for designing high functional catalysts. Probably, the systematic and consistent incorporation of hybrid methodologies should be the paradigm. For example, establishment of big data, material informatics based on neural network or machine learning technique, and multi-scale simulations such as molecular dynamics (MD) or Monte Carlo (MC) simulations can be the relevant procedure.

This presentation introduces machine learning driven computational framework for high throughput screen of electrocatalysts for hydrogen fuel generation.

Women claim discrimination at NIH's child health institute p. 21

Emotions in the faces of mice pp. 33 & 89

Hominin genera in South Africa 2 million years ago pp. 34 & 47

Science

$15
3 APRIL 2020
sciencemag.org

AAAS

VISIBLE VARIETY

Platinum nanocrystals are surprisingly nonuniform p. 60

595901 Direct Access to Functional Porous Materials for Energy Conversion and Storage

Abstract:

Multifunctional and hierarchical porous materials have attracted much attention as host electrode materials for electrochemical energy conversion and storages. Our research group has developed powerful methods to control multiscale porous inorganic nanostructurs via simple "one-pot method" by employing blends of block copolymers and homopolymers. The new approach allows access to a high degree of control over pore structure and size, particle shape, particle size and chemical composition including metal oxides, metal nitride and conductive carbon. Multiscale porous materials have been employed as a multifunctional sulfur host, integrating the advantages of multiscale porous architectures with high performance electrocatalytic property to achieve high-power and long-life lithium-sulfur batteries. A new and intuitive strategy for tuning and enhancing the kinetic activity of Fe-N_4 sites was designed by controlling electro-withdrawing/donating properties of carbon plane. Fe-N_4 integrated mesoporous materials showed a high catalytic performance comparable to that of Pt/C in anion exchange membrane (AEM) fuel cells.

600504 Preparation of Multifunctional Carbon-Supported Irru Catalyst By Novel Acidic Polyol Procedure for Fuel Cell Application

Abstract:

Fuel cell systems have recently attracted attention as eco-friendly devices that can be applied to eco-friendly power trains for fuel cell electric vehicles and to combined heat and power generation, which is considered the important measures to the hydrogen society [1,2]. Among them, proton exchange membrane electrolyte membrane fuel cells (PEMFCs) have been mostly developed due to their possibility of ambient operation and high efficiency under the 100°C. In addition, the improvement of anion exchange membrane fuel cells (AEMFCs) has been remarkable along with the enhancement of the membrane performance during recent years [3]. These fuel cells are based on two basic electrochemical reactions such as hydrogen oxidation reaction (HOR) and the oxygen reduction reaction (ORR). To date, much research has been focused on searching the new ORR catalyst to increase the activity of the catalyst and to reduce the cost of the fuel cell systems. In the other hand, novel HOR catalysts is also demanded to be developed to address the issues from the Pt catalyst for PEMFC and AEMFC. For example, the Pt/C catalyst is very vulnerable under the fuel starvation situations that can occur during FCEV operation, which used the PEMFC system. In addition, in the case of AEMFC, it is known that the HOR activity of the Pt/C catalyst is drastically reduced.

In this presentation, as a universal HOR catalyst under acidic and alkaline, the carbon-supported IrRu alloy catalysts will be introduced. Recently, the IrRu alloy catalyst has been developed for replacing the Pt catalyst in the anode of the automotive PEMFC [4-6]. As a continuous study, the carbon-supported IrRu alloy catalysts with different ratios of Ir and Ru were prepared by polyol process under acidic condition (pH < 7) for the first time and investigated as the universal HOR catalysts under acidic and alkaline condition. The performance of the IrRu alloy catalysts in the single-cell level will be discussed.

[1] R.-H. Lin, Y.-Y. Zhao, B.-D. Wu, Int. H. Hydrogen Energy, in press (2020), doi.org/10.1016/j.ijhydene.2020.01.047

[2] The Fuel Cell Industry Review 2019 at www.FuelCellIndustryReview.com, E4Tech, (2019).

[3] D. R. Dekel, J. Power Sources, 375, 158 (2018).

[4] E. You, M. Min, S.-A. Jin, T. Kim, C. Pak, J. Electrochem. Soc. 165, F3094 (2018).

[5] S. W. Lee, B. H. Lee, T.-Y. Kim, C. Baik, M. S. Kim, G. S. Chai, C. Pak, Catal. Commun. 130, 105758 (2019).

[6] T.-Y. Kim, S. W. Lee, C. Pak, J. Ind. Eng. Chem. 85, 87 (2020).

598452 Electronic and Electrochemical Characteristics of Pt Monolayer on Graphene and Its Electrochemical Properties

Abstract:

Epitaxially grown Pt thin film on a graphene template is found to have a few monolayer thickness with structural stability comparable to Pt(111) surface and promising electrochemical activity. The newly synthesized thin film shows a quadrilateral-polygon structure of Pt in monolayer on graphene surface. In this study, we present a computational research on the unique architecture of the graphene-templated epitaxial Pt monolayer. Recently proposed strongly constrained and appropriately normed (SCAN) density function study (DFT) is employed to investigate the characteristics of the system. Electrochemical activities of the system are evaluated in terms of the free energy variation in oxygen reduction reaction (ORR). In architectures, Pt exhibits registry with the C-C bridge sites along the armchair and zigzag directions forming strong covalent bonds. Here, the details of the atomistic/electronic structures and binding energies are discussed. Interestingly, the ORR can occur on both Pt and graphene surfaces. This would provide a good strategy to protect metallic catalyst and to tune electronic structure of catalyst.

598810 The Counterintuitive Dependency of Bifunctional Effect on the Degree of Ordering in Alloy Electrocatalysts for Methanol Oxidation Reaction

Abstract:

Bifunctional effect is commonly regarded as the reason why bimetallic catalysts outperform pure-metal catalysts in processes that include CO oxidation. It has been shown that alloy catalysts shorten the distance between sites for CO_{ad} and sites for OH_{ad} into sub-nano range and are able to oxidize CO at much lower overpotential than non-alloyed catalysts with similar compositions. To see whether the highest intimacy between Pt and secondary metal in the ordered alloy could provide the highest activity, we modify the degree of ordering of Pt_3Sn nanocubes (NCs) while maintaining the exposed {100} facets, and test them in methanol oxidation reaction and CO oxidation. Strikingly, the 60% ordered Pt_3Sn NCs outperform the 95% ordered Pt_3Sn NCs in methanol oxidation by almost 5 times in converted charge density. 60% ordered Pt_3Sn NCs also oxidize CO at lower potential in CO stripping and are more effective in bulk CO oxidation than the 95% ordered Pt_3Sn NCs and Pt NCs. Moreover, we demonstrate the outstanding CO tolerance of less ordered Pt_3Sn NCs comparing to 95% ordered Pt_3Sn NCs by saturating the electrolyte with CO during methanol oxidation. We reveal that the Sn atoms at the surface of the 60% ordered Pt_3Sn are more prone to be oxidized under oxidative potential whereas the Sn atoms at the surface of the 95% ordered Pt_3Sn NCs are resistant to oxidation. We propose that the formation of SnO_x in the vicinity of Pt sites create higher proximity between OH_{ad} and CO_{ad} in less ordered Pt_3Sn NCs, making less ordered alloys more active for methanol oxidation reactions, in which CO oxidation plays a crucial role.

605484 Safety Risks When Storing and Handling Bulk Solids

Abstract:

Each year hundreds if not thousands of individuals are severely injured -- and many are killed -- when working in and around bulk solids storage and handling facilities. While this represents but a small percentage of all fatal occupational accidents reported by OSHA and other government agencies, these accidents often receive more press coverage and, in many cases, are avoidable for reasons explained in this paper.

Numerous codes and standards specify means to calculate material-induced loads that are needed to design silos. Unfortunately, these codes do not provide users with consistent information, and many common silo design conditions are not covered.

In reviewing the several dozen litigations in which I have been personally involved as an expert witness that resulted in either serious injury or death plus an even larger number that, fortunately, did not, I have found several main causes of accidents that have occurred when storing, handling and processing bulk solids.

Safety awareness is not limited to the chemical processing and mining industries -- it is on the mind of workers and management in every industrial plant worldwide. While there should not be any disagreement that it is unacceptable in this day and age to have workplace losses caused by stupidity or carelessness, one is left wondering if it is possible that bulk solids are so pervasive that it is impossible to regulate workers into being safe.

Conducting a Job Hazard Analysis (JHA) before performing a task that might result in injury or death is always a good idea.

An extremely useful reference is E. Scott Geller's The Psychology of Safety. The author of this practical handbook teaches science-based and feasible approaches to change what people think (attitude) and do (behavior) in order to achieve what he calls "a Total Safety Culture". In this culture, everyone feels responsible for safety and pursues it on a daily basis. At work, employees go beyond "the call of duty" to identify environmental hazards and at-risk behaviors. Then, they intervene to correct them. Safe work practices are supported with proper recognition procedures.

Building a Total Safety Culture requires a long-term continuous improvement process. It involves cultivating constructive change in both the behaviors and attitudes of everyone in the organization.

More education of the dangers involved in the storage, handling and processing of bulk solids is required. This is a huge undertaking given that owners, operators, workers, contractors, designers, repair crews, trucking firms.... all need to be educated. Papers such as this one hopefully raise awareness of the problems that can occur, but this information needs to be distributed to a very wide audience.

607517 Understanding and Modeling Granular Segregation—Karl Jacob's Vision

Abstract:

As large and small particles flow, small particles fall between large particles to segregate in lower portions of a dense granular flow while displacing large particles upward, a process known as "percolation." Similarly, heavy particles segregate below light particles as they flow due to "buoyancy." While many phenomenological studies of segregation have been performed, most focused on pattern formation, which is of academic interest, but not immediately applicable to industrial problems. Furthermore, previous models of segregation depend on obliquely defined or difficult-to-measure empirical parameters, making them impractical for process engineering problems. Over a decade ago, Karl Jacob had the vision to connect the academic interest in pattern formation and modeling with practical segregation problems in industry. This started as simple experiments for quasi-2D heap flow to understand size segregation typical of hopper filling. However, this prescient choice of a flow geometry provides a test bed with steady flow and segregation as well as a simple velocity field and segregation that is easily characterized. Because the heap flow geometry provides a wide range of flow and segregation conditions that are easily simulated and measured, this opened the potential for modification of the standard advection-diffusion equation to account for segregation. The resulting segregation model is physics-based, but, due to Karl Jacob's insistence, it is readily applied to many practical flow geometries and conditions in industry. The model has been used to predict steady and transient segregation in chutes, rotating tumblers, hoppers, and shear cells and has been extended to density-disperse particles, multi- and polydisperse particles, non-spherical particles, and even particles varying simultaneously in size and density. Funded by The Dow Chemical Company, the Procter & Gamble Company, and the National Science Foundation.

608948 Reflections on Karl Jacob and Some New Results on Powder Flow Enhancement Using Magnetically Assisted Impact Coating

Abstract:

Personal reflections on Karl Jacob's contributions to powder technology, education in particle technology, and his dedication to being a life-long Packer Backer will be made. The main portion of this talk will be on new work using Magnetically Assisted Impact Coating (MAIC) with flow aids to enhance powder flow. Previous work will be reviewed and new results on the use of fumed silica and magnesium stearate as flow aids on acetominophen will be discussed. MAIC processed materials will be compared to samples prepared using v-blending and dual asymmetric centrifugal mixing methods. FT4, particle size distribution, and SEM characterizations methods will be used to describe the powder properties from each method. Results will be presented that show show enhanced powder flow using the MAIC process as opposed to the v-blending or centrifugal mixing methods with low levels of flow aids and converging powder flow enhancement of all application methods at higher flow aid application levels. Discussion of aging effects on flow aid usage will also be made.

614127 Reflections on a Career in Particle Technology

Abstract:

Karl will reflect on his career in particle technology. Most importantly, he will discuss the people who made it possible and how they shaped his career and life in the field, not only in solids processing but more broadly in chemical engineering.

605034 Numerical Modeling of Deagglomeration of Cohesive Particles By Turbulence

Abstract:

The transport and deposition of fine particulates in turbulent flows play important roles in many engineering and medical systems. Examples include dry powder inhalers for drug delivery and fluidized bed reactors. Micron-sized particles tend to form aggregates due to inter-particle cohesion. The dynamical evolution and morphology of these aggregates involve a complex interplay between turbulent stresses and inter-particle cohesive forces. As a result, particle clumping can arise under various circumstances, which is known to compromise the performance of the aforementioned systems. Of particular interest to the present study is turbulence-induced breakup of fine particulate aggregates. Solid particles with diameters smaller than the Kolmogorov length scale ($d_p < \eta$) are initially aggregated into a spherical 'clump' of diameter $D > \eta$ and placed in homogeneous isotropic turbulence. Parameters are chosen relevant to powder suspended in air such that cohesion due to van der Waals is important. Simulations are performed using a CFD-DEM framework that models two-way coupling between the fluid and solid phases and resolves particle-particle interactions. Aggregate breakup is investigated for different Adhesion numbers Ad, Taylor micro-scale Reynolds numbers Re_λ and nondimensional clump sizes D/d_p. The intermittency of turbulence is found to play a key role on the early-stage breakup process, which can be characterized by a turbulent Adhesion number Ad_η that relates the potential energy of the van der Waals force to turbulent shear stresses. A scaling analysis shows that the time rate of breakup for each case collapses when scaled by Ad_η and an aggregate Reynolds number proportional to D. A phenomenological model of the breakup process is proposed that acts as a granular counterpart to the Taylor Analogy Breakup (TAB) model commonly used for droplet breakup. Such a model is useful for predicting particle breakup in coarse-grained simulation frameworks, such as Reynolds-averaged Navier–Stokes, where relevant spatial- and temporal-scales are not resolved.

605325 Regimes of Impact-Induced Erosion of Particles of Varying Strength

Abstract:

Particle breakage is relevant to many applications, from pharmaceutical processing to jet printing to defense technologies. Fundamental understanding of particle behavior during and after impact as a function of particle strength and impact velocity is needed for process optimization. In addition, detailed information about the number, shape, size and velocity of resulting fragments is difficult to obtain, particularly at conditions of high impact velocity. Discrete element method (DEM) simulations provides the capability to simulate the physics of particle impact and breakup at the micro-scale. This work explores varying particle breakage behavior upon impact with a flat surface. Impacting particles are modeled as spherical agglomerates consisted of a number of smaller constituent particles held together via a constant adhesive force characterized by a surface energy. Under the influence of a wide range of impact velocities and particle surface energies, five distinct behavioral regimes – rebounding, resting, fragmentation, pancaking, and shattering – are identified. In the rebounding regime, the coefficient of restitution decreases linearly as impact velocity increases. In the fragmentation regime, the rebound velocity generally decreases with increasing fragment size. These trends are consistent with the experimental results obtained in the study of Hassani-Gangaraj *et al.* (2018). Some regimes are only present at certain velocities while the extent of these regimes varies with particle strength. The type of regime depends largely on how the impact force is transmitted and dissipated through particle-particle contacts within the agglomerate.

Hassani-Gangaraj, M.; Veysset, D.; Nelson, K.A.; Schuh, C.A. "Melt-Driven Erosion in Microparticle Impact," *Nature Communications* **2018**, *9*, 1-7

606339 Breakage of Structured Particles: A Comparison between Stress-Strain and Impact Events

Abstract:

The breakage of structured particles is important in the food, automotive, electronic, and ceramic industries. Many of the structured particles created consist of relatively uniform or well-graded primary powder particles designed to have minimal segregation that are then made into larger particles that have intricate structures and designs. The tendency for structured particles to break depends on the strength of the bonds formed between the primary particles as well as on the structure of the larger particles. Breakage also depends on the set of forces that act on a structured particle during a breakage event. A population balance model is a reasonable means to describe breakage events and can be used to examine how breakage might occur. However, current formulations of the population balance model do not include the effects of global structure on breakage. They also do not include the effect of particle scale bond strengths, nor do they include any description of the influence of the different types of stress-strain or impact-induced stresses that are present with the material during processing. There is a difference between the breakage mode when the primary influence of breakage events consists of impact events as compared to stress-strain events.

The goal of this paper is to look at particle scale strength as a function of particle size distribution and then examine the differences in breakage of structured particles during stress-strain events versus impact events. A population balance model is used to determine the breakage selectivity coefficients. These coefficients are then correlated with the mode of breakage (impact versus stress-strain) using FEM models of stress induced by each type of breakage. However, the breakage rate constants are correlated to the strength as a function of the particle size distribution of the primary particles. Combining the breakage pattern computed by FEM analysis of particles with particle scale strength models and the population balance model provides a rich toolset to understand and predict breakage. A set of structured particles (balls, rods, crosses, and grid-shaped) were created from a consistent set of cohesive particles. These particles were then subjected to stress-strain events and impact events. The breakage studied by the population balance model and FEM analysis of structured particles were used to determine the breakage selectivity behavior. A polydisperse particle scale strength law was used to determine strength as a function of the primary particle size which was correlated to the breakage rate constants.

608742 Abradable Particles in DEM through Spherical Harmonics

Abstract:

Abrasion is a particular form of particle attrition in which a particle changes in shape by becoming smoother over time through the preferential removal of protrusions on the particle's surface. Fines are produced by abrasion which are much smaller than the particle's size. Abrasion plays a significant role in industrial applications, as does attrition more broadly. Fines can reduce flowability which can impair processing operations such as conveying, blending or tableting [1]. Attrition affects the bulk density, specific surface area, segregation behaviour and dissolution rate which has major implications for the quality of pharmaceutical products [2, 3]. The rehydration characteristics of infant formula are significantly affected by attrition [4]. Abrasion of fluid cracking catalysts is a significant cost factor in these processes as it necessitates the periodic replacement of the catalysts [5].
Because of its importance to a wide range of industrial particulate processes, quantifying abrasion, and ultimately understanding abrasion in order to better control it, are of fundamental importance. DEM simulations have contributed greatly to our understanding of granular materials behaviour in recent years [6], as they give access to particle-scale information such as interparticle contact forces which are difficult, if not impossible, to acquire directly from experiments. Unfortunately, most DEM simulations are based on spherical particles, while our everyday life experience shows that abrasion depends heavily on the particle shape: angular particles are more susceptible to surface breakage than rounded ones. Additionally, simulations of abrasion in DEM are very few and are mostly application-based [7].
In view of these considerations, a novel approach is presented for modelling non-spherical, abradable particles in DEM. The particle's shape is described by an expansion in spherical harmonic functions [8], with the high-degree terms mainly responsible for microscopic details of the shape, e.g., the surface texture. Mathematical considerations [9], together with recent experimental results [10], show that higher harmonics, corresponding to the surface texture, are the first to be eroded, while ellipsoidal shapes, found in abundance in nature, take much longer to become spherical, i.e., the sphere is the equilibrium shape [9]. Therefore the abrasion process can be represented through the sequential removal of the highest spherical harmonics (see Fig. 1). During abrasion, mass is lost in the form of fines, and an additional set of scaling factors ensures that the high-degree spherical harmonic expansion, representing the original shape, bounds the expansion at any lower degree: the abraded shape. The sequence of spherical harmonic expansions with decreasing order allows prediction of a particle's shape evolution during the abrasion process. Finally, this abrasion is related to microscopic wearing laws, based on energy considerations. A comparison between the abraded shapes obtained through the microscopic wearing laws and the predicted shapes is discussed. The implementation of this novel method in the open-source LAMMPS code is presented along with possible sources of performance improvement.

[1] A. Lekhal, K. P. Girard, M. A. Brown, S. Kiang, B. J. Glasser, J. G. Khinast (2003), Impact of agitated drying on crystal morphology: KCl-water system, Powder Technol. 132(2-3): 119–130.
[2] C. Hare, M. Ghadiri, R. Dennehy (2011), Prediction of attrition in agitated particle beds, Chem. Eng. Sci. 66(20): 4757–4770.
[3] X. Hua, J. Curtis, Y. Guo, B. Hancock, W. Ketterhagen, C. Wassgren (2015), The internal loads, moments, and stresses in rod-like particles in a low speed, vertical axis mixer, Chem. Eng. Sci. 134: 581–598.
[4] K. J. Hanley, E. P. Byrne, K. Cronin, J. C. Oliveira, J. A. O'Mahony, M. A. Fenelon (2011), Effect

of pneumatic conveying parameters on physical quality characteristics of infant formula, J. Food Eng. 106(3): 566 236–244.
[5] J. Reppenhagen, J. Werther (2000), Catalyst attrition in cyclones, Powder Technol. 113(1-2): 55–69.
[6] G. Lu, J. R. Third, C. R. Müller (2015), Discrete element models for non-spherical particle systems: From theoretical developments to applications, Chem. Eng. Sci. 127: 425–465.
[7] P. W. Cleary, R. D. Morrison (2016), Comminution mechanisms, particle shape evolution and collision energy partitioning in tumbling mills, Minerals Eng. 86: 75–95.
[8] E. J. Garboczi (2002), Three-dimensional mathematical analysis of particle shape using X-ray tomography and spherical harmonics: Application to aggregates used in concrete, Cem. Concr. Res. 32(10): 1621–1638.
[9] J. F. Bloore (1977), The shape of pebbles, Math. Geol. 9: 113–122.
[10] I. Deiros Quintanilla, G. Combe, F. Emeriault, C. Voivret, J. Ferellec (2019), X-ray CT analysis of the evolution of ballast grain morphology along a Micro-Deval test: key role of the asperity scale, Granul. Matter 21: 30-1–30-12.

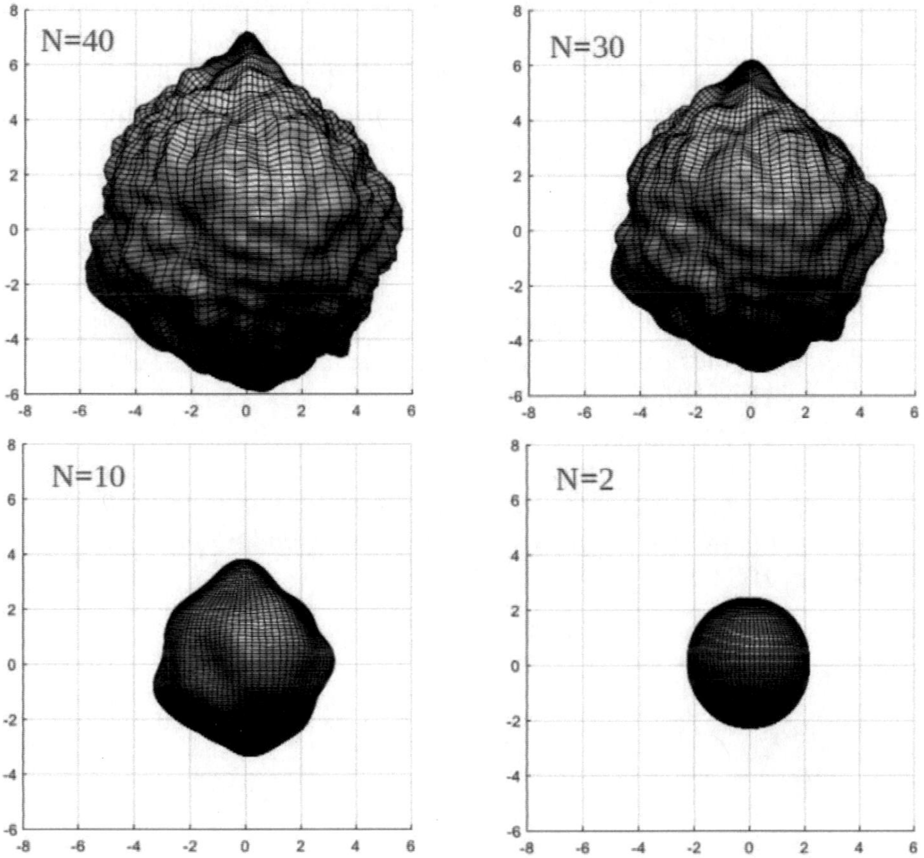

Figure 1. Four expansions for N=40, 30, 10 and 2 obtained by eliminating higher harmonic terms from the expansion and multiplying by suitable the scaling factors.

610419 Investigation into the Role of Holdup on the Gas and Particle Flow Pattern in a Spiral Jet Mill

Abstract:

Both breakage and classification occur within the main chamber of a spiral jet mill. The result is a complex fluid flow field that is difficult to analyse using traditional experimental techniques, whilst underlying mechanisms of particle size reduction remain poorly understood [1, 2]. Using Discrete Element Method (DEM) modelling, coupled with Computational Fluid Dynamics (CFD) the effect of hold-up has on both the fluid and particle phase within the mill. The design of the mill domain is based on the Hosokawa Micron AS-50 spiral jet mill (Runcorn, UK) and five mass loadings were used: 0.4 g, 0.8 g, 1.2 g, 1.6 g & 2.0 g. It was found that as the mass of material held within the bed increases, the average particle velocity decreased, along with the span of particle velocity distribution. As a result, the energy associated with an average collision decreased. It was also found that the bed has its own velocity gradient, and the solid particles travel with a greater velocity towards the surface. The result is that greatest amount of energy transfer takes place where the largest presence of shearing is found. This is located along the bed surface and on the back-face of each jet, as shown in Figure (1). These results are consistent with the experimental work of Kurten and Rumpf [3] and Luczak *et al.* [4]. Finally, it was found that the fluid velocity surrounding the classifier decreases with increasing particle load. The decrease in fluid velocity is due to damping of the fluid field, as momentum is transferred to stabilise the particle bed.

1. MacDonald, R., et al., *The spiral jet mill cut size equation.* Powder Technology, 2016. **299**: p. 26-40.
2. Bnà, S., et al., *Investigation of particle dynamics and classification mechanism in a spiral jet mill through computational fluid dynamics and discrete element methods.* Powder Technology, 2020. **364**: p. 746-773.
3. Kürten, H. and H. Rumpf, *Strömungsverlauf und Zerkleinerungsbedingungen in der Spiralstrahlmuhle.* Chemie Ingenieur Technik, 1966. **38**(11): p. 1187-1192.
4. Luczak, B., et al., *Visualization of flow conditions inside spiral jet mills with different nozzle numbers– Analysis of unloaded and loaded mills and correlation with grinding performance.* Powder Technology, 2019. **342**: p. 108-117.

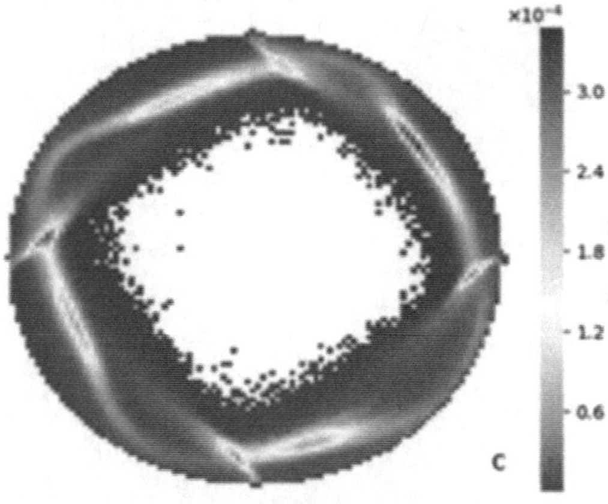

1: Particle contour plot highlighting total energy dissipated (J) from 1.6 g loading case recorded over 0.03 s.

611549 Breakage Functions for High Aspect Ratio Particles

Abstract:

Filtration of particles is influenced by the size-shape distribution of the particles. Therefore, it is necessary to determine how upstream processes affect the size-shape distributions in upstream processes before entering the filter. One such upstream process is particle breakage in stirred vessels such as crystallizers . Experiments on high aspect ratio particles such as paracetamol and urea demonstrate how the size-shape distribution changes with time. This work presents recent developments in modeling fragmentation of high aspect ratio crystals based on the experimental results in stirred suspensions.

602062 A Nonlinear Elasto-Plastic Bonded-Sphere Model for the Discrete Element Simulation of Biomass Particle Breakage

Abstract:

To enhance the current understanding of biomass fracturing mechanism so as to optimize the mechanical preprocessing techniques for biomass size reduction, a discrete element method (DEM) model has been proposed and developed with customized bonding laws designed specifically for biomass materials. The customized bonding laws are able to capture the elasto-plastic deformation within a single biomass particle under compression, bending, and twisting. To better match the experimental observations, two additional nonlinear force-displacement form for normal force contact freedom are proposed as well, i.e., quadratic form and square root form. The fracturing of biomass particles is simulated by allowing the bond to break based on the normal and shear strength criteria. Simulation cases of fracturing tests of loblolly pine cube have been developed and further validated by comparing to the physical experiments. It is shown the elasto-plastic DEM model with a square root force-displacement form can replicate the mechanical failure of pine cubes under shearing with a higher fidelity comparing to the linear elastic bonded-sphere model. By assigning different bonding strengths across the growth ring, the DEM model is able to simulate the anisotropic behavior of pine cubes with different fiber orientations, which is a key factor that should be considered in the biomass processing modeling. Future research will focus on understanding the grinding process of biomass particles for equipment optimization and energy consumption minimization.

594139 Mathematical Modeling of Continuous Milling Processes Using a Cell-Based PBM Approach: A Parametric Study

Abstract:

Particles with a high specific surface area usually exhibit desirable functional responses, such as high reactivity, fast dissolution, and good content uniformity. For this reason, various industries usually increase the specific surface area of particles, like pharmaceuticals, pigments, and minerals, using milling processes that decrease the size of particles for downstream processing or end use. However, current milling technologies are both energy-intensive and energy-inefficient [1]. Therefore, designing a milling process that can operate at or near an optimal condition would benefit the overall performance, but finding optimal milling operational conditions by trial and error is costly. For this reason, computer simulations of the process could provide significant operational insights while enhancing understanding of the milling process, thereby aiding in rational process design and optimization at a minimum cost.

Several modeling approaches, such as the discrete element method and the population balance model (PBM), have attempted to develop a fundamental understanding of the milling processes. Among these approaches, only the PBM provides a quantitative analysis of the spatio-temporal evolution of the particle size distribution (PSD) during milling. In this theoretical study, we developed a cell-based PBM that takes into account the complex nonlinear breakage kinetics and the degree of mixing in an open-circuit continuous dry mill with a discharge screen, unlike previous studies [2–6]. In the context of dry milling, the proposed mathematical model is the first in literature that accounts for (i) the elusive nonlinear breakage kinetics that originate from the multi-particle interactions, (ii) a non-ideal degree of mixedness, and (iii) particle separation via an ideal discharge screen. Various back-mixing ratios (R) and number of cells (n) were used to emulate different extents of particle mixing in the mill including the idealized cases of perfect mixing and no back-mixing (plug flow). Other simulations were performed to examine the impact of cushioning action of fine particles and the screen size. The simulation results were reported in terms of cumulative PSD at steady-state and during the transient period. Apart from the PSD of the product at the mill outlet, spatial variations of the PSD inside the mill were also examined.

For continuous mills without a screen, the cell-based PBM with $n = 1$ exactly emulated the PBM solution of an idealized perfect mixing scenario, whereas the cell model with $n = 60$ nearly represented the ideal plug-flow (no back-mixing) scenario, with an increase in n converging to the PBM solution for the plug-flow obtained previously in the literature. These findings demonstrated the consistency of the cell-based PBM. The simulation results also suggest that an increase in the number of cells and/or a decrease in the back-mixing ratio, emulating lower extent of back-mixing, led to finer product. While the cushioning action of finer particles led to coarser product PSD in the absence of a screen, this nonlinear impact becomes less important when a screen is present and especially when the screen opening size is small. This study has demonstrated for the first time in literature that the use of a discharge screen with a fine opening not only allows for passage of the desired fine particles and retention of the relatively coarse particles, but also it reduces the retardation impact of the fines on breakage kinetics, thus ultimately resulting in a finer product PSD.

References

[1] B.A. Wills, An introduction to the Practical Aspects of Ore Treatment and Mineral, Butterworth-Heinemann, Oxford, 2016.

[2] L.G. Austin, P.T. Luckie, D. Wightman, Steady-state simulation of a cement-milling circuit, Int. J. Miner. Process. 2 (1975) 127–150.

[3] H. Benzer, L. Ergun, M. Oner, A.J. Lynch, Simulation of open circuit clinker grinding, Miner. Eng. 14 (2001) 701–710.

[4] E. Bilgili, B. Scarlett, Numerical simulation of open-circuit continuous mills using a non-linear population balance framework: incorporation of non-first-order effects, Chem. Eng. Technol. 28 (2005) 153–159.

[5] O. Genc, S.L. Ergun, A.H. Benzer, The dependence of specific discharge and breakage rate functions on feed size distributions, operational and design parameters of industrial scale multi-compartment cement ball mills, Powder Technol. 239 (2013) 137–146.

[6] C. Mihalyko, T. Blickle, B.G. Lakatos, A simulation model for analysis and design of continuous grinding mills, Powder Technol. 97 (1998) 51–58.

602571 Computational Fluid Dynamic Simulations of Experimental and Natural Granular Flows: First Insights into the Flow-Wall Interaction Dynamics

Abstract:

Volcanic granular flows (some pyroclastic density currents, debris avalanches, debris flows) can be defined as gravity-driven currents of solid particles where the particle-particle interaction dominates the motion. The interaction with topography is another relevant factor controlling the dynamics and the evolution of these flows. In the last decade, granular flow modelling has become one of the main tools to replicate past events and to investigate potential hazards of volcanic granular flows. The research on the behaviour of volcanic granular flows is one of the main topics in present day geophysics and volcanology, due to the hazard that they pose in areas located in the vicinity of the volcanoes.

The aim of this research is to investigate the dynamics of channelised volcanic granular flows by combining large-scale experiments with multiphase computational fluid dynamic simulations using the MFIX software (http://mfix.netl.doe.gov/). In particular, we focus on the dynamics that regulate the flow-wall interactions. MFIX provides a suite of models that treat the carrier phase (fluid phase) and the dispersed phase (volcanic particles) as interpenetrating continua. Sensitivity analyses of the boundary conditions and of the frictional models implemented in the numerical code were carried out to search for the most appropriate numerical set-ups to simulate experimental granular flows. These analyses have shown how the coefficients, which rule granular energy and momentum diffusion equations, are crucial to model the physics of the granular flow close to the wall. The optimal MFIX configuration was applied to large-scale experiments of a monodisperse system and comparisons between simulated and observed flow thickness and instantaneous and average velocities were carried out. Preliminary results show that the model predictions are significantly affected by the diffusivity of granular energy and momentum at the wall and, therefore, a detailed study of the latter would improve knowledge of the boundary conditions and hazard control.

597290 CFD-Dpm Simulations of Spray Drying Process: Drying Kinetics of Individual Droplets in Water Droplet Group

Abstract:

Spray drying processes have played considerably essential roles in several industries, such as pharmaceutical, food, ceramic, and others. However, many factors affect the physical properties of dry particles, and the detailed mechanism of dry particle generation is not sufficiently clarified. Therefore, the operating conditions are determined by trial and error to obtain the desired product. A look at the work published so far reveals the existence of maps of velocity, temperature, humidity, and pressure within spray dryers obtained using computational fluid dynamics; CFD. However, the drying kinetics of individual droplets and their subsequent change in diameter as their trajectories are still missing.

In this study, we performed CFD-DPM simulations of droplet drying in a mini spray dryer. The developed model and calculation results were validated based on mini spray drying tower experiments. We focused on the drying behavior in the droplet group and investigated the changes in the droplet diameter and the driving force of evaporation. The results of the numerical simulation revealed that the driving force of evaporation, ΔT, changed every moment as the droplets traveled through different temperature distribution fields in the drying tower. Therefore, each droplet has different drying kinetics in the spray dryer. Also, it was suggested that the prediction of the drying kinetics of the droplet group was insufficient under the assumption of the steady condition, and the influence of the surrounding droplets should be considered.

Furthermore, we investigated the effects of operating conditions of the spray drying process (such as air temperature, airflow rate, liquid feed rate, and droplet diameter) on the drying kinetics of the droplet group. The numerical simulations were performed under various conditions to calculate the drying kinetics of the droplet group in the spray dry process. The calculated results revealed that the drying kinetics of the droplet group strongly depended on the driving force for the droplet evaporation(ΔT). Finally, we proposed a method to estimate the drying kinetics of the droplet group from the operating conditions by organizing with the thermal efficiency and Nusselt number.

601976 Deep Neural Network-Based View Factor Modelling of Radiative Heat Transfer between Particle-Particle and Particle-Wall

Abstract:

Introduction

Above 700 [°C] radiative heat transfer is the most dominating heat transfer mechanism. Such high temperature processes occur in a wide range of industrial applications. Examples are: pebble bed reactors, laser sintering and high-temperature particle oxidation or reduction processes [1]. The modelling of heat radiation is an ongoing challenge in many simulation fields, e.g. Computational Fluid Dynamics (CFD) and the Discrete Element Method (DEM).

In this study we investigate radiative heat modelling from a particle perspective. In principle the emitted heat flux $Q_{i\text{-}j}$ between two bodies (including walls) depends on the following factors: temperatures T, surfaces A, emissivity E and the view factor $\varepsilon_{i\text{-}j}$ between these objects, as well as the Boltzmann constant σ_S. The view factor is the ratio of the radiation leaving a surface i and the radiation that is striking surface j.

The analytical calculation or numerical integration of view factors cannot be performed for real-sized systems, that feature billions of particles, due to the astronomical computational cost. Methods to overcome this limitation have been presented in literature, based on the Monte Carlo algorithm by Walker et al. [2], or the Projection Method based on the projection of a defined number of test points that are distributed on particle surfaces by Forgber & Radl [3]. Both Methods reflect the widespread tradeoff in simulation science between accuracy and speed. The Monte Carlo raytracing achieves very accurate view factors but takes up high computational cost. The projection method provides precise view factors in specified systems at significantly less computational effort, although the demand is still too big to adopt the method in general for heat radiation modelling in DEM simulations.

Methods

To overcome this tradeoff in the presented approach we use Machine Learning (ML) techniques to create a pre-trained heat radiation model based on a deep neural net (DNN), that predicts accurate view factors at high speed. The training data for the DNN is generated with the Monte Carlo raytracing method from randomly-positioned particle beds, considering different particle volume fractions as well as the wall-view factors.

The neural network model parameters are formed by the hyperparameters, e.g. the learning rate, and the model design that consists of number of layers, number of nodes, activation function, optimizer and so on. The input layer of a DNN is defined by the features of the training data that are often called markers. A neural net with the surface distance cannot model the widely distributed view factors in a particle bed. Therefore, additional markers like the solid angle between interacting particles and derivations of a local particle fraction are needed. These markers are then consequently analyzed by a feature analysis. To achieve the best possible generalization with the DNN-based model these model parameters are optimized by the randomized search approach [4].

The outcome of the presented approach will be compared to the Projection Method and a simple regression analysis in terms of accuracy and computational cost. The overall quality of prediction is measured by the mean squared error (MSE). Because the MSE is significantly more influenced by big view factors and does not reflect the overall quality of the prediction, the calculated view factors are correlated to the view factors determined via Monte Carlo raytracing. The coefficient of determination (R^2) is then used to describe this correlation. Another quantitative measure is the coefficient of determination ($R^2_{spread}(x_u, x_o)$) of particle distances within an upper boundary x_o, that is the half domain size plus one particle diameter, and a lower boundary x_u, defined as the half domain size minus one particle diameter.

The investigated particle bed contained 587 particles and the total dataset therefore consists of 586*586 view factors values. The total dataset was divided into the training dataset, the test dataset and the validation dataset with the respective relative amounts of 0.8, 0.1 and 0.1. The top and the bottom walls of the particle bed were used for the creation of the particle-wall dataset. The particle positions were randomly changed to gain more data. 33 different particle bed configurations were used. The total particle-wall dataset then consists of 66*587 view factor values. The split ratios were the same as for particle-particle interactions.

Results

A simple linear regressor, which is derived from the training dataset, achieves an MSE of 4.3e-7 on the test dataset at a basically instantaneous speed. The R^2-value of the correlation is 0.981. For large surface distances the simple regressor cannot accurately predict the spread of the view factors. This results in an R^2_{spread}-value of -0.336. For particle-wall interaction a simple linear regressor achieved an MSE of 2.5e-6, an R^2_{wall}-value of 0.901 and $R^2_{spread,wall}$ is -2.133.

The Projection Method (number of test points = 200) achieves an MSE of 7.7e-7 on the test dataset. The overall coefficient of determination is 0.967, and R^2_{spread} is -2.530. Overall the computational demand of the Projection Method is too high, to be adopted in general to DEM simulations.

The DNN-based model achieves an MSE of around 2.5e-7 while taking around 0.30 [s] to calculate the view factors and 0.55 [s] to load the pretrained model. The R^2-value for the total correlation with the test dataset is 0.989. The R^2_{spread}-value is 0.424.

Particle-wall view factors of a particle bed show a different data structure than particle-particle interactions. More big view factors occur since the nearest particle bed layer is irradiated without any obstacles. Therefore, a separate DNN-based model for particle-wall interactions was trained. The MSE for wall interactions is then 7.9e-8, R^2_{wall} is 0.937 and $R^2_{spread,wall}$ is 0.491.

Conclusion

For particle-particle interaction it is demonstrated that a pretrained DNN can model view factors for any distance at significantly less computational effort, higher overall accuracy and reasonable prediction quality at certain distances.

The same conclusion can be made for particle-wall interactions. Thus, a combination of two DNN-based models, one for particle-particle interactions and one for particle-wall interactions, is recommended for general heat radiation modelling in DEM simulations.

References

[1] M.F. Modest, Radiative Heat Transfer - Second Edition, Academic Press, 2003. https://doi.org/10.1017/CBO9781107415324.004.

[2] T. Walker, S.C. Xue, G.W. Barton, Numerical determination of radiative view factors using ray tracing, J. Heat Transfer. 132 (2010) 1–6. https://doi.org/10.1115/1.4000974.

[3] T. Forgber, S. Radl, A novel approach to calculate radiative thermal exchange in coupled particle simulations, Powder Technol. 323 (2018) 24–44. https://doi.org/10.1016

/j.powtec.2017.09.014.

[4] J. Bergstra, Y. Bengio, Random search for hyper-parameter optimization, J. Mach. Learn. Res. 13 (2012) 281–305.

610648 Thermal Treatment of Granular Materials in a Rotary Drum Equipped with Baffles

Abstract:

Rotary drums play a vital role in the thermal treatment of granular materials in processes such as calcination and drying. A rotary drum equipped with baffles/flights is of interest as studies show that baffles can effectively decrease the required processing time and improve the uniformity of the final product. However, much uncertainty still exists about how baffles impact the rate of heat transfer in rotary drums for different operating conditions and particle properties. In order to understand how baffles affect heat transfer of granule materials in rotary drums, a series of simulations for different operating conditions and particle properties were carried out using the discrete element method (DEM). A number of different operating conditions were considered including: fill level, drum size, baffle size, and number of baffles. Different thermal properties of the particles were considered by varying the thermal conductivity and the effect of particle size was also considered. The simulation result show that the heat transfer initially improves significantly by increasing the number and size of the baffles and the decreasing the fill level. However, it should be noted that adding an excessive number of baffles can decrease the rate heat transfer of granule materials in rotary drums. Moreover, we observed, there is an optimum number and size of baffles for a given fill level. By introducing a new scaling parameter we were able to categorize all the simulation result for different operating conditions and material properties and introduce an optimum condition for adding baffles to the rotary drum. The results of the study have advanced our understanding of thermal treatment of granular materials in a rotary drum equipped with baffles.

609991 Simulations of Heat Transfer to Flowing Particles Used for Long Duration Thermal Energy Storage

Abstract:

Energy storage systems are crucial to transform the world's energy reliance from coal-based energy production to clean energy production. The National Renewable Energy Laboratory (NREL) is leading an effort to develop the **E**co**n**omic Long-**Dur**ation Electr**i**city Storage by Using Low-Cost Thermal E**n**er**g**y Storage and High-Efficiency Power Cycle (ENDURING) system that aims to provide clean energy at low cost. In the ENDURING system, silica sand is the heat transfer fluid as well as the thermal storage material because it is a relatively low-cost material that is reliable at high temperatures. To warm the particles, an array of electric heating elements is powered by excess energy from the transmission grid and used to heat the flowing sand particles. In our work, we study how to maximize the heat transfer from the electric heating elements to the sand particles. To investigate the heat transfer problem, we use MFiX, a Discrete Element Modeling (DEM) computational solver developed by the National Energy Technology Laboratory (NETL). The particle-particle and particle-wall friction coefficients, surface angle, and particle surface roughness were the key parameters investigated in this study. Also, an analytical model of the heat transfer process was developed to extrapolate the small-scale DEM results to a large-scale prototype. After exploring the parameter spice, we found that a relatively shallow surface angle of 20° increases the residence time and heating rate of particles without incurring jamming. The simulations also reveal that friction can enhance the heat transfer by improving thermal contact, but cases with high friction can restrict the flow and result in worse heat transfer. The simulation results are discussed in detail and configurations for optimal performance of the particle heater system are presented.

608815 Thermal Atomic Layer Deposition of Gold Nanoparticles for Catalysis: High-Precision & Scalable Production

Abstract:

Atomic layer deposition (ALD) has become an established technique for the growth of conformal, ultrathin films in, for example, semiconductor applications. However, it is also an attractive technique for the controlled growth of nanoparticles (NPs). As such, it can play an important role in the synthesis of nanostructured catalysts with high precision. Gold nanoparticles have been extensively studied for their applications in catalysis. For Au nanoparticles to be catalytically active, controlling the particle size is crucial [1,2]. While for several catalytic materials ALD precursors have been available already for a long time, only recently ALD precursors for gold have been developed [3, 4]

Here we present a low temperature (105 °C) thermal atomic layer deposition approach for depositing gold nanoparticles on TiO_2 with controlled size and loading using trimethylphosphino-trimethylgold (III) and two co-reactants (ozone and water) in a fluidized bed reactor. We show that the exposure time of the precursors is a variable that can be used to decouple the Au particle size from the Au loading. Longer exposures of water broaden the particle size distribution while longer exposures of ozone narrow it [5]. By studying the photocatalytic activity of Au/TiO2 nanocomposites we show how the ability to control particle size and loading independently can be used not only to enhance performance but also to investigate structure-property relationships. This study provides insights into the mechanism underlying the formation and evolution of Au nanoparticles prepared for the first time via vapor phase atomic layer deposition. Employing a vapor deposition technique for the synthesis of Au/TiO_2 nanocomposites eliminates the shortcomings of conventional liquid-based processes opening up the possibility of highly controlled synthesis of materials at large scale [6].

[1] Valden, M., Lai, X., & Goodman, D. W. (1998). Onset of catalytic activity of gold clusters on titania with the appearance of nonmetallic properties. science, 281(5383), 1647-1650.

[2] Li, L., Larsen, A. H., Romero, N. A., Morozov, V. A., Glinsvad, C., Abild-Pedersen, F., Greeley, J., Jacobsen, K.W & Nørskov, J. K. (2013). Investigation of catalytic finite-size-effects of platinum metal clusters. The journal of physical chemistry letters, 4(1), 222-226.

[3] Griffiths, M. B., Pallister, P. J., Mandia, D. J., & Barry, S. T. (2016). Atomic layer deposition of gold metal. Chemistry of Materials, 28(1), 44-46.

[4] Makela, M., Hatanpaa, T., Mizohata, K., Raisanen, J., Ritala, M., & Leskela, M. (2017). Thermal atomic layer deposition of continuous and highly conducting gold thin films. Chemistry of Materials, 29(14), 6130-6136.

[5] Hashemi, F. S. M., Grillo, F., Ravikumar, V., Benz, D., Shekhar, A., Griffiths, M., Barry, S. & van Ommen, J. R. (2020). Thermal Atomic Layer Deposition of Gold Nanoparticles: Controlled Growth and Size Selection for Photocatalysis. Nanoscale, accepted.

[6] van Ommen, J. R., & Goulas, A. (2019). Atomic layer deposition on particulate materials. Materials Today Chemistry, 14, 100183.

599758 Alloying and Ligand Effects in Supported Gold Nanoparticle Catalysts for CO Oxidation

Abstract:

Alloying metals are an interesting and important method to modulate the chemistry of active sites in nanoparticles. This effect of alloying is more pronounced as nanoparticles become small enough to be classified as nanoclusters (< 2 nm). In this size range, the dispersion significantly increases and the metal utilization improves as almost all atoms are present on the surface and are available as active sites. Furthermore, small particle sizes result in a greater fraction of undercoordinated sites which are often very active for catalysis.

Introducing alloying metals to gold has a few different effects, namely the ability to compress or to strain the structure of intermetallic bonds, inducing a charge transfer from one metal to another, or to change the composition of the active site surface. These effects all alter the energy of the transition state and introduce a new chemical structure moving from one state to another. In addition to composition, the environment surrounding the surface plays a role in the activity of the catalyst. There are many efforts which aim to generate a "naked" nanoparticle with an entirely ligand-free surface; however, in this work, the aim is to observe changes in catalytic performance with ligands attached. The presence of the ligands is important to alter both the steric demand of reactants that reach a catalytically active surface and the possible ligand electron donation/withdrawal with the metallic surface. Bearing this in mind, we select two preparation methods. Strong Electrostatic Adsorption (SEA) for the "naked" nanoparticles, and a Sterically Demanding Ligand Synthesis (SDLS) for a phosphine capped nanoparticle.

This work demonstrates the effects of alloying Au with Cu, Co, Ni, Pd and Ru metals on the active site and the impact of a sterically demanding ligand such as triphenylphosphine (TPP) on both the selectivity and activity for gas-phase CO oxidation. Differences in metallic composition as well as ligand-metal interactions show observable differences in activity compared to monometallic Au. Ligand-bound nanoparticles show activity for CO oxidation in the range of 340-420 K, while the "naked" nanoparticle synthesis illustrates activity at a temperatures of about 300 K, with the exception of AuRu showing activity at a temperature of 330 K. Interestingly, AuNi bimetallic particles show the lowest activity for SDLS relative to the other alloys but show the highest activity for SEA synthesis relative to the other bimetallic nanoparticle compositions. STEM confirms all alloyed particles are in the size range of 2-5 nm, which have high fractions of undercoordinated bimetallic surface atoms, except AuRu which have larger particle sizes of 6.8±1.7 nm. Alloys show a shift in Enthalpy-Entropy compensation with ligand presence. We also observe trends in the material structure cell parameter with the activation energy of the material.

DRIFTS allows for the elucidation relationship between CO chemisorption properties and the structure of the materials. CO vibrations are present at about 2000 cm^{-1} for Au-FCC metals while about 2060 cm^{-1} for Au-HCP metals. The lower IR frequency indicates a higher electron density of the material due to greater pi-backbonding. Ligand coverage is a reasonable explanation for decreased activity in SDLS. The trends in the location of the chemisorbed CO in the IR spectra are consistent with the lower electron density that is observed in the "naked" nanoparticles which suggest that the ligand interaction provides an electron-rich surface. In addition, the activation energy trends with both the alloy composition as well as the presence of the ligand.

604856 Thermally Induced Restructuring of Core@Shell Nanoparticles As a Strategy for Increasing Catalyst Stability and Activity

Abstract:

The restructuring of catalytic nanomaterials that occurs during prolonged operation at high temperature remains a significant challenge in catalysis. Typically, thermally induced restructuring results in the sintering of active metal and the loss of support porosity, which compromises both the structure and number of active sites that can be used for reaction. Such outcomes significantly inhibit both catalyst activity and selectivity. The challenges imposed by sintering processes are magnified by the growing cost and scarcity of many of the metals that comprise industrial catalysts.

In this work, we investigate how initial catalyst nanostructure affects the outcomes of high temperature restructuring.[1] By comparing core@shell nanoparticles that have active metal cores encapsulated by porous metal oxide shells, with catalysts prepared by wet impregnation, we show divergent restructuring outcomes that depend on the placement of the active metal.

Elevated temperature aging (800˚C) of a Pd@CeO_2 core@shell catalyst redisperses core Pd species throughout the encapsulating CeO_2, increasing active site dispersion from 33% to 88%. The highly dispersed restructured Pd clusters exhibit an average diameter of 1.3 nm, and coordinate strongly with the reducible CeO_2 support, as characterized by CO chemisorption, transmission electron microscopy, and x-ray energy dispersive and photoelectron spectroscopies. The Pd redispersion also appears to stabilize the CeO_2 crystallites that comprise the shell, making them less prone to high temperature sintering. These restructuring processes appreciably increase catalytic activity, as indicated by a 45˚C decrease in the temperature required for 90% conversion in a probe CO oxidation reaction. The turnover frequency for CO oxidation increases from 0.2 s^{-1} to 0.39 s^{-1} and the apparent activation energy decreases from 42 to 33 kJ/mol after aging. Such changes in intrinsic catalytic activity suggest that the enhanced activity seen in the restructured catalysts is not solely due to an increase in active site dispersion. Such outcomes are in stark contrast to those observed for Pd/CeO_2 catalysts prepared by impregnation, which exhibit appreciable pore closure, sintering and impeded activity after identical elevated temperature aging. By extending this investigation to a number of aging conditions and active metal/ support combinations, we hope to better understand how directed restructuring of carefully designed nanomaterials can be used as a novel strategy to enhance catalyst functionality.

[1] Hill, Schwank, et. al. ACS Catal. (2020).

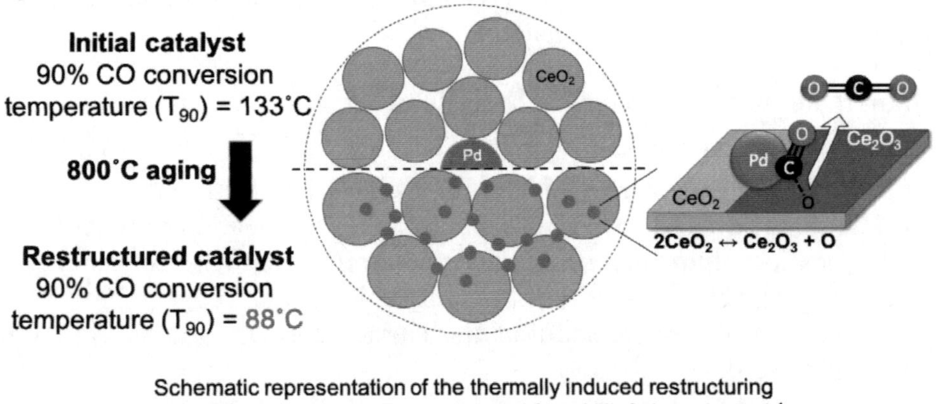

Schematic representation of the thermally induced restructuring of Pd@CeO_2 catalysts. Figure adapted from Hill, Schwank et. al.[1]

609152 Ultrasonic-Assisted Synthesis of Photoresponsive Silver Supported Nanoparticles in Continuous Flow

Abstract:

Graphitic carbon nitride (g-C_3N_4) photocatalyst, which is a novel metal-free polymer semiconductor with a band gap of about 2.7eV, has attracted growing attention in the photocatalyst for H_2 production. The unique electronic structure with π-conjugated graphitic provides special properties such as high stability and a visible light response. However, the low surface area and high recombination rate of photogenerated electron-hole pairs are still the barriers hindering the application of g-C_3N_4. Many strategies have been employed to increase the photocalytic activity. Among these methods, coupling with noble metal nanoparticles with surface plasmon resonance effect is a promising method. However, the coalescence and agglomeration of nanoparticles are great challenges for controlling particle size. Supported nanoparticles with small mean sizes can greatly improve the catalytic performance. Recently, size controlled supported nanoparticles have been synthesized continuously through microreactors. However, the inevitable use of organic steric capping agents hinders catalytic activity. Ultrasound, which can support cavitation and facilitate the disruption of particle agglomerates, is considered as an effective method for synthesis of small monodisperse nanoparticles.

Here in, we present a novel approach for ultrasonic-assisted continuous synthesis of Ag/g-C_3N_4 catalysts in the absence of organic steric capping agents. Supported nanoparticles were synthesized through mixing the support and reactants in a coiled flow inverter microreactor. Ultrasonic irradiation was applied in the nucleation and growth of nanoparticles in the microreactor. The synthesized Ag/g-C_3N_4 was characterized by TEM, ICP and PL techniques. The increase of ultrasonic power produces nanoparticles with smaller mean sizes, and the loading contents and photoelectron transfer performance of the supported catalyst are also affected. The catalytic performance of Ag/g-C_3N_4 was tested based on hydrogen-production from triethanolamine under visible light irradiation. The photocatalytic production H_2 rate with 5 wt% Ag/g-C_3N_4 prepared at an ultrasonic output of 60 W and at a residence time of 35 s was about 2.4 times as that of pure g-C_3N_4, respectively. This study not only provides a further understanding of the effect of the size of the supported catalyst on the performance, but also proposes an innovative method of continuous synthesis of the supported nanomaterials without organic steric capping agents, which provides a reference for the application of ultrasonic technology for industrial-scale production.

Reference:

1. Wu, KJ, Torrente-Murciano L. Continuous synthesis of tuneable sized silver nanoparticles via a tandem seed-mediated method in coiled flow inverter reactors. *React Chem Eng*. 2018;3:267-276.
2. Wu, KJ, Michet DVBG, Torrente-Murciano L. Synthesis of narrow sized silver nanoparticles in the absence of capping ligands in helical microreactors. *React Chem Eng*. 2017;2:116-128.
3. Cattaneo S, Althahban S, Freakley SJ. Synthesis of highly uniform and composition-controlled gold–palladium supported nanoparticles in continuous flow. *Nanoscale*. 2019;11:8247-8259.

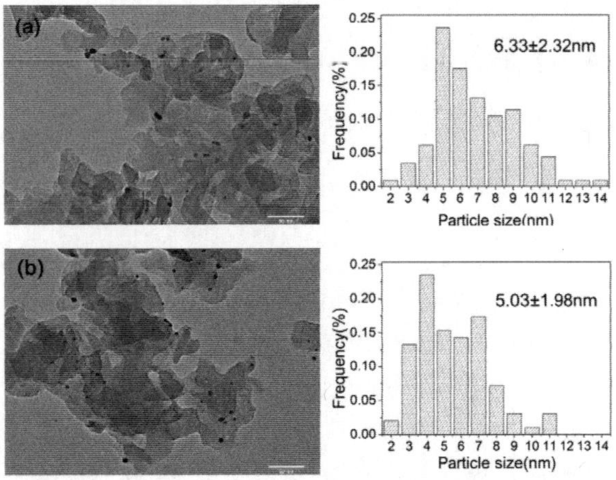

TEM images of 3 wt% Ag/g-C_3N_4 prepared at different ultrasonic power: (a) 0W and (b) 120W.

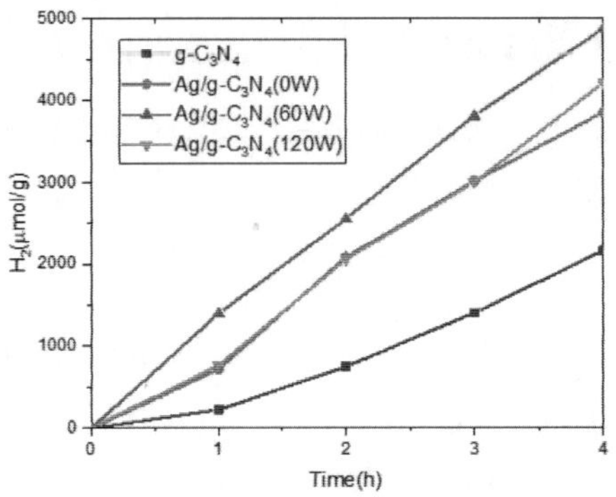

H_2 production of pure g-C_3N_4 and 5 wt% Ag/g-C_3N_4 prepared at different ultrasonic outputs at a flow rate of 2.4 mL/min.

602094 Granular Rheology with Sliding, Rolling and Twisting Friction

Abstract:

Intuition tells us that a rolling or spinning sphere on a table will eventually stop due to the presence of friction. Nonetheless, rolling and twisting friction are often neglected in particle simulations. Rolling and twisting friction are important in granular rheology, where the flow field induces rotations. Granular rheology is simulated using particle-based, discrete element simulations with sliding, rolling and twisting friction in different configurations, including chute flow and bulk-like stress-controlled shear. Stress differences, stress ratios and inertial numbers are used to characterize the rheological behavior for different friction states. The increase in critical stress to flow, for bulk-like rheology, and the angle of repose, for chute flow, due to rolling and twisting friction is measured. The fabric tensor is used to characterize the structure of the flowing states. Beyond the contact, normal and tangential fabric tensors, the two rotational fabric tensors, from rolling and twisting friction, are also calculated. Chute flows are further characterized by property profiles perpendicular to the plane.

Sandia National Laboratories is a multimission laboratory managed and operated by National Technology & Engineering Solutions of Sandia, LLC, a wholly owned subsidiary of Honeywell International Inc., for the U.S. Department of Energy's National Nuclear Security Administration under contract DE-NA0003525.

595896 A Multi-Level Coarse Graining Technique for Discrete Element Method (DEM) Simulation

Abstract:

Discrete Element Method (DEM) and coupled CFD-DEM are the two important techniques that can simulate flow of particulate matter accurately from first-principles. However, one of the major issues with such detailed models is that the computing power required to simulate industrial scale systems is impractically high. Even large-scale parallelization cannot always resolve this problem completely. Coarse graining techniques can be useful in this regard. In a standard coarse-graining technique (*Sakai and Koshizuka, Chemical Engineering Science, 2009, 533 -539*), a collection of original particles are represented by a coarse grained particle in such a way that the linear and rotational kinetic energy between the coarse grained and fine grain descriptions are equal.

Although*Sakai and Koshizuka*type coarse graining has been demonstrated successfully for many systems, the accuracy is good only until the diameter ratio between the coarse grain particle to the original particle is about two. If the coarse graining ratio is increased to a higher value, the accuracy of the simulation reduces substantially even for a simple hopper discharge problem. Beyond a certain threshold, the simulation artefacts become dominant at certain parts of the equipment. One such artefacts is choking in hopper flow.

In this work we demonstrate a new method which resolves this problem by using an adaptive coarse graining strategy. In this new method, the domain is divided into many sub domains each with its own coarse graining ratio. When a particle move from one zone to another, the degree of coarse graining changes by either breaking the particles into smaller one or gathering smaller particles into a bigger particle as shown in the attached figure.

We use this adaptive coarse graining technique to simulate hopper flow as shown in the attached figure and the results are compared with classical coarse graining technique of *Sakai and Koshizuka*. A simulation with originally 7,500,000 particles were taken and resolved into a simulation of 956,689 particles using a fixed coarse graining ratio of 2. Higher coarse graining ratio yielded inaccurate results. With adaptive coarse graining, a total of only 75,000 particles of varying sizes were used. The results obtained from the fixed coarse graining and adaptive coarse graining were comparable, but the fixed coarse graining took 180 hours to simulate the discharge while the adaptive coarse graining took only 7 hours to complete the simulation.

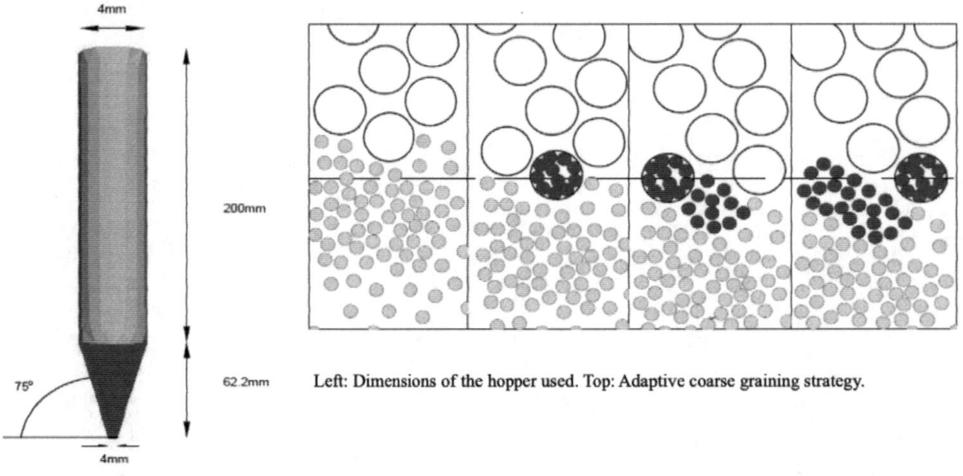

Left: Dimensions of the hopper used. Top: Adaptive coarse graining strategy.

600295 CFD-IBM-DEM Analysis of PM Filtration in Fluidized Bed Based on Force Balance to Evaluate Operation Conditions.

Abstract:

Particulate matter (PM) is a well-known air pollutant and most of them are soot emitted from fossil fuel combustion. PM diameter has been decreased with improvement of combustion technology. Especially, the concentration of $PM_{2.5}$ (diameter less than 2.5 μm) has increased. Fine PM has adverse effects on human body, and the highly efficient $PM_{2.5}$ filtration method is required. The granular bed filter is promising method for PM filtration, and our research group shows that this device can collect fine PM at highly efficiency with using gently fluidized bed [1]. It is important for practical use of this device to understand fundamental phenomena in PM filtration, because the performance of this device strongly depends on operating parameters, such as gas velocity, void fraction, and operating time. In previous study, CFD (computational fluid dynamics) and DEM (discrete element method) coupling simulation was conducted with using locally averaged Navier-Stokes equation [2]. PM-gas interaction is averaged, and micro flow, which may have large effect on PM filtration, is also averaged. It is difficult to accurately reproduce the motion of PM in bed particle layer.

In this study, numerical simulation of PM filtration was conducted via CFD-IBM (immersed boundary method)-DEM [3] to consider the effect of accurate flow around bed particle. The effects of operating conditions are assessed based on forces acting on PM and the ratio of captured PM number to total PM number. The schematic diagram of numerical method is shown in Figure. Since the method requires highly computational cost, a part of bed particle layer was demonstrated. In the following figure, grey particle is bed particle and PM is blue particle. The diameters of bed particle and PM are 420 and 2 μm, respectively. PM is supplied from bottom of bed particle layer at feed rate of 100 per 0.0001 seconds, and initial velocity is same as superficial velocity. The most PM collides with bed particle, and it is captured on bed particle surface by adhesion force. The collection efficiency is about over 90% at superficial velocity of 0.4 m/s. The highly collection efficiency is obtained even at higher gas velocity of 0.8 m/s when bed particle is about closed packed. This gas velocity is twice of 0.4 m/s at which collection efficiency of 100% is obtained in previous experimental work [1]. The cohesive interaction between PM-bed particle or PM-PM is dominant force for fine particle. It makes collection efficiency high, and PM is stably maintained on bed particle by adhesion force.

[1] T. Yamamoto, K. Hori, J. Tatebayashi, An experimental investigation of the PM adhesion characteristics in a fluidized bed type PM removal device, *Powder Technol.*, 289, 2016, 31-36.

[2] K. Yokoo, T. Yamamoto, H. Matsune, M. Kishida, J. Tatebayashi, CFD-DEM analysis of PM filtration due to adhesion force using fluidized bed, 18th Asian pacific confederation of chemical engineering, 2019, B321

[3] T. Kajishima, S. Takiguchi, Interaction between particle clusters and particle-induced turbulence, Int. J. Heat and Fluid Flow, 23, 2002, 639-646.

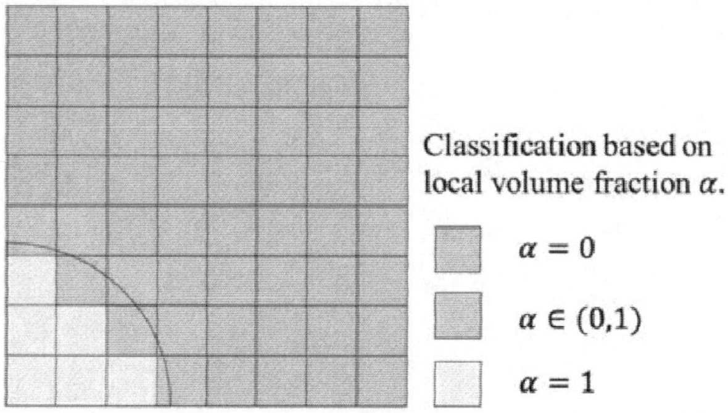

Particle volume-weighted velocity ; $u = \alpha u_p + (1-\alpha)u_f$

Gas-Particle momentum-exchange ; $f_p = \alpha \frac{u_p - u}{\Delta t}$

PM (blue) filtration in bed particle (grey) layer.
PM is supplied with gas from bottom side.

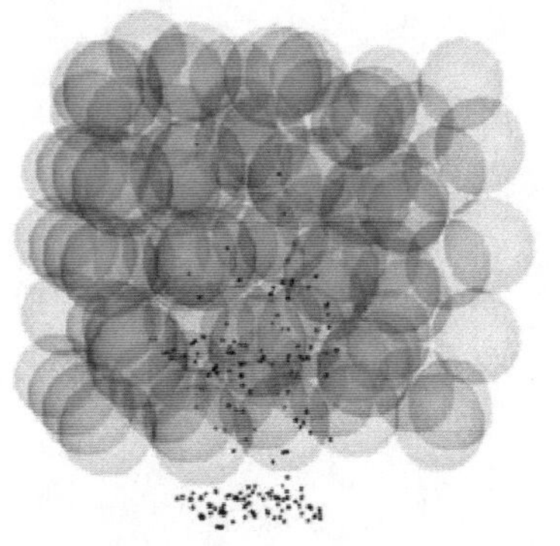

602403 Continuum Rheology of Dense Granular Flows Near the Flow-Arrest Transition

Abstract:

Granular materials exhibit a dynamical transition between arrested and steady flowing states at a critical ratio of shear stress and pressure. Although their simple-shear steady-state rheology is well characterized, the transition itself is accompanied by interesting dynamical phenomena such as transient dilatancy and shear jamming, which are not well understood. This transition is highly stochastic, which makes its continuum modeling quite challenging. Additional complications are introduced by a dependence on loading paths, with important differences between the rheology of shear-induced and compression/extension-induced flows. We demonstrate such complex rheological scenarios that emerge at the flow-arrest transition of granular materials using stress-controlled discrete element simulations. We introduce and calibrate a continuum rheological model, derived from a dissipative rheological theory, which is applicable to viscometric and extensional flow regimes, and incorporates important rheological features such as normal stress effects and flow-arrest transitions.

609957 Continuum Simulations of Dense Granular Flow Near the Maximum Packing Limit

Abstract:

The rheology and hydrodynamics of powder flow are critical in many industrial processes such as agricultural and pharmaceutical production, and petrochemical refining. Continuum simulations solve the averaged equations for conservation of mass and momentum by treating the granular medium as a continuum and can predict granular hydrodynamics. A comparison of continuum simulations of silo discharge (Vidyapati *et al.* 2013) with experimental data and discrete element model simulations indicates that they are limited in their accuracy of predicted discharge rates by the constitutive model for granular stress. Specifically, accurate representation of the intermediate regime of granular flow, which lies between the rapid granular regime and the quasi-static regime, is needed for predictive accuracy. One of the challenges in incorporating newly proposed constitutive models for the granular stress in continuum simulations, which account for intermediate regime behavior (Chialvo *et al.* 2012, Sun *et al.* 2011, Vidyapati *et al.* 2016) is ensuring their realizable behavior near the maximum packing limit. Realizability in this context means guaranteeing that the solid volume fraction is always non-negative and does not exceed its theoretical maximum value without *ad hoc* modifications. The behavior of continuum solvers near the packing limit depends on the constitutive model for the granular pressure as a function of solid volume fraction. As the solid volume fraction approaches the maximum packing limit, there is a threshold above which the granular material behaves like an incompressible fluid. The equations also change type from variable density to incompressible at this incompressible threshold. We report solutions to a one-dimensional (1-D) granular flow illustrating this transition and develop a computational scheme to implement this in 3-D continuum simulations.

611290 Model Powder Flow in an Oscillatory Rheometer Using DEM

Abstract:

Powder flow in an oscillatory rheometer was modeled using the discrete element method (DEM). In the simulation, non-cohesive spherical particles were sheared between two curved boundaries in an oscillatory motion. The study aims to study the relationship between the spatial distribution of normal stress on boundaries and the transient jamming structures, i.e. force chains. Specifically, the normal stress distributions on boundaries were visually presented with heat maps. The power spectra were computed to determine the periodical variation of the stress field. The force chain network within the powder bed was characterized and the evolution of the network was studied. Sensitivity analysis was conducted to investigate the effect of selected material properties (particle size, particle size distribution, particle-particle sliding friction coefficient, particle-boundary sliding friction coefficient, particle-particle rolling friction coefficient) and operational conditions (shear rate, inline angle of the boundaries, fill volume) on boundary stress field and force chain fluctuation.

606131 Effects of Particle Size and Shape on Granular Shear Flows in Shear Cells

Abstract:

In this study, shear flows of non-spherical particles of different sizes and shapes are investigated in both Jenike and Schulze Ring Shear Cells. The effects of particle size and particle shape are explored and the difference between these two shear cells are investigated. The discrete element method (DEM) is also applied to model the shear flows in the ring shear cell, in order to explore the underlying mechanism of the shear flows and the effect of the influential factors. Results show that shear stresses measured from Jenike and Schulze Ring Shear Cells are basically identical. Solid fraction and shear stress predicted from DEM simulations are consistent with those measured from shear cell experiments. It is found that for particles with same Aspect Ratio (AR), the solid fraction decreases with the increasing particle size, leading to a decreased shear stress with increasing particle size. It is also found that the shear stress decreases with increasing AR. Specifically for cylindrical particles with the same bottom diameter (0.5 mm × 2 mm and 0.5 mm × 4 mm), more elongated particles have a smaller stress ratio. It is believed that the packing structure and particle orientation play an important role on the shear stress.

598350 Spherical Agglomeration: Process Modelling and Validation for Kinetic Estimation

Abstract:

Background

Manufacturing particulate products with optimal performance requires tightly controlled processes across several industries such as pharmaceuticals. A key unit operation at the drug substance to drug product interface is crystallisation which is a critical separation and purification technique for the recovery of active pharmaceutical ingredients. It is therefore essential to understand and control crystal properties as they have shown to strongly impact downstream processes and the manufacture of pharmaceutical tablets [1].

Spherical agglomeration (post-crystallisation) is an emerging particle formation method which has the capability to improve problematic particle shapes (needles, rods, flakes) and increase particle size [2]. This involves the controlled addition of a partially-miscible solvent termed the bridging liquid to pre-suspended crystals through a single intensified unit operation. The bridging liquid should have a high affinity with the drug to agglomerate the crystals and under optimised conditions, compact and dense spherical agglomerates are produced. As a result, the produced agglomerates display high bulk densities, better flow properties, easier filtration and downstream processing in addition to suitability for direct compression of active pharmaceutical ingredients [3].

Process Modelling

Analogous to wet granulation mechanisms, spherical agglomeration exhibits different rate processes: wetting/nucleation of primary crystals by the bridging liquid, consolidation/growth of agglomerate nuclei, and attrition/breakage of the agglomerates through an immersion and distribution mechanism (Figure 1). A recent novel mathematical model was developed to predict kinetics and formation of agglomerate nuclei occurring in an immersion spherical agglomeration process [4]. Importantly, three different regimes for agglomeration nucleation were found; *immersion rate limited regime*, *collision rate limited regime* and *intermediate regime*. The mechanistic model was then successfully integrated along with key material properties and process parameters into a gPROMS FormulatedProducts® (PSE) flowsheet to develop a population balance model framework.

The aim of this work is to identify and assess the impact of material attributes, equipment and process parameters on product attributes through population balance modelling. The influence of key input model parameters such as the bridging liquid mean particle, liquid flow rate and addition time on the particle size, size distribution and porosity is investigated. Furthermore, characteristic process trends such as initiation and duration of nucleation are studied. As such, a comprehensive sensitivity analysis will be evaluated for risk mitigation approaches and greater process understanding of the predictive model. For kinetic estimation and model validation, carefully designed small-scale decoupled experiments will be performed to account for the different rate processes in spherical agglomeration.

References:

1. Variankaval, N., A.S. Cote, and M.F.J.A.J. Doherty, *From form to function: Crystallization of active pharmaceutical ingredients.* 2008. **54**(7): p. 1682-1688.

2. Pitt, K., et al., *Particle design via spherical agglomeration: A critical review of controlling parameters, rate processes and modelling.* 2018. **326**: p. 327-343.
3. Amaro-González, D. and B.J.P.t. Biscans, *Spherical agglomeration during crystallization of an active pharmaceutical ingredient.* 2002. **128**(2-3): p. 188-194.
4. Arjmandi-Tash, O., et al., *A new mathematical model for nucleation of spherical agglomerates by the immersion mechanism.* 2019. **4**: p. 100048.

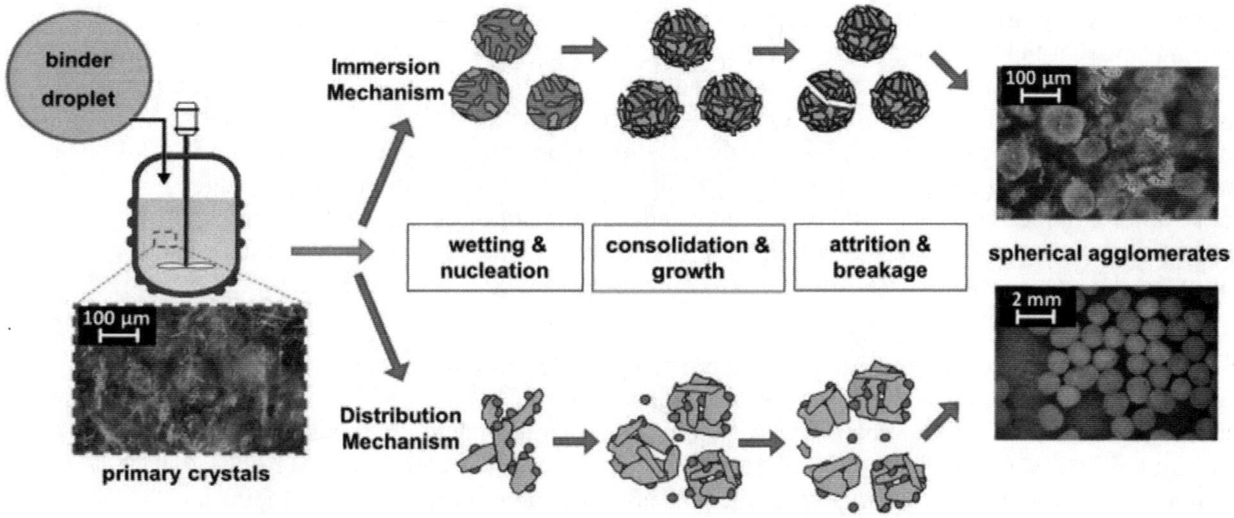

Figure 1: Schematic of rate processes and possible mechanisms occurring during spherical agglomeration [4].

599488 Understanding AUTO-Agglomeration of Dry ACTIVE Pharmaceutical Ingredients

Abstract:

Introduction:

Dry Active Pharmaceutical Ingredient (API) auto-agglomeration is a type of unwanted powder transformation that occurs without any additives. If uncontrolled, formation of hard, i.e. capable to withstand drug manufacturing process, API agglomerates may lead to an inadequate API distribution within the final formulation exhibiting the real risk to safety, efficacy and processability of drug products. Both, the fact that API auto-agglomeration may occur at, almost, any stage of drug manufacturing and/or handling processes as well as that a variety of general powder transformation mechanisms have been previously reported (for instance, mechanical, chemical, and triboelectric modes of bulk powder caking) make API auto-agglomeration phenomenon highly complex and challenging to anticipate and to prevent. However, despite the key role it has in the drug product performance and manufacturing, the number of studies addressing auto-agglomeration is currently limited and its underlying root-causes remain largely unknown. As a result, pharmaceutical industry relies mainly on the cost-ineffective trial and error approach when facing issues related to the unwanted API agglomeration.

The aim of this paper is to explore the correlation between particle properties, its processing/handling conditions and auto-agglomeration phenomenon as well as to establish a method/tool that would allow anticipation and, hence, better control of API auto-agglomeration leading to enhanced product quality, sustainability and cost-effectiveness. The experimental work was carried out to test the following hypothesis: API surface chemistry impacts its auto-agglomeration tendency. Ibuprofen recrystallised in different solvents was used as the main model API. Its auto-agglomeration tendency was tested using mechanical vibration that corresponds to drug handling conditions. Size distribution and particle shape of ibuprofen were analysed using SEM and G3 morphology, FTIR was used to confirm the identification of ibuprofen structure, and Instron to study the strength of the ibuprofen particles before and after the mechanical vibration.

Results and Discussion:

Four batches of ibuprofen of distinct morphologies were recrystallised in order to test the effect of crystal surface chemistry on its auto-agglomeration tendency. Collected FTIR data confirmed the ibuprofen structure in each case. SEM and morphology G3 analyses revealed that the shape regularity of recrystallised ibuprofen decreases with decreasing solvent polarity as follows: ibuprofen recrystalised from methanol (IbuMeth) > ibuprofen recrystalised from ethanol (IbuEth) > ibuprofen recrystalised from acetonitrile (IbuAce) > ibuprofen recrystalised from hexane (IbuHex), in line with previously reported data in the literature. Mechanical vibration experiments resulted in formation of visually detectable, relatively soft agglomerates that differ in size and shape across the batches. In addition, morphology G3 results indicated formation of agglomerates at microscale in the case of IbuEth and IbuHex being vibrated: d10, d50, and d90 of recrystallised IbuEth approximately doubled after mechanical vibration, whereas in the case of IbuHex this shift was less considerable. The strength of both IbuEth d90 as recrystallised and after mechanical vibration was tested using Instron revealing that the latter is approximately ¾ of the former.

Conclusion:

Detection of ibuprofen agglomerates upon mechanical vibration suggests there is a correlation between ibuprofen particle morphology, hence, particle surface chemistry and ibuprofen auto-agglomeration tendency. Further investigations of ibuprofen agglomerates as well as other model APIs agglomerates will include analysis of their strength (Instron, Nanoidenter), internal structure (X-ray micro computed tomography), and their surface chemistry (X-ray photoelectron spectroscopy).

595680 Modeling of Granule Size Distribution in Pharmaceutical Roller Compaction Process

Abstract:

Pharmaceutical roller compaction (RC) is a particle size enlargement process in which granules are produced by making ribbons from an active or placebo compound and milling them into smaller agglomerates. Determination of the granule size distribution (GSD) as the major product attribute requires extensive experimental efforts which are time consuming and material demanding. Modeling the impacts of process parameters and ribbon properties on granule attributes provides a quantitative insight into the RC granulation process. In this talk, we present a computational approach to simulate the size distribution of granules produced in the milling step of pharmaceutical roller compactors. RC process was emulated by making slugs and milling them with a lab hand-mill apparatus. The model was structured based on the GSD data obtained from different active compounds of Aspirin and Ibuprofen with various drug loads, including two commonly used excipients of Microcrystalline Cellulose PH102 and Lactose FastFlo 316. Experimental data were simulated using a Guassian distribution function, and the determined model parameters were correlated to ribbon Young's Modulus. Model was validated by data which were excluded from the training set. Results demonstrated the success of the model in predicting the GSD obtained from experimental measurements. This model holds only two fitting parameters which can be determined from regular material profiling techniques. This approach can be integrated with an RC model that can predict the ribbon attributes, and be utilized as a powerful computational tool for RC process space design.

608093 Infleunce of Natural and Synthetic Binders on Paracetamol Granules Properties

Abstract:

The role of binders in the formulation of tablets widely varies based on the intended application. Binders, or excipients, account for most drug formulations and provide enhanced properties that aid in drug delivery and bioavailability. Excipients are distinguished by their functionality, efficacy, safety, and qualify. Currently, scientists are seeking more naturally occurring sources (plant-based) to replace common synthetic binders such as polysorbates, povidone, polyvinylpyrrolidone (PVP) and polyethylene glycol (PEG). Natural excipients are considered non-toxic and more biocompatible. Overall, the tablet or granule cohesiveness, stability, manufacturability, and release characteristics will provide an indicator of the feasibility for using a given excipient in industrial-scale manufacturing applications. The purpose of this research is to investigate the influence various natural and semi-synthetic excipients on granule properties.

Paracetamol granules were prepared using the previously proven method of binder dropping. All excipients were dissolved into deionized water prior to placement on the powder bed. Analysis of the granules included weight and size analysis using Image-Pro Premier software, friability testing, compression testing, disintegration testing, and dissolution testing. Size measurements included diameter, aspect ratio, and projected area. Dissolution and disintegration tests were performed at a dip rate of 10 DPM in phosphate buffer solution at a pH of 7.0. The concentration of paracetamol in the solution was determined using UV-Vis spectrometry. Friability testing occurred for 100, 150, and 200 rotations. With the exception of PEG, the concentration of the excipient influenced the behavior of the paracetamol granules during dissolution and friability testing. However, the extent of the influence varied between the excipients. For most excipients, the increase in concentration produced larger granules that were more susceptible to breakage during friability testing. The bonding of the excipients determined whether the formed paracetamol granules immediately disintegrated or dissolved over time.

602835 Identifying Optimum Liquid-to-Solid Ratio Range for Microcrystalline Cellulose Based Formulations in Twin Screw Wet Granulation

Abstract:

Twin screw wet granulation (TSWG) is gaining increasing interest in the pharmaceutical industry as an effective continuous manufacturing process for solid oral dosage forms. Although multiple variables have been shown to impact various characteristics of granules produced by this technique, liquid-to-solid ratio is known to be one of the primary factors affecting granule attributes. Hence, identifying an optimum range of liquid-to-solid ratio can help streamline formulation development and minimize material requirements for TSWG process development. Here we present the results of TSWG studies conducted with microcrystalline cellulose-based (MCC) formulations of varying drug loads, using micronized acetaminophen as a model active compound. The blends were granulated using a GEA ConSigma-1 25 mm twin screw wet granulator with a standard kneading element screw configuration at a wide range of liquid-to-solid ratios until pasting was visually determined. It was observed that the paste points for different granulations were dependent on the level of MCC in the formulation. We couple this process-focused analysis with a lab-scale approach to establish viable ranges of liquid-to-solid ratios in advance of actual process development. Comparisons of lab-scale mechanical properties of wet granular material, various process parameters, and material and granule characteristics will be discussed, with guidelines for optimizing liquid addition levels proposed.

601969 Parameter Sensitivity Analysis of a High-Shear Wet Granulation Model for Experimental Design and Parameter Estimation.

Abstract:

A typical workflow in developing a new product via high shear wet granulation (HSWG) involves experimentally testing every combination of input variables, across 3 to 5 scale of operation. When done using traditional DoE approaches, the experimental effort increases as 2^n where n is the number of process and formulation parameters. This makes HSWG a costly process to formulate a design of experiments in time, money, and materials. Therefore, this work aims to apply a model-driven design workflow to identify the most critical process parameters and thus reduce and better target experiments to be performed.

The model-driven design focuses on developing a predictive and well-calibrated process model. The process model is based on a population balance modelling framework. Mechanistic understanding of the rate processes is incorporated through appropriate kernels. The most impactful modelling parameters of these kernels must be identified in addition to the operating parameters for the process. To do this, we adopt a Gaussian Process (GP) surrogate modelling approach to directly interrogate the wet granulation computational model. The GP is trained from an input space of the wet granulation computer model produced from randomly sampling the model parameters, normally distributed about the mean. Predicting the model output using a GP enables a reduction in the considerable computational effort required to analytically calculate the Sobol' indices for a Global Sensitivity Analysis (GSA).

Cross-validation ensures the GP surrogate model is capable of predicting the model outputs given all twenty modelling parameters. Subsequently, calculations of the Sobol' indices demonstrate that only four of the twenty parameters have sufficient impact to influence the DoE. The most critical process parameter is the liquid spray rate, dominating all four models outputs. The collision coefficient and the critical pore saturation are together the most important modelling parameters with respect to the fine fraction and D50, with each variable ascribing to 20% of the output variance alone and above 60% when including cross-effects. Whereas, for fine fraction the breakage coefficient is the most significant modelling parameter, resulting in a first-order Sobol' indices of 25% and negligible interactions. Interestingly, the total Sobol' index of the nuclei-to-drop diameter ratio is 50% making it the most dominant variable for granule porosity but it has little relevance towards the other outputs. Overall, the GSA has identified the critical process parameter and the impactful modelling parameters, and has enabled a proposal of an appropriate experimental design and model calibration workflow. By applying this workflow within model-driven design, industrial development is improved by determining more beneficial production conditions and reducing experimental effort by 30-60% compared to conventional approaches.

604611 Analysis and Evaluating Design Space of Continuous High Shear Granulation and Drying Process Using Gproms

Abstract:

Continuous wet granulation (high shear granulation) process is widely used in the chemical and pharmaceutical industries. This work aims to investigate the product properties and production efficiency of the continuous high shear granulation with experimental studies and simulation using gPROMS. In the continuous high shear granulation process, powder and liquid binder contact in a short resident time, where particles experience nucleation, growth and breakage. This work is also the first time to examine the effects of varying process parameters in a Schugi mixer (continuous high shear granulator). At the meantime, gPROMS is used to simulate the processing efficiency (yield) and the product properties (PSD, porosity, etc.) with the different formulation and process parameters in the granulation process.

603359 Modeling Segregation in Modulated Granular Heap Flow

Abstract:

While granular segregation in steady flows has been investigated and successfully modeled for various operating conditions and geometries in previous studies, segregation in unsteady granular flows, though present in many industrial applications, remains largely unexplored. In this study, we focus on stratification in unsteady size-bidisperse bounded heap flow. Experiments show that periodically changing the feed rate of particles falling onto the heap results in stratified segregation similar to that which occurs at low feed rates, but with more regular stratified layers of large and small particles and a higher overall feed rate. We model this stratification in a bounded heap using the unsteady form of the advection-diffusion-segregation equation. Simply repeatedly switching the model from a low volume flow rate to a high volume flow rate instantaneously along the entire length of the flowing layer results in stratification patterns similar to those observed in experiments. Higher fidelity is achieved with a flow kinematics model that describes the downstream-moving front of particles after a change in the flow rate. Quantitative agreement of continuum model predictions with experiment over a range of operating conditions validates our modeling approach.

The research is funded by NSF Grant #CBET-1511450.

598677 Linear Vibration and Mixing in Continuous Spatial Particle ALD

Abstract:

Continuous spatial Particle ALD reactors enable powders to be continuously coated with sub nanometer films by sending the powder through alternating regions of precursor gas using a process called linear vibration. Linear vibration systems use a self-reversing actuator to vibrate the bed, producing steady and uniform powder flow. Some materials do not flow as easily using linear vibration, particularly as the powder diameter decreases. Fine powders are frequently used in Particle ALD, so understanding and predicting their flow behavior is essential for process control. Another important aspect of process control is guaranteeing equal exposure time of each particle to the precursor gas. For thick powder beds, effective powder flow and top-bottom mixing must occur simultaneously to guarantee equal exposure time with high powder throughput. In this work, the relationship between linear vibration and mixing in a continuous spatial Particle ALD reactor is investigated. An MFIX-DEM model was developed and the results were compared to in-situ images of dyed glass microsphere mixing during continuous bed vibration.

597259 Combined Size and Density Segregation in Granular Materials

Abstract:

In flowing granular mixtures of particles varying in either size or density alone, small particles sink (driven by percolation) and light particles rise (driven by buoyancy). However, when the constituent species vary from each other in both size and density, which particles will rise or sink remains unclear. In particular, modeling the segregation of mixtures of large, heavy particles and small, light particles is a challenge due to the opposing effects of the two segregation mechanisms. Here we use DEM simulations to study combined size and density segregation for a wide range of particle properties and flow conditions. Based on the simulation results, the segregation velocity is accurately modeled by particle size and density ratios through a function that is quadratic in concentration. Concentration profiles predicted by substituting this segregation velocity model into a continuum transport equation match DEM simulation results for quasi-2D bounded heap flows with different combinations of particle size and density ratios. Most surprisingly, the direction of segregation depends on the local concentration for a range of mixtures of large, heavy particles and small, light particles. Funded by NSF Grant CBET-1929265.

604728 Particle Mixing Measures in a Vertical Bladed Mixer Using X-Ray Computed Tomography (CT)

Abstract:

Vertically bladed mixers are widely used in various solids-handling industries such as cosmetics, food, and pharmaceutical manufacturing. Difference in the particle size, particle density and volume fraction of the different species in a binary mixture can affect the location and velocity profiles of these particles. These effects in turn, influence the time scales and behavior of the mixing/segregation process. In this study, mixing of binary mixtures was investigated in a bench scale vertically bladed mixer fitted with two 45° blades. Wood and Delrin® particles of different sizes, masses and in different volume fractions were used to experimentally simulate the effects of these parameters on the mixing/segregation process of the binary mixture.

X-ray computed tomography (CT) was performed on the bladed mixer to determine the 3D location of all particles in the mixer. Since X-rays are attenuated differently by materials of different densities, the different constituents in the binary mixture can be easily distinguished in the CT reconstructions. The dispersion of the two species in different regions of the mixer vessel at different mixing times was investigated and was used to provide an accurate representation of the mixing/segregation process. The magnitude of mixing was quantified using Lacey's Index [1], a particle scale mixing index (PSMI) [2] and a grid-independent segregation index [3].

References

[1] P.M.C. Lacey, The mixing of solid particles, Chem. Eng. Res. Des. 75 Supplem (1997) S49–S55.

[2] G.R. Chandratilleke, A.B. Yu, K. Shinohara, A particle-scale index in the quantification of mixing of particles, AIChE J. 58 (2012) 1099–1118.

[3] M.S. Siraj, S. Radl, B.J. Glasser, J.G. Khinast, Effect of blade angle and particle size on powder mixing performance in a rectangular box, Powder Technol. 211 (2011) 100–113.

597268 A Segregation Force Model for Mixtures

Abstract:

Segregation of bidisperse particle mixtures is characterized by species concentration and particle size and density ratios. As the concentration of one species approaches zero, corresponding to the single particle limit, the problem reduces to that of an intruder in a homogenous bed, for which previous studies have successfully modeled the segregation driving force as a function of size and density ratios. This model consists of two multiplicative terms that account for percolation and buoyancy, respectively. However, a general segregation force model is still lacking for mixtures having non-zero species concentrations. Here we extend the virtual-spring-based force measurement approach used for single intruders in DEM simulations to bidisperse mixtures for a wide range of particle species concentrations, properties, and flow conditions. Based on the simulation data, we propose a segregation force model that depends on species concentration, thus bridging the gap between single intruder motion and mixture segregation. Our results show that the segregation force varies smoothly with concentration. For large-particle concentrations less than 0.5, the scaled segregation force is nearly equal to that for a single intruder. Funded by NSF Grant CBET-1929265.

609442 Segregation of Competing Mechanisms

Abstract:

Segregation is a mechanistically driven event. That means it has several causes. It is important that formulators understand these causes so as to know how or what to change in the formulation to facilitate smart product design. Without this knowledge, formulators are left with a trial-and-error approach, and often get lost in conducting a vast amount of testing and characterization using multi-factor experiments. Fines sift through a matrix of coarse particles, air currents present during filling carry fines to different parts of the process, particles with different frictional properties or shape roll or slide down piles at different rates, particles with different coefficients of restitution bounce down the pile with different trajectories, or entrained gas within the particle bed can be released upward and carry fines to the top surface. Each segregation mechanism induces the separation of particles during processing. Each segregation mechanism also induces a pattern of segregation based on key particle scale properties that can be easily characterized. One could create a vast computational method using DEM and FEM models and then let the models calculate to generate the overall segregation pattern and magnitude. But that brute force approach still does not give the formulator any means of knowing how to change the formulation to prevent or minimize segregation. Nor does the brute force method provide process engineers with any means of knowing what to change in the process to mitigate segregation. A superposition principle may help. Each segregation cause (mechanism) will induce a unique pattern where the magnitude of segregation of the various components is dependent on the key particle scale properties inducing the segregation (i.e. particle size, surface friction, coefficient of restitution, or aerodynamic radius). If the overall segregation pattern and magnitude are known, and the segregation pattern for each mechanism is understood, then that information can be combined to yield a means of determining the root causes of segregation on a component-by-component basis – thereby giving formulators and process engineers tools to estimate how to change the process or the product to mitigate segregation. This paper looks at an ideal system with three active mechanisms (sifting, repose, and rebound) and presents a method of determining the root causes of segregation for each component in the system.

597737 Investigating Indirect Ultrasound on Particle Size, Shape and Nucleation of APIs through Process Intensification

Abstract:

Background

Particle engineering of active pharmaceutical ingredients (API) is an increasingly studied area of research & development. During drug substance manufacturing, particle formation typically originates from crystallisation which is a critical multiphase unit operation for the purification and separation of APIs. It is imperative that product properties such as particle size, shape and polymorphic form are well defined and well controlled as they affect a drug's bioavailability, shelf-life and to effectively formulate the API into a finished drug product. [1]

However, achieving control over these properties is not necessarily met at the point of API crystallisation and isolation. Process intensified strategies through incorporation of ultrasound devices has shown wide interest in the last decade for manipulating particle properties during or immediately after the crystallisation step. For instance, the integration of ultrasound with crystallization allows for multipurpose functionality such as particle size & shape modification, [2] de-agglomeration, [3] scale-up, [4] accelerating the rate of crystallization kinetics and presents an overall more flexible approach to particle engineering. [5]

Ultrasound technology is commonly applied across several industries through insertion of a probe horn in direct contact with a liquid medium. For pharmaceutical crystallisation, this can be advantageous in process development especially when small seed particles $<15\ \mu m$ are preferred for controlling the crystal growth and thus, final product properties. [6] Whilst still a popular method, direct contact of the probe horn with particles has well reported limitations such as surface erosion, [7] substantial temperature deviations, [8] excessive noise in the vicinity of health and safety, rapid dissipation of energy transferred per unit volume and continuous operability. Therefore, it is worthwhile exploring alternative particle engineering strategies which can avoid the aforementioned limitations whilst ensuring the manufacturability and quality of the drug substance.

Case-Study

In this work, indirect ultrasound is explored as a viable and robust means for particle engineering of APIs. A semi-continuous platform which applies indirect ultrasound for inducing nucleation, particle size reduction and shape modification is experimentally investigated via process intensification. The process consists of an inline flow-cell which generates intense ultrasonic vibrations via electrical stimulation (Figure 1). These mechanical vibrations are then transferred to a glass tube which in turn sonicate the incoming process stream without direct contact.

Two APIs are selected for this study both of which have been found hard to nucleate with standard operating conditions under current industrial manufacturing processes. To establish key relationships between process parameters and product attributes, a series of isothermal desupersaturation experiments were conducted which monitored the impact of power amplitude (0, 20%, 40% and 80%) on inline count profiles, chord length distribution (CLD), particle size distributions (PSD) and solution concentrations. Nucleation rates affected from ultrasound impact were then extracted through population balance modelling as a proof-of-concept demonstration. To our knowledge, this is the first study to test and implement this indirect ultrasound flow-cell strategy on APIs for nucleation acceleration and control of particle attributes. The overall platform integration and ease of operation were additionally assessed for potential application in continuous manufacturing of pharmaceuticals.

References:

1. Variankaval, N., A.S. Cote, and M.F.J.A.J. Doherty, *From form to function: Crystallization of active pharmaceutical ingredients.* 2008. **54**(7): p. 1682-1688.
2. Zeiger, B.W. and K.S.J.J.o.t.A.C.S. Suslick, *Sonofragmentation of molecular crystals.* 2011. **133**(37): p. 14530-14533.
3. Cote, A. and E.J.A.P.R. Sirota, *CRYSTALLIZATION: the pursuit of a robust approach for growing crystals directly to target size.* 2010. **13**(7): p. 46.
4. Kim, S., et al., *Crystallization process development of an active pharmaceutical ingredient and particle engineering via the use of ultrasonics and temperature cycling.* 2003. **7**(6): p. 997-1001.
5. Ruecroft, G., et al., *Sonocrystallization: the use of ultrasound for improved industrial crystallization.* 2005. **9**(6): p. 923-932.
6. Jiang, M., et al., *Indirect ultrasonication in continuous slug-flow crystallization.* 2015. **15**(5): p. 2486-2492.
7. Price, C.J., *Application of Ultrasound in Crystallization (Sonocrystallization)*, in *Engineering Crystallography: From Molecule to Crystal to Functional Form.* 2017, Springer. p. 301-313.
8. Zhang, Z., et al., *Enhancement of crystallization processes by power ultrasound: current state-of-the-art and research advances.* 2015. **14**(4): p. 303-316.

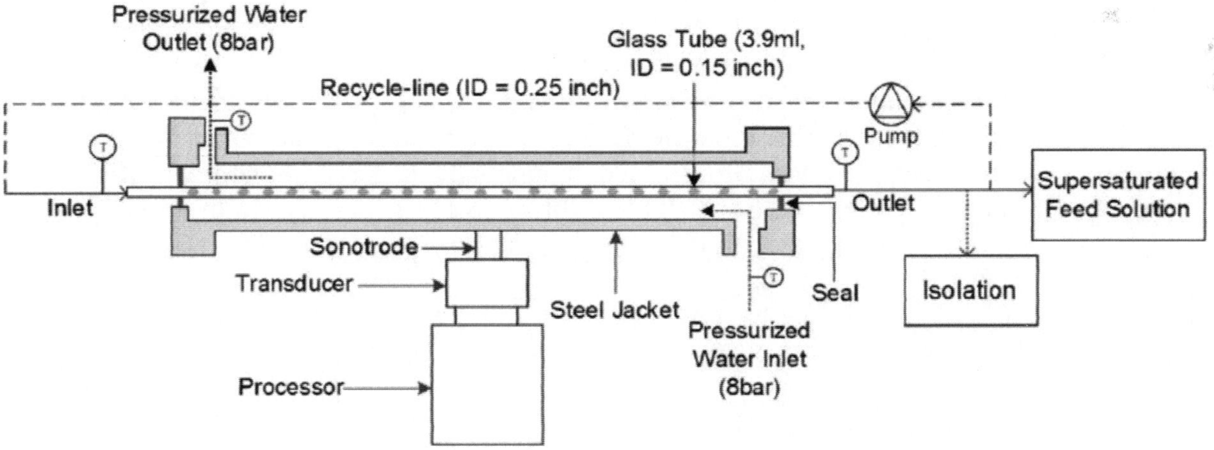

Figure 1: Sketch of the implemented indirect ultrasound flow-cell used in this study

607549 Experimental Investigation of Wet-Mill Integrated Crystallization with Temperature Cycles for Size and Aspect Ratio Control of a Commercial Pharmaceutical Ingredient

Abstract:

The control of the crystal properties has been in the focus of pharmaceutical industries and it represents a challenging aspect of the active pharmaceutical ingredient (API) production processes. To ensure that the desired critical quality attributes (CQAs) of the product are met, such as size, shape or polymorphic form, as well as to ease the downstream operations such as filtration or tableting, a careful design of the crystallization processes is critical. In this work we demonstrate the systematic design of a crystallization system for a commercial API, (compound A) from Takeda Pharmaceuticals International. The challenges related to the crystallization of the API is that the process is nucleation dominated and crystals are difficult to grow to the desired particle size (150-250 μm) by a simple cooling process. Moreover, the compound A tends to form high aspect ratio (AR) crystals, which can yield manufacturability problems. The aim of the crystallization design is to produce low AR and sufficiently large crystals with narrow distribution. Two methods were applied to reach these goals: (1) implementing temperature cycles to internally remove the fines, and (2) application of immersion milling to further control the shape and size of crystals.[1,2,3] These approaches were suggested in the literature based on the results of comprehensive model-based optimization studies.[4]

The work shows the implementation of a Quality-by-Control (QbC) guided wet-mill integrated crystallization design to produce large crystals with low AR. In this work, nine batch cooling crystallization experiments were performed. The first experiment was a seeded crystallization with very slow linear cooling rate. The remaining eight experiments were divided into two sets: (i) temperature cycling using simultaneous internal wet-mill; and (ii) crystallizations using temperature cycling but without milling. All experiments were performed with similar batch times but increasing the temperature cycle number from zero to four. Then, by using the same temperature profiles and operating procedure, simultaneous milling was applied for 75 % of the duration of the first temperature cycle as suggested in the literature, by using an immersion mill.[3]

A batch seeded cooling crystallization with very slow cooling (-0.02 °C/min) was performed, which showed that even this very slow cooling was insufficient to obtain the desired large crystals (> 150 μm) because of strong secondary nucleation in the process. Instead, applying internal fines removal by temperature cycles led to considerable increase of the product crystal size. Increasing the number of temperature cycles had a significant effect on producing large crystals (with mean size of 150-250 μm) by eliminating the fines produced by nucleation. These conclusions were also evidenced by using FBRM and in-line particle images. The results suggested that at least 2 temperature cycles were required to reach the desired product size for compound A. In the second part of the study, experiments with the same temperature profiles with wet-milling applied during the first cycle were performed. In these experiments, large crystals with low AR (>150 μm with mean AR around 2.5) could be successfully produced. The reason for using the wet-mill only at the beginning of the process is that it helps create large amount of small, low AR particles, which are partly removed by the first heat-up stage. In addition, it was shown with the 4 cycle-experiment that the milling rate does not have a significant effect on the CSD within the range applied (4000 and 8000 RPM) due to the achievable smallest size limit that characterizes the milling equipment. Therefore, the milling intensity was set to 4,000 RPM for all experiments. The experiments demonstrated that wet-mill contributed significantly to reaching the desired product crystal size and shape. This was quantified by 2D size distribution measurements based on optical microscopy images. Additionally, unseeded experiments were also performed, in which in situ seed generation was achieved by (1) cooling or (2) application of the immersion-mill as a high share nucleator. Both of these experiments demonstrated that applying wet-milling during the first temperature cycle not only significantly improves the CSD and AR of the product, but also considerably decreases the uncertainty caused by variation in the seed quality, and in combination with internal seeding via cooling or using the immersion mill as a nucleator can increase the overall efficiency and robustness of the process.

References:

[1] D. Ramkrishna and M. R. Singh, "Population Balance Modeling: Current Status and Future Prospects," Annu. Rev. Chem. Biomol. Eng., vol. 5, no. 1, pp. 123–146, 2014.

[2] D. Acevedo, V. K. Kamaraju, B. Glennon, and Z. K. Nagy, "Modeling and Characterization of an in Situ Wet Mill Operation," Org. Process Res. Dev., vol. 21, no. 7, pp. 1069–1079, 2017.

[3] Salvatori, F., Binel, P., & Mazzotti, M. (2019). Efficient assessment of combined crystallization, milling, and dissolution cycles for crystal size and shape manipulation. Chemical Engineering Science: X, 1, 100004.

[4] B. Szilagyi and Z. K. Nagy, "Model-based analysis and quality-by-design framework for high aspect ratio crystals in crystallizer-wet mill systems using GPU acceleration enabled optimization," Comput. Chem. Eng., vol. 126, pp. 421–433, 2019.

608902 Beyond Structure Determination: Crystallisation As a Purification Unit Process for Proteins and Peptides

Abstract:

The purification of high molecular weight proteins and peptides, for their use in biopharmaceutical drug therapeutics, can still account for a significant proportion of the drug development cost, which averages at between $1-6 billion. Crystallisation of these materials has mainly been used only in the field of structural determination, using small scale methods such as hanging drop vapour diffusion. However, advances in the optimisation of crystallisation conditions has seen significant increase in the scalability of protein crystallisation, including the development of continuous crystallisation approaches. This has seen crystallisation emerge as a more economic and greener alternative to the existing chromatographic methods typically utilised in peptide and protein purification.

The Made Smarter review of 2017 identified that manufacturing industries should embrace transformative digital technologies to streamline their unit processes. Mechanistic process modelling, based on population balance equations to represent the underpinning physical phenomena, has proved a computationally efficient method of predicting optimal crystallisation conditions of the purification of small molecule active pharmaceutical ingredients. However, this has yet to be expanded into the crystallisation of biopharmaceutical proteins.

Here, we utilise a combined process modelling and experimental approach to scale up the crystallisation of lysozyme and insulin from the μl to the tens of ml scale. We successfully optimise conditions, such as stirring rate and operation, pH, concentration and templating additives to not only scale up, but also increase the achieved crystallisation yield.

Insulin, being more susceptible to denaturing, utilised shaking to induce mixing of the protein rich buffer and salt precipitant. Small scale hanging drop screening experiments found that the addition of amino acids could reproducibly encourage the crystallisation of insulin, offering the promise of control over the nucleation. Scale up experiments to 15ml allowed the investigation of the impact of shaker mixing from 50-150 rpm. The insulin case study also highlights how to circumvent common issues with protein crystallisation, such as working with extremely low solubility materials and identifying the nucleation and growth points during the crystallisation process.

The lysozyme protein is more resistant to denature than insulin, whereby it can be subjected to mechanical stirring. Here we show successful batch and continuous crystallisations, utilising both shaking and mechanical stirring, up to the 30ml scale. Induction times were measured and classical nucleation kinetics parameters were derived from this data, with the pre-exponential factor and the surface energies being consistent with previous studies. Further optimisation of stirring methods and initial protein concentration was investigated to optimise the size and shape of the produced lysozyme crystals, whereby their purity was compared to the as purchased lipolyzed powder.

Finally, process modelling techniques were applied to predict the crystallisation of lysozyme and insulin in the batch reactor conditions. Such models are well parameterised to describe the crystallisation of small molecule active pharmaceutical materials. However, the classical nucleation and growth equations, as well as the models to describe the liquid state, have been less commonly applied to biological materials. Kinetic parameters to describe the processes that were extracted from the literature were used as a starting point for the model, where careful optimisation of these parameters were used to identify problems and where discrepancies between modelled and experimental particle size distributions and concentration vs time data originated. The kinetic parameters that were optimised against reliable experimental data were confronted with further experimental data to examine the wider applicability of protein crystallisation process model.

It was found that the classical Mullin nucleation and power law growth kinetic equations could be successfully regressed against experimental concentration vs time data, even for long nucleation times (Figure 1).

Figure 1: Concentration vs time for a 30ml batch crystallisation of lysozyme, using a sodium acetate buffer and sodium chloride precipitant. Blue dots are experimental data with approximate 10% error from nanodrop UV vis measurements and the red line in the regressed fit using Mullin classical nucleation kinetics and power law growth kinetics

Figure 1 shows that a good fit of the process model to the experimental concentrations vs time data was achieved. It is interesting to observe that the classical two step growth kinetics failed to find a fit, where we speculate that the diffusion term associated with the large lysozyme molecule limits the growth of this material in the model. Further investigation of the exact parameters in the equations will be presented.

This study demonstrates a novel experimental and process modelling workflow for optimising the crystallisation of large biological molecules, paving a path for this to be a viable unit process in biopharmaceutical drug development. This study gives guidance and insights into the challenges of protein crystallisation and modelling of this process, along with what are the next steps to make both processes more robust for further scale-up.

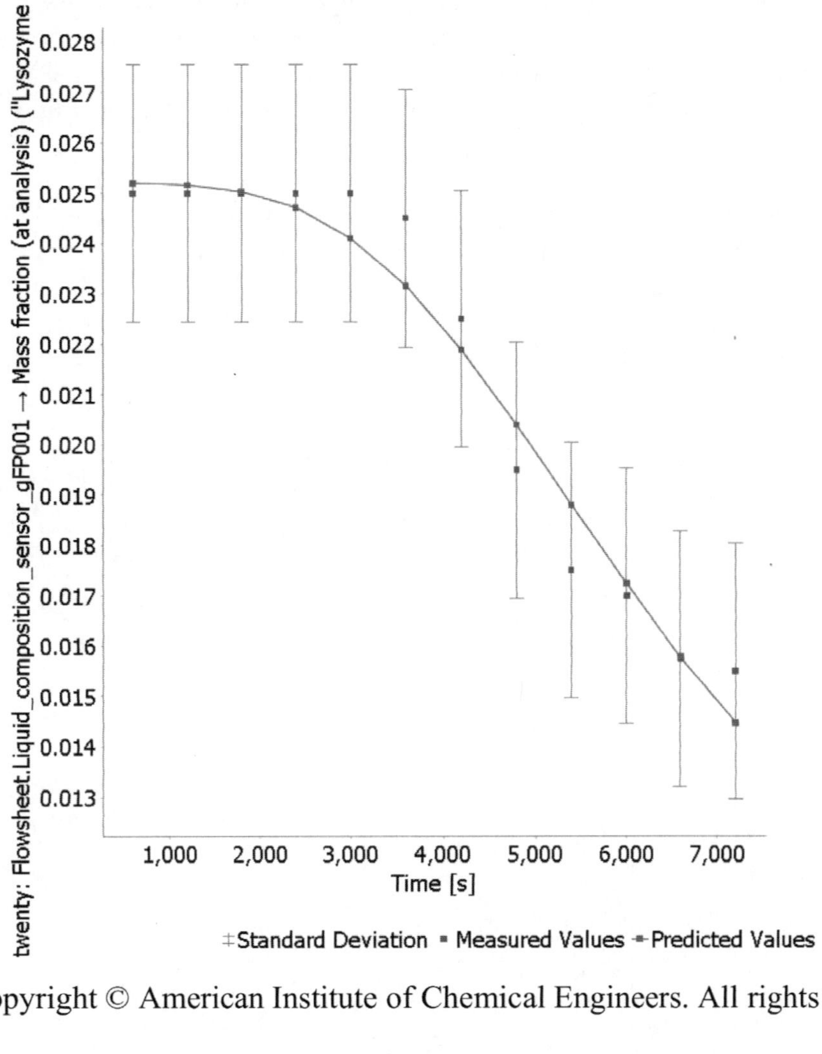

605706 Maneuvering Around Liquid-Liquid Phase Separation (LLPS) to Enable Development of Robust Crystallization Processes.

Abstract:

Anti-solvent and/or cooling crystallizations are commonly employed in purification of pharmaceutically relevant molecules. In addition, the crystallizations are also typically tasked with isolating solids with controlled physical properties. During development of multi-solvent crystallizations, it is not unusual to observe Liquid-Liquid Phase Separation (LLPS), commonly referred to as "oiling out". LLPS is generally undesirable due to its detrimental impact on purity and physical properties as well as due to the challenges it brings in during scale-up. LLPS is impacted by multiple parameters such as solute concentration, solvent composition, and temperature. Understanding of such a dynamic phase diagram, in addition to the kinetic parameters, is paramount to achieve a predictable process scale-up.

In this presentation, we detail a methodology to develop a hybrid anti-solvent / cooling crystallization process maneuvering around the LLPS boundaries. Use of automated Lab reactors (ALR) equipped with Process Analytical Technologies (PAT) is highlighted for performing data-rich experiments quickly. The generated data is further modelled with thermodynamic equations for enhanced process understanding. Finally, we will present a case study to showcase how the above-referenced methodology was utilized for enhanced process understanding and for development of a robust and scalable crystallization process.

All authors are employees of AbbVie and may own AbbVie stock. AbbVie sponsored and funded the study; contributed to the design; participated in the collection, analysis, and interpretation of data, and in writing, reviewing, and approval of the final publication.

603539 Utilizing Mechanistic Modelling to Support Crystallization Seeding Design

Abstract:

Introduction

Digital design of manufacturing processes using mechanistic models is fast becoming an essential tool during Active Pharmaceutical Ingredient (API) early process development activities. It enables rapid and effective exploration of the decision space for Critical Process Parameters (CPP), helping to reduce risk and product time-to-market, and aiding in the effective and safe production of high quality pharmaceutical products. In this work, we demonstrate how mechanistic modelling can be utilised to support the early stage design of seeding strategy for a batch cooling crystallization process. We outline a step-wise workflow consisting of model validation from which primary nucleation and growth kinetics are estimated, as a minimum list of crystallization mechanisms. The data used during the bespoken parameter estimation was experimentally measured in an unseeded small-scale crystallization process.

As part of this work, the model predictions for the seeded crystallization process was subsequently verified experimentally. This model-based approach along with rich experimental data yielded a more efficient workflow for seeding design, with reduced experimental effort compared to a more traditional purely empirical approach. A range of process data was utilized for the purposes of qualitative mechanism discrimination and quantitative model validation as follows:

- Qualitative mechanism & model discrimination
  - FBRM data to inform active crystallization mechanisms to consider
  - SEM and/or optical images of product crystals to inform crystallization mechanisms to consider within the mechanistic model
- Quantitative model validation
  - FBRM data to provide indication of onset of nucleation and access to the rate of nucleation in the unseeded crystallization runs.
  - Solute concentration data over time both online (IR/Raman) &/or offline (HPLC)
  - Particle size measurements of the product crystals from laser diffraction

The validated mechanistic model was subsequently used for model-based design of seeding policy, to assist in the seeding design (seed mass, seed addition point, and/or seed particle size distribution) required to achieve the target PSD specifications (defined by the volume-based quantiles d10, d50 and d90 of the product PSD). Uncertainty and sensitivity analysis were performed to understand the attainable region for the product PSD in terms of quantiles for the crystallization process and to identify the relative importance of process parameters, including seed mass and seed PSD.

Conclusions

The main conclusions of the work include the following:

- It has been shown that a model-based approach can be utilised to support the early stage development of crystallization processes to achieve the desired quality attributes of the product particles.
- A model-based approach has been shown to help to reduce the degree of experimental effort

required compared to a purely empirical approach.

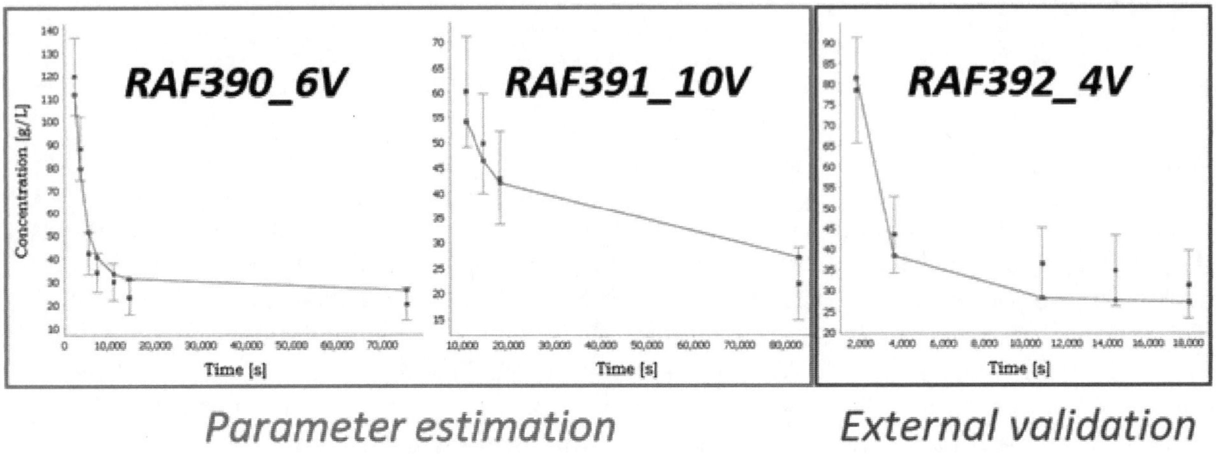

(a) (b)

Figure 1: Model validation results using de-supersaturation profiles from unseeded experiments. Data sets used for (a) parameter estimation and (b) external model validation.

Figure 2: Sensitivity analysis results showcasing the impact of mass of seed added and seed mean particle size on the D50 quantile of the crystallized material.

611464 Process Optimization for Controlling the Morphology of Cof-5 Crystals

Abstract:

Covalent Organic Frameworks (COFs) have gained significant attention in the field of material science over last two decades. 2D boronate-ester linked frameworks like COF-5, are porous, crystalline, organic polymers with low density, that can be used for numerous applications such as gas storage and as membranes for various industrial purification applications. COF-5 exhibits high thermal stability, high surface area and permanent porosity widening the scope of its applicability. To maximize its applicability, it is essential to optimize and control the morphology of the crystallites by tweaking appropriate processing conditions. COF-5 can be synthesized by evaporative crystallization using 1,4-diboronic acid (PBBA) and Hexahydroxy Triphenylene (HHTP), in an equimolar solvent mixture of Dioxane and Mesitylene. This process is heterogenous, uncatalyzed and slow, needing three days to reach a considerable and acceptable yield. Researchers have previously made efforts to reduce the reaction barrier and reaction time to 20 hrs. However, a systematic understanding of the effects of alterations in processing conditions on crystallite yield and morphology remain unanswered. Effective and accelerated crystallization of COF-5 while controlling the structure of these crystals is the prime focus of our research. This study has addressed the optimization of processing conditions, catalyst content, and solvent environment to synthesize and control the morphology of COF-5 at a higher rate. Additionally, these conditions were utilized to grow COF-5 films on substrates, while ensuring that their crystallite properties remained intact. Experimentally derived kinetic models using the optimized conditions, have been coupled with population balance equations to get a deeper insight on the crystal growth mechanisms of COF-5.

608089 Crystallization Model Discrimination Using in Situ Chord Length Monitoring and a Size Conversion Model (CLD2PSD model).

Abstract:

A new framework has been recently introduced to correlate in situ chord length distribution (CLD) measurements and offline particle size distribution (PSD) data (Irizarry et al, 2017) and applied to different particle morphologies (Schoell et al, 2019). Subsequently the model was expanded to consider bimodal distributions and inferences of modality and morphology changes (Irizarry et al, 2020). This work applies the proposed approach to estimate the kinetics parameters for seeded batch crystallization processes of different APIs in combination with population balance modelling. First, a data-driven CLD2PSD model was built for the studied pharmaceutical compounds since the differences in optical properties and particle morphology require a new model for every single substance. Then, time-resolved PSDs have been extracted from in situ CLD data of seeded batch processes and the CLD2PSD model. Finally, the online PSDs have been combined with online solute concentration data to allow for the estimation of crystal growth, secondary nucleation and agglomeration kinetics, depending on the supersaturation level encountered during the process. The validity of this approach has been confirmed by comparison to kinetic data determined using a traditional approach relying on offline data. The main advantage of the proposed approach is that all data is generated online and in situ, thus avoiding sampling errors.

References

- Irizarry, R. et al., 2017. Data-driven model and model paradigm to predict 1D and 2D particle size distribution from measured chord-length distribution. Chemical Engineering Science, 164, pp. 202-218.
- Schoell, J. et al., 2019. Determining particle-size distributions from chord length measurements for different particle morphologies. AIChE Journal, 65, e16560

606939 Continuous Solid-State Deracemization Via Temperature Cycles – a Model-Based Process Development

Abstract:

Introduction

Although preferential crystallization is a standard industrial process to obtain enantiopure crystals of conglomerate forming compounds, its maximum theoretical yield is 50%, unless one adds to it recycle and racemization steps [1]. On the contrary, one can produce an enantiopure powder from an initially racemic one with 100% yield through solid-state deracemization by exploiting several techniques [1,2]. Among such techniques, one based on temperature cycling is a promising candidate for industrial applications due its simplicity of operation [1,3,4]. However, the process itself is complex and hence it warrants the need for a mathematical model to understand the effect of initial conditions and of operating parameters on the process outcome, i.e. the production of the desired enantiomer, the process time, etc.

Continuous processes have been shown to outperform batch operation. In addition, they can recover from process disturbances, because they run at stable steady state [5]. Continuous crystallization processes have been adapted for the separation of conglomerate forming enantiomers, in order to exploit the advantage of continuous crystallization techniques. In case racemization of enantiomers in the solution proves difficult, continuous preferential crystallization without a racemization reaction is a widely studied choice [6]. If the racemization reaction is available, continuous preferential crystallization with racemization reaction or continuous Viedma ripening can be performed. Alternatively, deracemization via temperature cycles in mixed suspension mixed product removal crystallizers (MSMPRC) can also be performed, but this is an unexplored method of choice.

Mathematical model

A population balance equation (PBE) -based mathematical model has been developed to provide a clear explanation of the phenomena involved in the deracemization process. The model describes the interplay of several phenomena involved in a temperature cycling process (size-dependent solubility, crystal growth and dissolution, agglomeration, attrition/secondary nucleation and racemization) accounting for the dependence of their thermodynamic and kinetic parameters on the crystal size and on the temperature [5].

Results and discussion- batch

Using the PBE-model that we have introduced, first, we have simulated temperature cycles in batch crystallizers. A mathematical model of deracemization is a useful tool to perform process analysis: it provides information on all the relevant physical quantities during the whole process and enables decoupling of the individual phenomena, and investigating the effect of variations in e.g., the breakage rate intensity, the agglomeration intensity, etc. After identifying the necessary phenomena to deracemize using temperature cycles, process characterization was performed using the developed model, in order to identify an optimal process. Parametric studies varying operating parameters like the temperature and cooling rate were performed and based on the outcome it is shown that the simulation outcome qualitatively describes experimental results [1,4].

Moreover, by investigating the effect of slight variation in the initial conditions, an explanation to the large variations in deracemization time and in process outcome — that are typically observed in experiments — is provided using the mathematical model.

Results and discussion- continuous

The inherent disadvantage of the batch crystallization process is that it suffers from batch-to-batch product variability and requires control techniques to obtain the product with the specific quality of interest. As tuning the many degrees of freedom blindly while performing the actual experiments is ill-advised [7], this contribution aims at providing a mathematical tool that can be used to design robust continuous processes.

Therefore, we have extended our batch non-isothermal population balance-based model of solid-state deracemization via temperature cycles to simulate the proposed continuous configuration. To avoid undesired nucleation of the counter enantiomer, different process variants are considered, e.g., utilization of recycle loop, wet mill, fines dissolution loop. We have investigated the steady state operation of temperature cycles carried out in MSMPRC. Periodic operation of the continuous crystallizer via temperature cycling affects the transient behaviour of the process and thereby the attainable steady state. We have run simulations to study the interplay of the cycle time and the residence time with respect to the physical properties of the system, i.e., how the process characteristic times compare to the characteristic times of growth, dissolution and racemization.

The aim of this analysis was to verify whether one of the continuous configurations envisaged performs better than the batch alternative. The effect of the initial conditions (variations in the initial enantiomeric excess, in the suspension density, etc.) and of the operating parameters on the periodically perturbed process and its steady state in terms of productivity and attainable enantiomeric excess are identified. Together with this assessment, we aim to provide strategies for the optimal implementation of the process.

Acknowledgments

This research has received funding as part of the CORE project (October 2016 – September 2020) from the European Union's Horizon 2020 research and innovation programme under the Marie Sklodowska-Curie grant agreement No 722456 CORE ITN.

References

[1] Breveglieri, G. M. Maggioni, M. Mazzotti, Cryst. Growth Des. 18 (2018) 1873-1881.
[2] L. Noorduin, W. J. P van Enckevort, H. Meekes, B. Kaptein, R. M. Kellogg, J. C. Tully, J. M. McBride, E. Vlieg, Angew. Chem., Int. Ed. 49 (2010) 8435-8438.
[3] Iggland and M. Mazzotti, Cryst. Growth Des. 18 (2011) 4611-4622.
[4] Suwannasang, A. E. Flood, C. Rougeot, G. Coquerel, Org. Process Res. Dev. 21 (2017) 623-630.
[5] Köllges and T. Vetter, Cryst. Growth Des. 17 (2017) 233-247.
[6] Vetter, C. L. Burcham, M. F. Doherty, AICHE J. 61 (2015) 2810-2823.
[7] Bodák, G.M. Maggioni, M. Mazzotti, Cryst. Growth Des. 18 (2018) 7122-7131.

$$\frac{\partial f_i}{\partial t} + \frac{\partial\left(G_i\left(S_{\infty,i}, T, L\right) f_i\right)}{\partial L} = B_i(L, c_i, T) - D_i(L, c_i, T) + \frac{\dot{m}_{in} f_{in,i} - \dot{m}_{out} f_i}{m}$$

$$\frac{dc_i}{dt} = -k_v \rho \frac{d\phi_{3,i}}{dt} + R_i\left(c_i, c_j, T\right) + \frac{\dot{m}_{in} c_{in,i} - \dot{m}_{out} c_i + \dot{m}_{rec} c_i}{m} \qquad (i = D, L)$$

- $\phi_{n,i} = \int_0^{+\infty} L^n f_i(L,t) dL$
- $ee = \frac{\phi_{3,D} - \phi_{3,L}}{\phi_{3,D} + \phi_{3,L}}$

Initial and boundary conditions:

$$f_D(L,0) = f_D^0(L)$$
$$f_L(L,0) = f_L^0(L)$$
$$f_i(0,t) = 0$$
$$c_D(0) = c_L(0) = c_\infty(0)$$

Figure 1. Summary of the model equations that are the basis for the model of our continuous configuration. L: size of the crystal, T: temperature, S_∞: supersaturation, f_i: particle size distributions, G_i : growth rates, B_i and D_i: birth and death terms, $\dot{m}_{in}$, $\dot{m}_{out}$, $\dot{m}_{rec}$: mass flow rates, c_i: concentrations, R_i: racemization term

613536 Making Connections – Reflections on Stuart Daw as a Collaborative Innovator

Abstract:

Stuart Daw's life and career were marked by making connections between different disciplines of ideas and people. His pioneering work in nonlinear dynamics and chaos in multiphase reacting-flow systems such as combustion and fluidization paved the way for new opportunities in several engineering disciplines. He was a strong believer in increasing collaborative efforts, and this led to his leadership in two major government research consortia in a framework that is a model for the Department of Energy today. Stuart also approached many aspects in his personal life the same way among his varied interests. In this talk, I will give highlights of Stuart's life and career and the legacy he left with his colleagues and friends.

613537 Embracing and Controlling Chaos in Fluidization – in Honor of Stuart Daw

Abstract:

Thirty years ago, Stuart Daw's pioneering work, recognizing fluidized beds as deterministic chaotic systems, profoundly changed the way we consider their complex hydrodynamics. The ensuing work at ORNL and Delft allowed for refreshing approaches to embrace this complexity in fluidization, building on chaotic attractor properties, and use them to characterize fluidized beds, facilitate scale-up, provide an early warning of agglomeration, and, ultimately, open up avenues to control their behavior. I will briefly review this path, as a renewed look at holistic, "complex systems" behavior is timely, to complement bottom-up multiscale approaches, leveraging superior computational resources and algorithms, and informed by ever improving characterization methods. I will illustrate how fluidized beds could be intensified and structured, by controlling their chaotic hydrodynamics.

613538 Performance Improvement by Mitigating Gas Bypassing in Fluidized Beds of Group A Materials

Abstract:

Gas-solids fluidized beds have been extensively studied. Most of the gas in bubbling beds flows up the bed as bubbles with the remaining gas going into the emulsion phase. The rising bubbles cause rigorous solids mixing which leads to a uniform temperature in the bed and improves solids-gas contacting. Turbulent beds operate at higher gas velocities and have much fewer discrete bubbles than the bubbling beds resulting in higher heat and mass transfer.

This picture of a fluidized bed is, however, not always correct. Deep beds of Group A materials can fluidize poorly, even though all the "criteria" for good fluidization are met. Fluidization tests of FCC catalyst particles by Knowlton [1] in a 0.3-m-diameter column and a static bed height of 1.83 m and by Wells [2] in a 2.44-m-diameter semicircular column and a static bed height of 4.88 m, both units being transparent, showed a significant bypassing of the fluidizing gas occurring in the beds when at superficial gas velocities that are typical of the bubbling bed fluidization regime. Knowlton [1] observed that the fluidizing gas preferentially flowed through one side of the bed whereas Wells [2] found that the streaming gas moved about the center near the faceplate, occasionally splitting and passing up the sides of the unit. In both studies, gas-solids contacting was extremely poor. The remainder of the bed was a mass of nearly stagnant, poorly fluidized solids.

The gas bypassing phenomenon has rarely been described in the literature perhaps because most laboratory beds are often not deep enough to cause bypassing, or because the steel construction of large commercial beds makes visual observation of gas bypassing impossible. It appears that the reason for the bypassing is that the pressure head generated by a deep fluidized bed causes gas compression significant enough to cause defluidization of the solids. If gas bypassing occurs in industrial beds it can result in poor yields, afterburning in the freeboards of combustion reactors, poorly fluidized entrances to standpipes, and poorly fluidized discharge regions for cyclone diplegs and poor solids flows around CFB loop seals. It can also compromise scale-up work. A small-scale unit might not have gas bypassing while the larger unit can have it. Over the last decade, PSRI conducted considerable work on gas bypassing in in 0.6-m, 0.9-m and 1.52-m-diameter FCC fluidized beds. This paper reviews that work and presents measures to mitigate gas bypassing in deep fluidized beds of Group A materials.

613541 Simulation-Based Engineering of Biomass Fast Pyrolysis Reactors

Abstract:

Stuart Daw initiated the Computational Pyrolysis Consortium (CPC) over seven years ago (1) to help address the challenges posed by fast pyrolysis of biomass feedstocks and the catalytic upgrading of the bio-oil product. The multi-national laboratory consortium, with support from EERE's Biomass Energy Technologies office, combined experimental and computational techniques to improve the fast pyrolysis process, improve bio-oil properties and component utilization, and develop tools and understanding of the scale-up and integration of component reactors supporting commercialization of catalytic fast pyrolysis systems. Since its inception, the CPC has broadened its membership and scope to address a range of biomass energy challenges as represented by its new name: Consortium for Computational Physics and Chemistry (CCPC) (2). At the invitation of Dr. Daw, NETL joined the CCPC in 2017 to assist in the application of multiphase computational fluid dynamics (CFD) tools to study pyrolysis and catalytic upgrading reactors. In this work, in collaboration with the CCPC, NETL uses its MFiX Suite of multiphase flow CFD tools to study two experimental pyrolysis systems being operated at DOE's National Renewable Energy Laboratory. One reactor, a laboratory-scale bubbling fluid bed of biomass and sand, provides a platform to study fundamental pyrolysis over a range of biomass feedstocks and operating conditions. The second reactor is an entrained flow device for supplying pyrolysis gases for catalytic processing at the near-pilot scale. Particle drag models applicable to biomass, sand, and char were developed to accurately model solids hydrodynamics. A detailed pyrolysis chemistry mechanism (3) has been incorporated in the simulations to predict a broad range of product species that comprise the bio-oil, bio-gas, and bio-char products. Simulation results and comparison to experimental data are presented. Excellent agreement for both reactor systems was obtained. These tools are now being used to support development of pyrolysis reactor systems at larger scales using a broad range of forest product feedstocks.

1. https://www.energy.gov/sites/prod/files/2016/05/f31/bio_oil_daw_3617_3618_3619_36110_36111.pdf
2. https://www.cpcbiomass.org/
3. Debiagi, P., et al. (2018). A predictive model of biochar formation and characterization. Journal of Analytical and Applied Pyrolysis, 134, 326-335. doi: 10.1016/j.jaap.2018.06.022

605303 Christine Hrenya's Career Highlights

Abstract:

Professor Christine Hrenya's research has encompassed investigations into a variety of complex fluids, such as gas-solid systems, granular materials, and aerosols. Her work has been applied to a wide range of applications: improving the efficiency of gasifiers used for energy production, developing novel heat transfer methods for concentrating solar power plants, mitigating the ejection of lunar soil upon spacecraft landing, and predicting the agglomeration of cohesive particles found in nature or industry. She is the recipient of the AIChE PTF Lectureship Award in Fluidization, the AIChE PTF Service Award, a AIChE PTF Best Paper Award and AIChE PTF Best PhD Award.

596795 Scale-up of Heat Transfer in an Agitated Dryer: Simulations and Experiments

Abstract:

Processing of particulates plays an important role in a wide variety of industries, including the pharmaceutical industry, the bulk chemical industry, and the food industry, to name a few [1]. Yet, despite the ubiquity of particulate systems, a strong fundamental understanding of their behavior is lacking. We have focused on heat transfer in granular materials in order to better understand the drying of particulates. Drying is an important manufacturing step in the production of active pharmaceutical ingredients (APIs) [2]. Several approaches exist to carry out the drying procedure, but a common method used by the pharmaceutical industry is agitated drying. During the process, a wet bed of API is heated in a jacketed cylindrical vessel while being mixed by a rotating impeller until the moisture content is reduced to a desired level. The complexity of the process stems from the fact that heat transfer, mass transfer, and changes in physicochemical properties can occur simultaneously throughout drying. Complications often plague the procedure, including issues such as lengthy drying times, over-drying, nonuniform drying, agglomeration, attrition, and form changes. These circumstances make agitated drying a complicated process to understand and control. When considering scale up, these challenges are coupled with the difficulties typically associated with transferring knowledge from lab scale to pilot or manufacturing scale. As a result, it can be difficult to design an appropriate drying protocol that optimizes heat transfer and can be translated from scale to scale while minimizing the risk for adverse conditions. In this work, we decouple the problem and focus on studying the heat transfer aspect of agitated drying. More specifically, we investigate the influence of process and material parameters on heat transfer in an agitated bed of particle. Our approach consists of employing a combination of experiments and discrete element method (DEM) modeling techniques to analyze how heat transfer scales with bed parameters. We have carried out a design of experiments and computed heating times as well as heat transfer coefficients for the different conditions. We find that scaling influences the flow and compressibility of the bed and therefore creates a balance between conduction and granular convection as the dominant mode of heat transfer.

REFERENCES

1. Nedderman RM. Statics and Kinematics of Granular Materials. 2005.

2. Sahni EK, Chaudhuri B. Contact drying: A review of experimental and mechanistic modeling approaches. International Journal of Pharmaceutics. 2012;434, pp 334-348.

598740 Improved Understanding of the Role of Ultra-Thin Particle-ALD Films on Lithium Metal Oxide Cathode Particles

Abstract:

The surface engineering of Li-ion battery cathode materials has been of interest for over a decade. Large capacity fading, voltage decay and low rate capability are observed upon cycling of many otherwise promising cathode materials. These failures stem from dissolution of metals into liquid electrolytes and structural instability of cathode materials caused by lattice strain induced by Li+ intercalation upon each discharge. In general, surface coatings act as a protective layer to prevent direct contact of the electrochemically active cathode material with the electrolyte and thereby decrease the dissolution of transition metal ions. Additionally, these coatings suppress phase transitions, improve structural stability and decrease disorder of cations in crystalline lattices. As a result, the surface modification improves the structural stability of cathode materials with minimal loss of available specific capacity compared to other coating methods that deposit significantly more material. Consequently, marked improvement of the electrochemical performance of coated electrode materials includes their coulomb efficiency in the first cycle, cycling behavior, rate capability, and overcharge tolerance.

The deposition of < 6 alumina ALD nanofilms on Li ion battery cathodes enhances the cycling stability of lithium ion batteries. It is commonly theorized that ALD results in a uniform film that optimally is thin enough to facilitate lithium diffusion while blocking side reactions of the electrolyte with the cathode material. Here, we elucidate the true nature of low-cycle number ALD films on lithium nickel manganese cobalt oxide (NMC) cathode materials. Cathode particles were coated with alumina ALD films of 2,4,6,8,10, 12, and 15 cycle numbers, and then studied using EDS mapping, HRTEM/FIB milling, XPS, low energy ion scattering (LEIS) and secondary ion mass spectroscopy (SIMS). Surface analysis shows that low-cycle ALD films are not uniform over the surface of the NMC particles and that alumina ALD preferentially deposits on transition metal sites on the cathode surface and coats Li on the surface to a lesser degree, leaving some Li exposed. Contrary to current supposition, low-cycle ALD improves the cycling stability of battery cathodes through this preferential growth that stabilizes the transition metal oxides in the presence of electrolyte without blocking lithium intercalation pathways. The ALD film coating on the surface for < 6 ALD cycles is not uniform, but better represents a film that is nucleating on substrate sites prior to forming a uniform film. This is expected. The unexpected result is that the film preferentially coats transition metal sites to a greater degree than the coating of Li sites. An environment has been created that facilitates the preferential ALD growth of Al2O3 on Mn, Co, and Ni.

601379 The Past, Present, and Future of Commercial-Scale Fluidized Bed Modeling

Abstract:

For the last 30 years, engineers have been modeling fluidized beds using computational fluid dynamics. It has been an interesting journey with a strong relationship between fundamental research and applied development. Challenges defining solids pressure and viscosity to subgrid models for allowing drag models describe mesoscale effects. Today, CFD models have become a required tool with the scale-up of fluidized bed and circulating fluidized bed operations. Still, there are our challenges. Particle properties and changes in particle properties is needed to better understand attrition, catalytic deactivation, redox, and poisoning. Fast reactions, typical of fluidized bed and circulating fluidized bed operations, also need to be better captured. Such reactions are driven by micro-mixing, which is not complete capture with the broad gridding of a CFD model for a commercial unit.

"Dr. Charles Stuart Daw: Fluidization Visionary Yes, but More Importantly, Mentor and Friend to All"

By

Thomas J. Flynn, PE
Technical Consultant
The Babcock & Wilcox Company
tjflynn@babcock.com
330-280-0567

Paper 416e
Session Particle Technology Forum (03)
Thursday, November 19, 2020, 2:00 - 3:00 PM

Many of you probably thought you were attending a virtual tribute to Dr. Charles Stuart Daw. Well this talk is not about Stuart. It's about me! It's always about me! Of course, I am kidding. I have been blessed by a career of technology development activities with Stuart, and he has had a significant impact on my personal development, as both a mentor and friend. I only wish I could be with you in person to share this tribute to Stuart.

I first encounter Stuart during the early '80s, when the Babcock & Wilcox Research Center was the recipient of a large Electric Power Research Institute project to develop fluid bed technology for coal combustion. The project included pilot-scale testing on a 1-foot by 1-foot BFB facility, and a 6 foot by 6-foot facility. At first, Stuart was part of an advisory team on the project. Soon, however, it became obvious to all that he had much more to contribute especially with respect to the combustion modeling effort on the project. The combustion model was a reduced-order model based on characterization of fuel and limestone properties that were obtained from a 2-inch fixed bed reactor and 6-inch BFB reactor. Stuart provided guidance to coordinate the model and bench-scale tests. I remember Tom Morris, my colleague at the B&W Research Center, and who was working with Stuart on the project, commenting to me that Stuart was very, very smart. I barely knew Stuart at the time as I was in a different group at the Research Center but would dine with him and other members of the BFB team from time to time. I remember thinking, "How could such a humble, unassuming individual really be that smart?" Later, I would grow to appreciate Stuart's depth of knowledge, and just how much Stuart's humble disposition belied his intelligence. A characteristic all very accomplished and intelligent individuals should aspire to attain.

Years later in 1991, I attended the International Fluid Bed Conference in Montreal. From my earlier encounter with Stuart, I thought I would attend the session in which Stuart was presenting a paper. Stuart's paper, and a paper by Jaap Schouten at the University of Delft were influential in B&W's entry into the field of chaos theory and nonlinear signal analysis. At the time, I was involved in the operation and testing on our 2.5 MWt circulating fluid-bed facility. As an operator, I was struggling to understand some of the long-term dynamics of the CFB given its extremely long time constants. We needed a quantitative means to detect and predict changes in operating conditions to guide control. The short-term predictability of chaos theory was just what we needed to predict the onset of agglomeration or shifts in particle size distribution as they impacted fluidization quality.

At the same time, Jack Halow of DOE-Morgantown and Stuart were developing applications of chaos theory to fluidized beds with John Stringer of EPRI. Together they envisioned so many potential uses of chaos theory and principles of advanced signal analysis. They recognized that chaos analysis can tease out changes in fluidization quality. Using high-speed pressure fluctuations as an indication of a shift from a well-fluidized condition as indicated by a Gaussian distribution to a slugging regime as indicated by a periodic distribution. Together they analyzed fluidized bed capacitance imaging data that Jack had collected in HIS garage! Unbelievable! Jack was an incredible talent and deserves a tribute unto himself.

B&W has utilized these techniques to assess fluidization quality in pilot facilities and to detect the onset of agglomeration. In fact, we have developed a proprietary bench-scale technical procedure to determine the agglomeration potential of fuels in BFBs using these techniques. We also used these techniques in a project with Dr. Leon Glicksman and his students at MIT to verify that a physical flow model of the TIDD Pressurized Fluidized Bed faithfully reproduced the fluidization dynamics in the TIDD unit. We verified the match between the two systems using chaos parameters, such as entropy, mutual information and symbol sequence analysis.

During the late 90s, the Electric Power Research Institute approached Oak Ridge National Lab and B&W to extend our fluid-bed chaos analysis techniques to pulverized-coal burners to improve NOx control. Stuart and Charles Finney collaborated with Tim Fuller of B&W and me to develop combustion stability diagnostic techniques. Stuart and Charles screened many analysis techniques and identified the most effective ones to discriminate the various combustion states of PC flames. Tim Fuller and I interpreted the analysis results in terms of the physical inputs driving the various flame states. We also backed the analysis of the flame states with visual observations of flames on B&W's 100 MBtu/hr pulverized-coal facility. Stuart's insight into combustion was instrumental in the development effort. Stuart even suggested the name Flame Doctor for the diagnostic product. The effort culminated in B&W's commercial Flame Doctor® Burner Monitoring System that is still in use today. In 2005, Oak Ridge National Lab and B&W received the Federal Laboratory Consortium Southeast Region Excellence in Technology Transfer Award for Flame Doctor®. Although extremely technically deep, Stuart always had his eye on the practical application of what he was working on. In my opinion, his ability to transfer from technically deep theory to commercial practice was unparalleled.

In 2000, Stuart and I were guest lecturers at the Chaos in Manufacturing Conference in Santa Fe, New Mexico. As we were on different flights to Santa Fe, Stuart and I had agreed to meet at the shuttle terminal to share a car ride to the conference center. As I approached Stuart, he was crouched over his backpack and suitcase reading an article entitled "Ergodic theory of chaos and strange attractors" by Jean-Pierre Eckmann and David Ruelle. I could not even understand the title of the paper, and here Stuart was making use of a free moment to read AND comprehend a complicated article. His ability to concentrate was amazing to me.

During the conference Stuart gave a presentation on perturbation control as applied to fluid-bed systems. The application was for a beneficiation reactor at the Oak Ridge Y-12 plant. The reaction being carried out in the reactor was highly exothermic, so the fluidization regime could quickly shift from well-fluidized to slugging resulting in off-spec material. Perturbation control was used to maintain stable fluidization conditions. During Stuart's presentation he used a pointer to illustrate the principles of perturbation control. You are all aware of the trick we would practice as children to balance a broom handle vertically on the palm of our hand. The key to the trick is to know where to focus your attention…the top of handle, and where and how much to apply the perturbation control….at the bottom of the handle. This illustrates one of Stuart's strengths, that is, to be able to explain the most complicated principles simply enough for all to understand and appreciate.

After dinner that evening, we went on a walk around the conference center and Stuart strongly recommended that I seriously investigate the concept of chemical looping. This was years before chemical looping had gained the popularity it has today built on the many efforts of universities and national labs around the world. This is another example of Stuart's vision. He could recognize a paradigm shift offered by a technology, such as chaos theory or chemical looping, long before it was recognized in general.

Most recently before he passed, Stuart served as the leader of the Computational Pyrolysis Consortium program, which brings together the skills and facilities of five national laboratories. Stuart was a logical choice for this position because he was so well respected by personnel within all the national laboratories, and he had the breadth of technical knowledge to bring together the efforts of the individual laboratories into a coherent program. Stuart had a way of making each individual feel that their effort was important and appreciated. The CPCP is a model for national laboratory programs that I hold up as an example of how national laboratories and universities can efficiently collaborate on technology development.

To cite a tired example, Stuart's depth of knowledge was like the layers of an onion. He had an uncanny knack of knowing just how deeply to draw on his knowledge and insight to help a team address a development issue. I never really experience the bottom of his depth of knowledge. But I took from this experience a valuable lesson to know my audience and draw on my personal knowledge to communicate effectively the essentials of a concept. Of course, my depth of knowledge is no match to Stuart's, but his example often helped me to foster a collaborative environment between team members throughout my career.

Those that have all benefited from their collaboration with Stuart would acknowledge he was a joy to work with, and it was always a rewarding experience.

Finally, I extend a special thanks to Lynn Daw, Stuart's wife. Thanks Lynn so much for sharing Stuart with us, and personally, making my life and career so much richer having known Stuart and benefitting from his guidance.

Thank you for your attention and join me in recognizing an accomplished fellow colleague who had a significant impact on many of our professional and personal lives. I am grateful to the conference organizers for the opportunity to honor Dr. Charles Stuart Daw.

613544 Application of Quadrature-Based Moment Methods to Turbulent Mixing

Abstract:

In the context of turbulent reacting flows, modeling of molecular mixing (i.e. micromixing) has long been a challenge. Specifically, the composition probability density function (PDF) has a transport equation wherein chemical reactions are closed but the effect of micromixing, which involves spatial gradients, is not. While the physics are deceptively simple, finding appropriate closures that can account for it is not. For example, coupling the mixing rate to chemistry through enhanced gradients, or simply accounting for chemical species with very different molecular diffusivities are difficult to model in a general way.

In the PDF transport equation, molecular mixing appears in terms of the conditional scalar dissipation rate (CSDR), a square, symmetric matrix of functions depending on the composition variables. For a statistically homogeneous system (i.e. the mixing statistics are independent of the spatial location), a Fokker-Planck (FP) equation describes the evolution of the PDF in composition space with the CSDR as the diffusion matrix. In principle, for a given CSDR, one can solve for the composition PDF given its initial condition (e.g. completely segregated feed streams). In practice, this must be done with a Monte-Carlo (MC) code, leading to statistical noise; or by making use of simplifications such as conditioning on the mixture fraction whose univariate PDF has an assumed form (e.g. a beta PDF).

Recently we have developed an alternative solution method for the FP equation using quadrature-based moment methods (QBMM). Given a mathematically consistent form for the CSDR, the basic idea is to solve the moment equations corresponding to the FP equation, and then to use QBMM to find quadrature weights and abscissas for evaluation of the chemical source terms. Conceptually, the deterministic equations for the moments (and, hence, the abscissas) replace the stochastic differential equations for the MC notional particles. Remarkably, the moment equations for molecular mixing are closed when the CSDR is chosen to relax towards a beta PDF. Thus, the MC algorithm is replaced by a high-order ODE solver for the set of coupled moments, thereby eliminating the statistical noise.

In this presentation, a general overview of QBMM for turbulent mixing is provided, followed by a specific application using moments up to first order for reacting species (i.e. equivalent to the conditional moment closure). The implementation of QBMM for turbulent reacting flows in CFD codes at the level of RANS and LES will be discussed. Specific issues related to solving for moments, such as realizability, will also be covered.

613545 On CFD-DEM modeling of fluidized beds for bioenergy applications

Abstract:

Thermochemical processing of biomass in fluidized beds is of pivotal importance to displacing substantial quantities of fossil-based fuel and product consumption. One key issue with this technology is the limitations of empirical insight to support robust performance optimization and scale-up, necessitating the development of process-scale models. In this context, coupled Computational Fluid Dynamics and Discrete Element Method (CFD–DEM) is essential to understanding the complex interaction between the hydrodynamics and chemical processes involved in the thermochemical processing of biomass in fluidized beds. This talk will focus on our work on computational modeling of biomass gasification and pyrolysis in fluidized beds. We will showcase how we utilized the CFD–DEM model at the Biomass Conversion and Modeling Laboratory at the University of Tennessee to accurately capture the formation of inorganic contaminants during biomass gasification. We will also discuss our work in the Consortium for Computational Physics and Chemistry at Oak Ridge National Laboratory focused on simulating autothermal pyrolysis using CFD–DEM model coupled with particle-scale simulation.

598416 (Invited Plenary Talk) Wearable Electrochemical Sensors

Abstract:

Wearable sensors have received a major recent attention owing to their considerable promise for monitoring the wearer's health and wellness. The medical interest for wearable systems arises from the need for monitoring patients over long periods of time. These devices have the potential to continuously collect vital health information from a person's body and provide this information to them or their healthcare provider in a timely fashion. Such sensing platforms provide new avenues to continuously and non invasively monitor individuals and can thus tender crucial real-time information regarding a wearer's health. Our research over the past decade has aimed toward filling the gaps toward providing biochemical information, beyond that given by common wrist-watch mobility trackers. This presentation will discuss recent developments in the field of wearable electrochemical sensors integrated directly on the epidermis or within the mouth for various non-invasive biomedical monitoring applications. Particular attention will be given to non-invasive and minimally-invasive monitoring of metabolites and electrolytes using skin-worn and microneedle amperometric and potentiometric sensors, respectively, along with related materials, energy and integration considerations. The preparation and characterization of such wearable electrochemical sensors will be described, along with their current status and future prospects and challenges.

610185 (Invited Plenary Talk) an Integrated Paradigm for Precision Exposure to Airborne Chemical and Biological Stressors Based on Personal Sensing

Abstract:

Introduction

The current manuscript outlines the fundamentals of a methodological framework for precision exposure assessment and gives concrete examples of its application in different European cities to assess the health risk associated with exposure to fine and ultra fine particles of chemical and biological origin. The work described herein involves a multi-city personal sensor campaign aiming to refine PM exposure using low-cost portable sensor data, exposure and human respiratory tract deposition modelling to estimate the biologically effective internal daily intake on an individual basis and agent-based modelling to estimate exposure and consequent health risk to the community.

Methodology

A prototype monitoring device (IcaruSense) was developed for measuring three size fractions of PM (1, 2.5 and 10 μm), enabling direct assessment of personal exposure. The device is based on an Arduino microcontroller where miniaturized electrochemical sensor-modules are connected. In order to improve sensor performance, correction factors were used to account for diverse environmental conditions and they were calibrated with the help of a well-validated light scattering optical sensor. Study participants wore a physical activity wristband (Garmin Vivosmart 3) that records steps, distance, type of activity, heartbeat and sleeping patterns. Finally, participant positions were recorded using a GPS sensor, integrated into the PM sensor kit. After calibration and validation, the IcaruSense devices were used to capture daily variability of PM exposure. Exposure was further refined by estimating inhalation adjusted intake rate from the continuous time-activity data collected from the physical activity and the GPS sensors. Using the inhalation adjusted intake rate PM deposition across the human respiratory tract (HRT) was readily estimated using the Multiple Path Particle Deposition (MPPD) model.

A city scale agent-based model (ABM) was developed to support this work by expanding the precision exposure estimates from the individual to the community level. The model feeds into population-based exposure assessment without imposing prior bias, basing its estimations onto emergent properties of the behaviour of the computerised autonomous decision makers (agents) that compose the city-system. Population statistics, road and buildings network data were transformed into human, road and building agents, respectively. Informed by literature-extracted relationships between Socio-Economic Status (SES) and human behavioural patterns, the established ABM simulates societal dynamics. Time-use survey outputs were associated with human agent behavioural rules, aiming to model representative to real-world routines. Moreover, real time-geography of exposure data, extracted from the precision sensors campaigns in nine European cities were used to parametrise and enhance the model.

Mobility behaviour and eventually exposure to pollutants change as a result of a human agent-specific decision making. Virtual individuals of different sociodemographic backgrounds use different means of transportation and follow different sequence and types of activities. Behaviours that were not explicitly programmed, arise through the human agents' interactions, enabling the examination of emergent behaviours from the bottom-up. Spatiotemporal trajectories are coupled with spatially resolved Particulate Matter (PM) levels, enabling the assessment of personal and inhalation adjusted exposure, based on type of location and intensity of enrolled activities.

Results and Discussion

The above methodology was applied in the personal sensors campaign of the HORIZON2020 EU Project ICARUS, where exposure and intake to PM of almost 100 individuals in each out of nine European cities were monitored. The integrated hybrid approach outlined above, which couples multi-sensor datasets on an individual basis with advanced respiratory tract modeling, allowed us to significantly differentiate the actual intake of the participants, highlighting larger differences than the ones attributed to the spatial differentiation of air pollution based on fixed stations (difference of ambient PM levels of 50% were translated in intake differences up to 110%). These differences are the result of the differences in PM size fractions that are captured by the sensors and the capability of the HRT model to translate these differences in PM size distribution accounting for variation in respiratory tract physiology among the participants, which is ultimately reflected in intake estimates.

ABM findings indicated that inhalation adjusted exposure differences between 2 individuals living in the same neighbourhood can vary by as much as 87%, due to radically different spatiotemporal behaviours. This major difference was observed for cases where one human agent has a full-time job and often exercises in outdoor courts whereas his/her neighbour is a retired homemaker. For the same reason, ABM runs confirmed that even people who reside in the same dwelling might not necessarily be exposed to similar levels of pollution. Model results indicated that PM inhalation adjusted exposure between flatmates can differ by 61%. Evidence was, also, provided for vulnerable subgroups of population that are disproportionately subjected to high levels of exposure. ABM-retrieved population exposure findings indicated that subgroup median exposure for children, adult males with a full-time job and low income and human agents of a lower educational level was notably higher than the total population median value. The validated model can assess exposure for the entire sociodemographic spectrum of a simulated region, without the constant need for repeated individual monitoring which is often not feasible due to high costs, ethical constraints and complex organisation that is associated with such measurements. By modelling the heterogeneous routines of human agents, the ABM produces detailed information related to the societal system examined and generates data that could be used to fill in the gaps that exist in traditional datasets.

Conclusions

This study presents a new methodological approach to estimate the external exposome, which encompasses the totality of human environmental exposures at an individual and community level. This method helps to integrate substantive considerations of individual status into long-term urban planning and management; it would also facilitate the identification of both physically and socially practical means for reducing life-threatening exposure levels. The methodological paradigm for precision assessment of exposure at the individual and community levels laid out in this work can be used for evaluating the impacts of different public health policies prior to implementation.

614114 Investigation of Fluidized Particles for Use in Thermal Energy Storage and As a Heat Transfer Fluid in a Novel Energy System

Abstract:

Solid particles perform as a heat transfer fluid in novel energy technologies because they enable efficient thermal energy storage, are stable at high temperatures, and are inexpensive. Despite the positive attributes of solid particles, the energy transfer to and from the particles is a key technological challenge that must be overcome before particles can reliably be used in energy systems. In a currently developed energy system, particles flow over hexagonal heater elements and are warmed to high temperatures around 1200°C. The particles are then fed into an insulated silo. When the energy is needed, the particles are fed into a pressurized fluidized bed heat exchanger, where the fluidizing gas is warmed and used to drive a turbine. The current work presented uses discrete element simulations and experiments to investigate how energy is transferred to the particles in the charger, and how energy is extracted from the particles within the fluidized bed. The balance between particle fluidization and the heat transfer efficiency is investigated and the simulation results are compared to experimental data.

604821 Heat Transfer in Clustered Gas-Solids Flows

Abstract:

At sufficient mass loading, gas-solids flows exhibit strong two-way coupling where the dynamic evolution of the particles is intimately linked to a carrier gas. This strongly-coupled case can give rise to large-scale coherent structures in the particulate phase (clusters), which effectively 'de-mix' the underlying flow. Inhomogeneous particle distributions can lead to significantly reduced heat transfer due to poor gas-solids contact. Here, we utilize a volume-filtered Eulerian-Lagrangian (EL) method to examine the role of clustering on heat transfer. Specifically, the role of particle structuring on the thermal developmental length scale in a riser is quantified through a novel two-step simulation approach. The degree of particle clustering is first established with cold flow simulations of cluster-induced turbulence (CIT). At statistical steady-state, CIT outputs are fed into a secondary simulation with a defined particle-fluid thermal gradient. Thus, the thermal conditions in the second simulation evolve over the length of the riser and allow the effect of clustering on the thermal development length scale to be rigorously quantified.

600282 CO_2 Methanation Via Chemical Conversion in Bubbling Fluidized Bed and Biological Conversion in Stirred Bubble Column: Reactor Design Using CFD

Abstract:

The power-to-gas (PtG) technology has been known as one of the most viable solutions for storing intermittent renewable energy sources such as solar and wind to the existing gas grid. In this study, the chemical CO_2 methanation (CCM) is experimentally and numerically investigated for a bubbling fluidized-bed (BFB) reactor at 350 °C and 5 bar, whereas the biological CO_2 methanation (BCM) is operated in a stirred slurry bubble column (SBC) reactor at 60 °C and 8 bar. In CCM, due to the strong exothermic reactions, heat management is an important factor in improving reactor energy efficiency. In BCM, the mass transfer, which is dependent on the interfacial area between phases, is a crucial parameter for investigating the process performance. A two-fluid Eulerian computational fluid dynamics (CFD) model with an empirical correlation of gas-solid heat transfer and a reaction rate model was proposed for the CCM in BFB. A Eulerian CFD coupled with a population balance model (PBM) was developed for the BCM in SBC. The impeller-influenced regions were considered as moving frame reference zones under a fixed speed of 360 rpm. The reactor performance from CFD results for CCM and BCM such as the gas molar fraction of species, CO_2 conversion, CH_4 yield and producer gas purity along the reactor height was evaluated. By comparing CCM and BCM, the important hydrodynamics parameters of BFB and SSBC were highlighted for the CO_2 methanation reactor design.

611665 Fluidization with Supercritical Fluids

Abstract:

Supercritical fluids offer unique opportunities for process intensification including as fluidizing media. The sensitivity of transport and thermodynamic properties at such conditions bring forth unique possibilities limited to only fluidization with such fluids. This presentation aims to illustrate the potential through both experimental and modeling of select examples in pharmaceutical and energy domains.

610063 Development of a Coupled CFD-DEM Multiphase Model for a Fluidized Bed Dryer Process

Abstract:

Purpose: Drying is an essential unit operation in pharmaceutical dosage development process and is extensively used after wet granulation. Among all the drying processes, fluidized bed drying is the most popular technique because of the advantages that it provides such as excellent solid-gas mixing and enhanced heat - mass transfer which leads to efficient drying. However, a lack of fundamental understanding of the drying process of wet granules can potentially hinder a manufacturing process. A fluidized bed has multiphase flow in the system which can be difficult to control/ scale up. Thus, it is critical to understand the effect of process parameters on the quality attributes of the granules such as the moisture content. Understanding these parameters using experiments can be demanding, in terms of time and money. Therefore, studying such effects with a numerical model can prove to be useful and economical. Thus, in this study, a 3D coupled model has been proposed to study the effect of process parameters on the flow and moisture content of granules. This model would be validated with experimental data collected from a Glatt fluidized dryer.

Methods: Drying process of wet granules contains a continuous phase (air), modeled by Computational Fluid Dynamics (CFD) and a discrete phase (solid), modeled by Discrete Element Method (DEM). Two commercial software, ANSYS FLUENT for CFD and EDEM for DEM were coupled for the CFD-DEM approach. Geometry and meshing were implemented using ANSYS SpaceClaim and FLUENT, respectively. Heat transfer was implemented to look at the particle-particle and particle-fluid interaction. Different air flow velocities were implemented to study the pressure and velocity contours of the fluid phase and solid phase.

Results: Preliminary study was conducted using dimensions from a reference paper. Different air temperatures and velocities were used. Once the coupling was confirmed, dimensions of the actual Glatt equipment to be used was created. Different velocities were used in the new geometry. The effect of heat transfer was studied through the velocity and pressure contours and product temperature. After studying the flow patterns, the next step was to create wet particles. Liquid bridge model was used for this purpose to create particles with varying liquid contents. The mass transfer ability through the 'Species Transport' capability is currently being explored to look at the transfer of species between two phases to look at the change in moisture content. The developed model would be validated against the experimental data carried out in Glatt dryer with pharmaceutical raw materials.

Conclusions: Such a 3D coupled CFD-DEM approach can provide a good predictive model for studying the multiphase fluidized drying process. It can contribute towards faster optimization of manufacturing process with lesser batch failures.

614329 PSRI's Web-Based Benchmark Data Repository for Validating Computational Fluid Dynamics

Abstract:

Modelling fluid-solid-particle multiphase flow, such as fluidized beds, with computational fluid dynamics (CFD), is the focus of significant research as the availability of reliable model predictions would impact increase revenue with better designed, optimized designs, and better troubleshooting tools. Currently, the lack of consensus on the validity of physical models is one of the biggest hurdles with the application of modeling fluid-solid-particle multiphase flow unit operations.

This study reviews the current state of CFD as applied to multiphase flow and the experimental data available that has been available for validation. Previous studies have demonstrated reasonably accurate validation of CFD models; but, under a specific set of conditions (material properties, boundary conditions, flow regime, etc.). These same models have provided less accurate results for a different set of material properties or operating conditions. Specifically, changes to various physical models, such as the drag model formulation, were necessary to achieve an accurate comparison against measurements reported under varied flow regimes or using different material properties. A possible reason for the inconsistency is many commercial CFD tools lack of available comprehensive validation data sets need to move such modeling efforts from interpolation to extrapolation. Nor should it, as little data exist that include a wide range of system sizes, and fluidization regimes for a gamut of materials. What is missing is a validation study that compares the accuracy of CFD simulations with a specific set of physical models to predict the flow of the same materials in different experimental systems and/or different materials in the same experimental system.

To fill this data gap, PSRI is launching a new, publicly accessible, web-based repository for measurements from PSRI fluidized bed experiments. The main goal of this work is to provide experimental data such that CFD models can be validated against the measurements for the same material in various experimental units and various materials in the same experimental unit. Users will be able to access experimental data for more than 30 different materials with at least five flow regimes (operating conditions) per material. Instructions to access the experimental data will be provided. Moreover, details of the experimental units, materials, and operating conditions used to generate the benchmark data will be presented.

608962 Validation of a ConsiGma® Tablet Coater Digital Twin

Abstract:

The goal of this work was to validate our simulation model of the GEA ConsiGma® tablet coater. GEA developed the ConsiGma® coater to be integrated into the ConsiGma® continuous production line. This was done to simplify the switch from the traditional batch production to a continuous mode. Continuous production offers some advantages when compared to batch production, including flexibility, increased process/product quality and decreased cost.

Process understanding is the key element for effective process control. Especially for new processes, having a limited experience with process charactersitics, computational tools can help to increase the understanding, and thus, process performance. Moreover, development costs can be minimized.

In the last years, computational fluid dynamics (CFD) and the discrete element method (DEM) have been increasingly used to analyze and to understand granular processes in the pharmaceutical and various other industries. In our collaboration, we developed a CFD-DEM model to simulate the tablet coating process in the GEA ConsiGma® coater. The model includes the momentum, heat and mass transfer between the fluid gas phase and the bi-convex tablets. Due to the nonsphericity of the biconvex tablets, a drag model was developed that takes orientation and shape of the tablets into account.

Tablet movement, coating quality, heat- and mass-transfer during the coating process were recorded in the simulations and validated against experimental results. The experimental and simulation results agree well on all measured accounts and support the use of this simulation approach to investigate and expand the process space of the GEA ConsiGma® coater.

600982 Nucleation and Growth of Solid-Electrolyte Interphase Components at Electrode Surfaces: Role of Electrolyte Solution Chemistry and Concentration

Abstract:

Solid Electrolyte Interphases (SEI) grow at the electrode surface because of electrolyte chemical and electrochemical instability. The SEI nucleation and growth start even at open circuit conditions, especially with anode surfaces as reactive as Li, Na, and other metals, when they are in contact with liquid or solid electrolytes. At open circuit conditions, the reaction can be described as Li oxidation and electrolyte reduction. In this talk I will discuss this spontaneous SEI formation and especially the role of the electrolyte chemistry and concentration on the initial nucleation and posterior growth. Our methods include density functional theory, classical molecular dynamics, ab initio molecular dynamics, and constrained molecular dynamics simulations. The computational results are discussed in the context of experimental observations.

597901 Inorganic-Organic Composite Electrodes for Supercapacitor Application

Abstract:

Nickel cobaltite composite electrodes were fabricated by a fast, scalable, and cost-effective electrophoretic deposition for supercapacitor applications. NiCo2O4/PANI/rGO composite electrode went through heat treatment under nitrogen to carbonize PANI after electrophoretic deposition. NiCo2O4 platelets, distributed on carbonized polyaniline–rGO network, were active in charge storage with high capacitance, excellent rate capability (1235 F g−1 at 60 A g−1), and acceptable cycling stability (3000 cycles at 10 A g−1) in a three-electrode assembly. The practical application of the composite electrode was investigated by an all-solid-state asymmetric supercapacitor cell using NiCo2O4/C-PANI/rGO as the cathode and activated carbon as the anode with specific capacitance of 262.5 F g−1 at 1 A g−1 and a good capacitance retention of 78% after 3500 cycles at an expanded working potential of 1.5 V. This work shows the importance of the composite assembly process that governs the microstructure of the composite. In addition, the effect of oxide pore tuning on charge storage is investigated.

605989 Overcoming Intrinsic Energy Density Deficiencies in Supercapacitors By Embedded Electrode Doping

Abstract:

Developing new materials to enhance the energy storage capabilities of supercapacitors is pertinent to the advancement of the energy industry. Supercapacitors have been successfully developed to increase the energy density of capacitors by adding Faradaic components to the electrodes, either by the inclusion of conducting polymers or metal oxides. Faradaic energy storage is superior in energy density to charge separation based energy storage, although by incorporating both mechanisms, supercapacitors can store energy that can be released at high powers with improved energy densities. Cycle life for surface-bonded Faradaic components is a prominent hurdle, where the energy density advantages are only sustainable for ~50 – 100 cycles. Therefore in this work we incorporate our Faradaic components within the electrodes, trapping them sterically within. This permits the augmented energy density to persist for 1000 cycles and due to the porosity of the electrodes the specific capacitance is still > 300 F/g in multi-walled carbon nanotube based electrodes. These electrodes have additional benefits in enhanced heterogeneous catalysis, which has been utilized to increase the power density of redox flow batteries as well. Further development of the energy density of supercapacitors in important for future energy storage markets as the world enters a new phase of energy.

608322 Unique Mixing Behaviors of Nanoscale Fluid-Based Electrolytes for Sustainable Energy Storage

Abstract:

Although the costs associated with renewable energy technologies continue to decrease at a fast rate, they continue to face intermittency and storage challenges. It is now possible to store this excess renewable energy in the form of chemicals and fuels through electrochemical CO2 reduction or in the form of redox flow batteries (RFBs). The electrochemical conversion of CO2 remains challenged by the CO2 solubility in aqueous solutions and the energy density of redox flow batteries is limited by redox species solubility. Therefore, novel reactive materials are currently being explored as additives to aqueous and non-aqueous electrolytes in order to improve target species solubility and/or enhance reaction rates. Polymer functionalized nanoparticles are a promising class of electrolyte additives due to their thermal stability and chemical tunability. Interestingly, grafted polymers have been shown to change their conformations in response to their environment (temperature, pH, salt concentration), thus allowing them to be engineered to capture and subsequently release target species. In this study, Nanoparticle Organic Hybrid Materials (NOHMs)-based fluids have been designed as hybrid electrolyte systems for energy storage applications. The development of NOHMs-based electrolytes requires a fundamental investigation into the structuring and organization of these complex nanoscale fluids. A study of the mixing behaviors of NOHMs with a variety of secondary fluids has revealed that physical properties such as hydrodynamic size, viscosity and diffusion coefficient are highly dependent on the chemical nature and size of the secondary fluid molecule. Additionally, it was discovered that the viscosity and the diffusion coefficient of the untethered polymer were found to scale according to theoretical predictions for the semi-dilute unentangled regime, while this was not observed for NOHMs-based fluids. Lastly, the CO2 solubility of NOHMs-based electrolytes has been measured and a change in the selectivity of the CO2 reduction reaction has been identified in the presence of these nanoscale electrolytes. The results of this work provide important insights into the design and development of a novel reactive medium for energy storage applications.

600178 Facile and Ultrafast Microwave Synthesis of $NiCo_2S_4$/Graphene Nanocomposite for High Performance Supercapacitor

Abstract:

Binary metal sulfides, especially nickel-cobalt sulfides ($NiCo_2S_4$), and their composites have attracted considerable attention as electrode materials for supercapacitors because of their excellent electronic conductivity and high capacitance compared to mono-metal sulfides and oxides. A two-step hydrothermal process is widely used to synthesize $NiCo_2S_4$. However, the traditional hydrothermal method has several drawbacks such as low energy efficiency, long reaction time and complex synthesis steps. Herein, we present an ultrafast microwave approach for fabricating $NiCo_2S_4$/graphene nanocomposite used for supercapacitor electrodes with excellent electrochemical performance. The morphologies, crystalline structures and elementary composition of the obtained nanocomposites were characterized by SEM, XRD, FTIR, and EDS. It is found that $NiCo_2S_4$ nanoparticles can uniformly grow on the surface of graphene after tens of seconds microwave irradiation. The presence of graphene not merely enhances the conductivity of nanocomposites but also provide the transport paths for ions in electrolytes and help to reduce energy loss, thus leading to enhanced electrochemical performance. Electrochemical results indicate that the specific capacitance of as-prepared $NiCo_2S_4$/graphene nanocomposite can reach up to 656 F g^{-1} at 0.5 A g^{-1}. In addition, the $NiCo_2S_4$/graphene nanocomposite exhibits excellent cycling stability with capacitance retention over 80% even after 10000 cycles at 4 A g^{-1}. Thus, the current work provides an ultrafast and energy-efficient microwave method to synthesize high-performance $NiCo_2S_4$/graphene nanocomposite for supercapacitor applications.

610567 Investigating Ion Intercalation Mechanisms in Mxene Structures through Molecular Dynamics Simulations

Abstract:

Supercapacitors are electrical energy storage devices that have garnered much interest in recent decades due to their high power density in comparison to conventional batteries[1]. Though they play important roles in various fields, the lower energy density of supercapacitors limit their widespread application. In an effort to enhance the energy density of supercapacitors, various novel materials have been synthesized and studied as electrodes. MXenes, a new family of 2D transition metal carbides, have attracted particular interest due to their excellent pseudocapacitance originating from redox reactions at the interface.[2] Studies have shown that the process of ion intercalation can significantly enhance the specific capacitance of MXenes[3]. As a result, it is of fundamental importance to understand the microscopic behavior of ions under the confinement of MXenes.

Through a series of molecular dynamics simulations, we study the atomistic interactions between ions and the MXene surface throughout the process of ion intercalation. By analyzing various structural and transport properties, such as the coordination number and the dynamics of the solvation cages, we aim to understand how ion intercalation affects the interlayer spacings of MXenes. Additionally, the constant potential method is utilized to study how ion behavior is impacted by potentials of varying magnitude. Overall, a detailed understanding of ion solvation behavior at a wide range of potentials can further guide the design of MXene supercapacitors that display simultaneous high power density and energy density.

[1] Miller J R, Outlaw R A, Holloway B C. Graphene Double-Layer Capacitor with Ac Line-Filtering Performance. Science, 2010, 329(5999):1637-1639

[2] Lukatskaya M R, Mashtalir O, Ren C E, et al. Cation Intercalation and High Volumetric Capacitance of Two-Dimensional Titanium Carbide. Science, 2013, 341(6153):1502-1505

[3] Li J, Yuan X, Lin C, et al. Achieving High Pseudocapacitance of 2d Titanium Carbide (Mxene) by Cation Intercalation and Surface Modification. Advanced Energy Materials, 2017, 7(15):1602725

600181 Highly Flexible, Strong and Conductive Cellulose Nanofibrils/PEDOT:PSS Nanopapers for All-Solid-State Supercapacitors

Abstract:

Recently, cellulose nanofibrils (CNF) produced from cellulosic materials (e.g. wood, agricultural and industrial waste, etc.) have received increasing attention due to their unique properties such as high specific surface area, high elastic modulus, high thermal stability, as well as renewability and biodegradability. These unique properties make CNF as promising building blocks for the manufacture of many sustainable nanomaterials.

In this work, we firstly prepared CNF from pulp mill sludge by a sustainable process of formic acid (FA) hydrolysis pretreatment and the followed microfluidization. It was found that the mild FA hydrolysis (at 95 °C for 3-6 h) pretreatment could hydrolyze most of hemicellulose, swell and break down the cellulose fibers, and the collected cellulosic solid residue with a high yield (over 75%) could be further converted to CNF with relatively low-intensity microfluidization (only two passes). Then, the CNF was coated with PEDOT:PSS *via in situ* polymerization. Afterwards, flexible and conductive CNF/PEDOT:PSS nanopapers were prepared from the above obtained PEDOT:PSS-coated CNF suspension by a simple vacuum filtration approach. Furthermore, CNF/PEDOT:PSS nanopapers were treated with dimethyl sulfoxide (DMSO) to increase the conductivity. Results showed that the DMSO-treated CNF/PEDOT:PSS nanopapers had high tensile strength (over 60 MPa), high thermal stability, and low sheet resistance (18 Ω sq^{-1}). Finally, a symmetric all-solid-state supercapacitor was assembled using the DMSO-treated CNF/PEDOT:PSS nanopapers as electrodes. It was found that the assembled supercapacitor could deliver the maximum areal specific capacitance of 888.7 mF cm^{-2} (corresponding to 111.1 F cm^{-3}) and offer the highest areal energy density of 79.0 μWh cm^{-2} (corresponding to 9.9 mWh cm^{-3}), which are among the highest values reported for PEDOT:PSS based supercapacitors. More importantly, the assembled supercapacitor showed remarkable cycling stability with the capacitance retention of 109.5% after 10,000 charge/discharge cycles at a current density of 20 mA cm^{-2}. Considering the sustainable and facile preparation process and excellent electrochemical performance, the obtained CNF/PEDOT:PSS nanopapers could be a promising candidate for flexible energy storage devices.

608197 Electrospun Infrared-Transparent Solar-Reflecting Nanofibers for Direct Radiative Cooling

Abstract:

Radiative cooling is a promising way to alleviate urban heat island effects and decrease the energy required for occupant thermal comfort. Among many emerging approaches, the use of infrared-transparent films and fabrics that allow objects to directly radiate heat through bands of atmospheric transparency while blocking solar irradiation are particularly well suited for seasonal and other temporary uses. These covers have been widely based on polyethylene due to its simple chemistry and consequently intrinsically low IR absorption. Here, we investigate the effects of polymer nanofiber morphology on the scattering and transmission properties of polyacrylonitrile (PAN) based films. By controlling the electrospinning process, a beaded fiber morphology can be fabricated which features ellipsoidal beads connected via thin cylindrical fibers. These beaded fibers exhibit greater solar scattering than either spherical or cylindrical polymer structures, as confirmed using electromagnetic simulations. These noteworthy scattering properties decrease the amount of material needed to reach above 95% solar-weighted total reflectance, enabling high infrared transmittance (> 70%) despite PAN's intrinsic absorption in the 8-13 μm range. When scaled up and tested outdoors, the beaded nanofiber PAN film (nanoPAN) decreases the temperature of an unpolished PDMS-coated aluminum sheet by 7°C by shielding it solar radiation while allowing it to radiate heat to space. Heat transfer modeling agrees with the outdoor experiments and predicts temperature drops exceeding 10°C below ambient under standard sky conditions with an optimized beaded nanofiber film. This work on tailoring the morphology of polymer nanofibers paves the way for unconventional materials and lower-purity feedstocks to be used as solar filtering covers in direct radiative cooling applications.

612443 Understanding the Role of Small Molecules in Mineral Oxide Processing By Atomic Layer Deposited Titanium Dioxide Surfaces and Colloidal Probe Atomic Force Microscopy

Abstract:

Small molecule polyelectrolytes such as polyphosphates, citric acid and Dispex N40 are widely used in mineral processing. These polyelectrolyte salts are reasoned to provide electrostatic stabilisation (possessing high surface charge densities); however this is mediated by the solution conductivity (Debye screening length). Moreover, high charge densities ($\zeta \sim 50$ mV) on mineral oxide surfaces can be achieved through pH moderation, therefore the question remains as to why small molecule polyelectrolytes are so popular within mineral processing operations?

Atomic layer deposition (ALD) was used to fabricate flat and smooth (RMS 0.5 nm) TiO_2 surfaces. With surface charges varying from 0 to -45 mV (corresponding to pH 4.7 to pH 10.5 respectively), the interaction force between TiO_2-TiO_2 (colloid probe and surface) showed jump-in to contact at all pH values. When coated with SHMP, the interaction forces were purely repulsive, with no jump-in on approach or pull-off during retraction. Such behavior was consistent over a wide pH range which included the $pH_{i.e.p.}$ for SHMP (pH 3.4) and highly charged surfaces (pH 11).

The approach data was modelled by both off-set electrical double layer forces (DLVO) and steric forces (Alexander-de Gennes model). The steric interaction was found to depend on pH with a step-change behavior in the region of pH 7.5. Below pH 7.5 the size of the SHMP molecule was ~1.0 nm and above pH 7.5 the size slightly increased to ~1.5 nm, see Fig. 1. Molecular modelling (using Avagadro software ©) estimated the length and width of SHMP to be 1.5 nm and 0.3 nm respectively. Therefore the change in steric contribution likely relates to a conformational change of the SHMP molecule at the metal oxide-liquid interface. At high pH, the hydroxyl groups of SHMP are dissociated and the molecule becomes increasingly negatively charged (confirmed by the pKa values of SHMP hydroxyl groups), with the molecule attached end-on and perpendicular to the surface. At low pH, the molecule can interact via hydrogen bonding which leads to a slight tilt in the molecule orientation.

The presentation also highlights the effect of ion-concentration (10^{-3} to 10^{-1} M), ion-type (following a Hofmeister series – CsCl, NaCl, and $CaCl_2$), and pH on interaction forces and molecular orientation. The underlying mechanism of the small polyelectrolyte to provide excellent lubricity will be concluded.

Fig. 1. SHMP AdG modelled steric size as a function of pH (in 1 mM NaCl). Inset show the conformational change on SHMP when adopting an end-on and perpendicular orientation at high pH (dominated by electrostatic repulsion between molecules), to a slightly tilted orientation where molecules interact via hydrogen bonding when the pH is below 7.5.

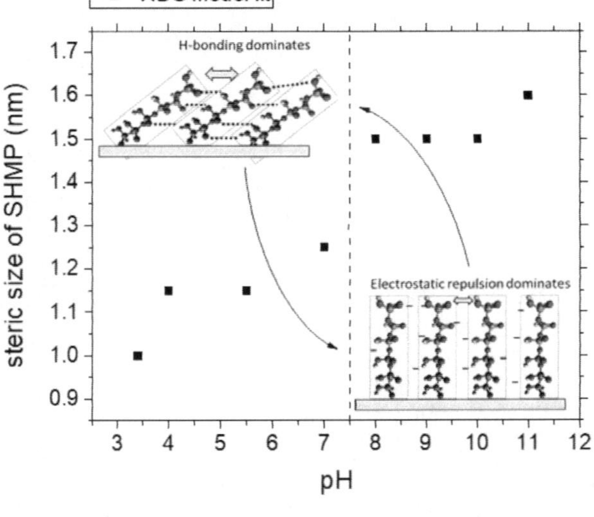

597563 Atomic Layer Deposition of Boron Nitride As a Hydrogen Environmental Barrier Coating

Abstract:

Hydrogen environmental barrier coatings can be used to reduce hydrogen embrittlement of structural materials by reducing diffusion into susceptible substrates. This protection is especially important for hydrogen pipelines and in fields such as nuclear and fuel cell applications where hydrogen is used as a fuel source. The higher the operational temperature, though, the fewer options for material choice. A nanoscale film of the current state-of-the-art material for nuclear applications, tungsten, was previously shown to increase the temperature at which hydrogen reacts with the substrate. However, the increase was still far from the operational temperature of these nuclear applications. In this work, density functional theory (DFT) calculations for energy barriers of atomic hydrogen diffusion were used to predict h-BN as a favorable material for preventing hydrogen diffusion. In addition, the diffusion pathway and the effects of hydrogen in the BN lattice were characterized. To experimentally confirm the effectiveness of BN as a barrier film, particles were coated with nanoscale BN environmental barrier coatings via atomic layer deposition (ALD). The films were then characterized for composition and thickness, and the samples were tested in a high temperature hydrogen environment using differential thermal analysis. Results showed that BN performed as predicted, preventing reaction of hydrogen with the coated sample beyond the maximum temperature of the current state-of-the-art, tungsten.

601486 Predicting Biological Interactions to Monolayer-Protected Gold Nanoparticles Using Molecular Simulation-Derived Descriptors

Abstract:

Gold nanoparticles (GNPs) have attracted significant interest as materials for biomedical applications because they can be detected *in vivo*, deliver drugs, and target specific sites in the body (*e.g.* cancer cells). GNPs are often coated by self-assembled monolayers (SAMs) consisting of alkanethiol ligands with varying backbone and terminal end groups. SAM-protected GNPs are advantageous because of their ease of fabrication with tunable parameters, such as the gold core size, the ligand chain length, and ligand end group chemistry. However, a challenge that inhibits the use of GNPs in biomedical applications is the lack of understanding in how changing these parameters affects the interactions between GNPs and biomolecules, such as proteins or lipid bilayers. The experimental trial-and-error exploration of GNP-biomolecule interactions is cost-prohibitive, motivating the use of alternative techniques to narrow the GNP design space. One promising method to guide GNP design is to relate ligand-specific descriptors to experimental outcomes. For example, recent work has related GNP hydrophobicity – quantified by the octanol-water partition coefficient (logP) of ligand end groups – to phenomena including bilayer adsorption, antimicrobial behavior, and cellular uptake. This approach thus provides some ability to select GNP coatings *a priori* that achieve desired behavior. However, a single ligand end group descriptor is insufficient to characterize changes to GNP surface properties that emerge from variations in GNP size or collective interactions between surface-grafted ligands.

In this work, we use classical atomistic molecular simulations to model SAM-protected GNPs and calculate molecular descriptors capable of characterizing GNP properties while accounting for variations in ligand chemistry, backbone structure, size, and ligand-ligand interactions. We first develop a generalized system preparation workflow that can model a range of SAM-protected GNPs. We then model GNPs in different solvation environments and calculate molecular descriptors that encode information about the GNP surface chemical (*e.g.* hydrophobicity, charge, hydrogen-bonding capabilities, *etc.*) and structural (*e.g.*, eccentricity, ligand fluctuations, solvent-accessible surface area, *etc.*) properties. In particular, we show that GNP hydrophobicity – previously quantified by logP – also depends on SAM, which itself depends on NP core size. We expand the molecular descriptor set to capture GNP-lipid membrane interactions by modeling GNPs in the presence of a lipid bilayer and show that descriptors can predict biological outcomes. The molecular descriptors developed in this work will help predict the effects of GNP functionalization on GNP-biomolecule interactions, opening avenues towards efficient screening of GNPs for selective interactions in the biological environment.

608018 A Cellulosic Shell Installed Onto Nanoparticles Protects Cores from Organic Solvents

Abstract:

We here describe the discovery and characterization of a cellulosic layer on nanoparticle surfaces that protects particle cores from organic solvents. Core-shell nanoparticles 150nm in diameter are formulated using Flash NanoPrecipitation (FNP) and contain a core made of lumefantrine, a model hydrophobic antimalarial drug, and a shell made of hydroxypropyl methylcellulose acetate succinate (HPMCAS), an amphiphilic polymer commonly used as a pharmaceutical excipient. Following FNP, NaOH is added to the nanoparticle suspension, which is then spray dried. During drying, the caustic environment de-esterifies the HPMCAS layer coating the nanoparticles, removing the acetyl and succinyl substitutions from the backbone. The de-esterified HPMCAS – now hydroxypropyl methylcellulose (HPMC) – then vitrifies to form a glassy layer around the NP cores. The HPMC shell layer is impermeable to many organic solvents, and with it in place, less than 10% of the lumefantrine core can be extracted using THF (in which it is readily soluble). Without the NaOH treatment, the THF-soluble HPMCAS shell is dissolved and 100% of the lumefantrine is extracted.

The HPMC shell is characterized by NMR and DSC to confirm its chemical and thermal properties. Necessary conditions for the NaOH treatment and drying are identified, and a method for washing the spray dried powders to remove excess dry NaOH is described. The layer's ability to protect the core is then tested against a variety of common organic solvents. We then describe possible uses for this technique in catalyst and personal care applications, such as temperature-triggered catalyst release or crush-induced fragrance delivery.

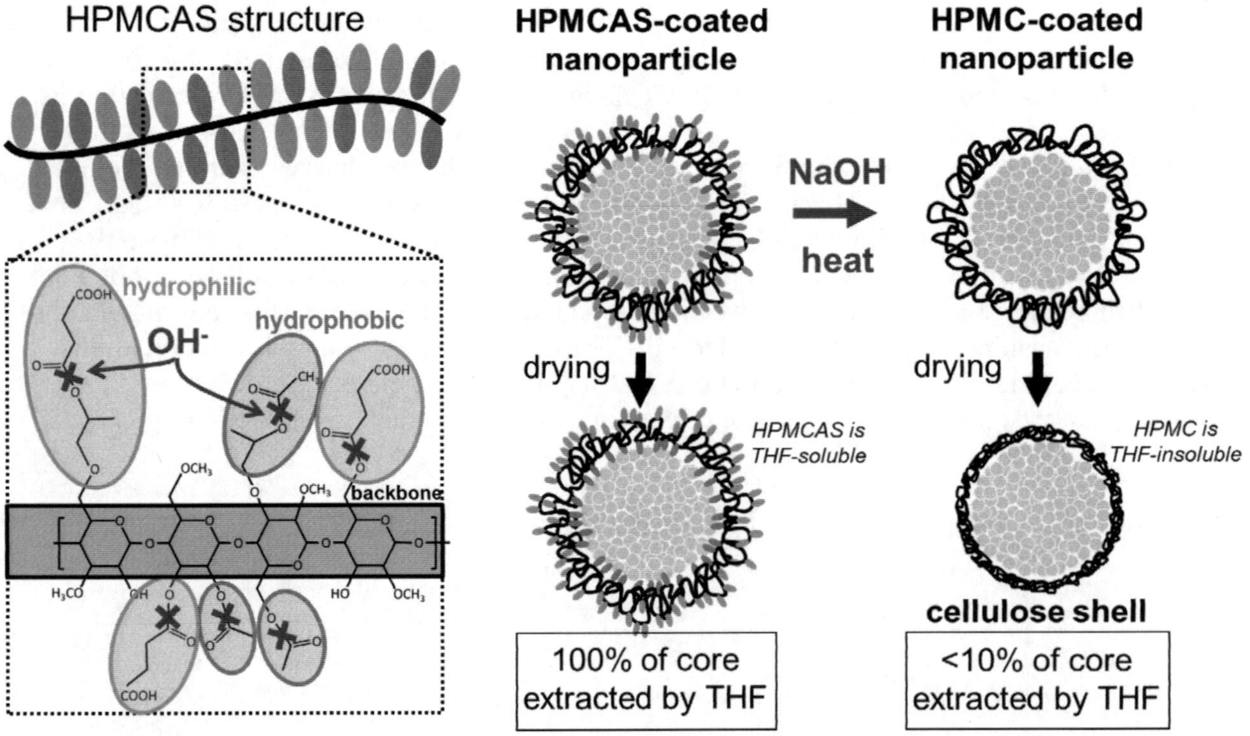

608831 Temperature-Controlled Nanomechanical Shear-Thinning between Polymer Nanoparticle Surfaces during Shear-Induced Aggregation.

Abstract:

Temperature-controlled nanomechanical shear-thinning between polymer nanoparticle surfaces during shear-induced aggregation.

José Francisco Wilson[a, [1]](wilsonj@vscht.cz), Miroslav Soos[a] (soosm@vscht.cz).

[a] University of Chemistry and Technology Prague, Technicka 5, 16628 Prague 6, CZECH REPUBLIC

Abstract

In this research, we studied the impact of material fusion as adhesion mechanism on the size and structure of fractal aggregates formed during shear-induced aggregation of fully destabilized NPs. To precisely control particle adhesion, we used nanoparticles (NPs) with a Core-Shell (C-S) structure, where the core is composed of polymethyl methacrylate and the shell is composed of a combination of polymethyl methacrylate and polybutylacrylate. The core acts as a hard-sphere with a high glass transition temperature (T_g) to maintain the structure of the NPs during changes in temperature. On the other hand, the composition of the shell was designed to target a T_g of 50°C , giving the surface of the NPs the capability to soften upon increasing temperature. During aggregation experiments we observed that the size of aggregates grow significantly at temperatures above T_g of the shell, indicating an increase of adhesive force between the NPs. Additionally, at temperatures higher than $T_{g,}$ it was observed that the increase in adhesion is countered by a reduction of aggregate size when higher shear rates are applied. This suggest that the bond between nanoparticles weakens upon increase of shear rate when the material on the surface experiences a transition from soft to a viscous polymer. This finding indicates that polymer NPs surfaces at temperatures above T_g might experience a local shear thinning effect. To prove this hypothesis, we used DEM simulations and study the interplay between the increase of adhesion due to material softness and the decrease of bonding strength upon increase of shear rate. Obtained results support the hypothesis that these two contradicting mechanisms determine the final mechanical properties or produced material.

[1]

603340 Nano-Coating Stabilized Particulate Photocatalysts for H_2 Production

Abstract:

Photocatalytic nanoparticles that are suspended in water or sprayed onto a panel promise solar-to-fuel energy conversion at scale. The fabrication processes for these photocatalysts to achieve efficient energy conversion should be simpler than solar cells and photoelectrodes. However, the design principles for reaching their performance limit are still elusive. We use few-nanometer thin stabilization coatings to enable an entire class of otherwise unstable particulate semiconductors for coevolution photocatalysis. CdS particles that are sprayed into a film and stabilized with an ALD coating coevolve H_2 at a record rate of 48.5 mmol·h^{-1}·g^{-1} or 2.5 mL H_2·h^{-1}·cm^{-2} under 1-sun solar illumination in ambient air. Moreover, we demonstrate a redox-mediated particulate reactor that separates the H_2 and O_2 products, which is safe. We systematically advance towards the solar-to-fuel efficiency limit of for photocatalysts, instead of trial-and-error that is often practiced in this field. To justify our approach, we will discuss the holistic understanding for the photophysical, electrocatalytic, and transport processes that are coupled at the nanoscale.

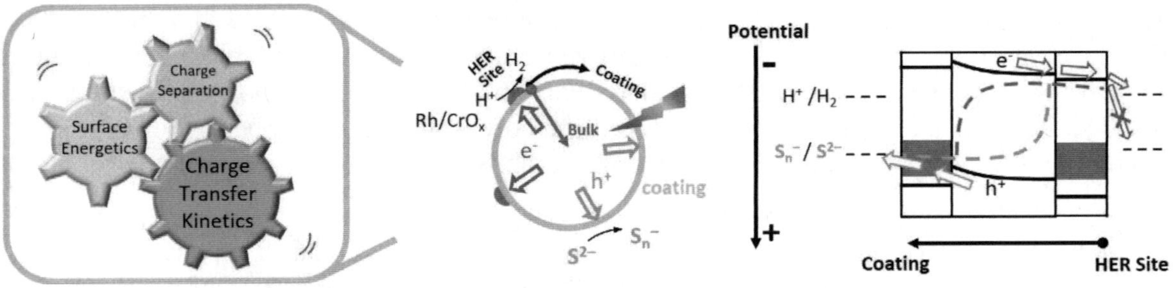

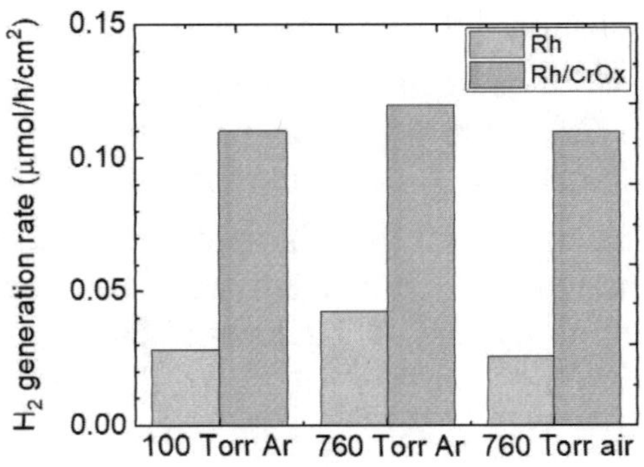

605674 Understanding the Unique Transport, Chelating and Electrochemical Properties of Nanoscale - Based Electrolytes for Sustainable Energy Storage

Abstract:

The recent growth in the worldwide installation of renewable energy has allowed the first steps towards decarbonization of power sector. However, further penetration of renewable energy into the grid is challenged due to the limitations of storage of intermittent renewable electricity. Thus, vast research efforts are on-going to develop electrochemical energy storage systems such as batteries and dense energy carriers. Electrolyte design and selection is a key component of these electrochemical energy systems, mediating transport and solubility of the reactive species to ultimately deliver high energy, current and power density systems. An understanding of stability, electrochemical and physicochemical properties, as well as functionality, is necessary to guide the design of next-generation electrolyte materials for sustainable energy storage systems.

A novel class of nanoparticle organic hybrid materials (NOHMs) has been developed that can serve as breakthrough electrolytes with concentrations of electrochemically active species not achievable in conventional solvents. NOHMs consist of polymer-tethered nanoparticles, where the polymeric canopy is grafted to an inorganic core either ionically or covalently. Due to the polymer tunability, NOHMs are versatile electrolyte media for redox flow batteries and other electrochemical systems, such as electrochemical CO_2 conversion to chemicals and fuels.

NOHMs have been explored as complexing agents of redox active species, increasing their effective concentration in solution. They have been found to effectively chelate copper and zinc redox active species of interest in electrochemical applications, and complexation behavior has been studied via various spectroscopy tools. The electrochemical behavior of NOHMs-complexed copper species has been explored, indicating changes of the overall conversion mechanism through stabilization of the copper (I) state. The degree of chelation of different redox active species has been found to be easily controlled by the solution pH. Tailoring transport properties is also of particular importance in NOHMs-based electrolytes mixtures, as they are challenged by inherently high viscosities, impacting charge transport critical in electrochemical performance. These properties can be tuned by intermolecular interactions between the polymeric canopy and the surrounding fluid. In particular, NOHMs are highly responsive to ionic stimuli, with the addition of even low salt concentrations inducing large reductions in the viscosity of NOHMs-based electrolyte mixtures. Alterations in the degree of polymer swelling and the conformational structure of the NOHMs polymer canopy with ionicity have been probed to explain measured bulk physicochemical properties.

Overall, NOHMs have been found to be electrolyte materials enabling high concentrations of electrochemically active species and with highly tunable transport properties, making them versatile electrolytes for a variety of electrochemical energy storage systems.

593515 Correlating Polyamide Powder Flow Properties to Mechanical and Physical Properties of 3D Printed Parts

Abstract:

Selective Laser Sintering (SLS) is an Additive Manufacturing technique typically used for rapid prototyping and low volume production of functional components. The process employs a laser beam to sinter powdered material, binding it together to create a solid structure. The laser selectively fuses pre-defined areas of a powder bed by scanning cross-sections generated from a 3D digital description of the required part. After each cross-section is scanned, a new layer of material is applied on top, and the process is repeated until the part is complete.

Generating the layers of powder is a precision process and requires a feedstock that can be reliably distributed by the delivery system which is deposited on to the fabrication bed in a consistent manner without agglomerates or voids. Intermittent flow, or agglomerates within the bulk, will cause non-uniform deposition, adversely affecting the efficiency of the process and the properties of the final product. Identifying which powder properties are conducive with the formation of uniform, repeatable layers allows new formulations to be optimised, and suitable raw materials identified, without incurring the significant financial and time overheads associated with running materials through a process to assess compatibility. This approach also helps reduce the occurrence of final products that are out of specification.

This study investigates the relationships between the flow properties of two different grades of Polyamide 12 (PA12) powder and mixtures of, with the mechanical and physical characteristics of the printed parts. The powder bulk properties including flowability were characterised using powder rheology (FT4 Powder Rheometer®, Freeman Technology Ltd, UK). Absolute, or skeletal volume of 3D printed cubes were measured using helium pycnometry (AccuPyc, Micromeritics Instrument Corporation, USA). Tensile specimens were also fabricated, and their mechanical properties were correlated with the initial powder's flow properties.

By comparing the powder flow properties and the properties of the built parts it was possible to identify strong relationships between various parameters, thereby allowing prediction of final product quality prior to printing. Understanding these relationships helps to identify and develop powders that enhance both process efficiency and the mechanical properties of the final product.

606590 Comparative Study of the Time Dependent Rheological Behavior of Particulate Suspensions - Cement Paste and Silicon Carbide Paste

Abstract:

Understanding the rheology of particulate suspension is important in many industrial applications, including coatings, paints, food, pharmaceuticals, and cements and where pumping and extrusion are used, among others. For cement paste suspensions, it is rheology that is critical to its suitability for use in construction applications, including 3D printing. Rheology depends on a number of physical and chemical properties of the cement and the paste formed thereof. Furthermore, the way that the paste was processed matter, including how it was mixed and even the measuring conditions. Critically important is the need to separate the effects of these factors on the rheology of cement paste, which has been shown to depend not only on the shear history of the paste, but also on the age of the paste i.e., properties evolving with time, thus exhibiting time-dependent rheological behavior.

This work attempt to separate the contribution of one of the contributing chemical factors, hydration, from the overall evolution of time-dependent rheological behaviors in cement-based systems. The effect of shear history and apparent particle size distribution of two solids, a Type I-II hydraulic portland cement and non-hydraulic silicon carbide surrogate, was studied. The rheological behavior of their respective pastes in deionized water, 0.15% superplasticizer solution, and 0.15% ammonium polymethacrylate solution was determined and compared. Contrary to the earlier suggestions that the thixotropic to anti-thixotropic transition of cement paste might be due largely to hydration related structural buildup, these comparative results show that particle-association or flocculation-de-flocculation related structural changes might be equally as significant.

608723 Thc Influence of Temperature and Moisture on Polymer and Metal Materials for Additive Manufacturing

Abstract:

Polymeric materials in additive manufacturing add a unique difficulty in characterization to a field that is used to deal with metal powders.
Due to the molecular structure of these polymer powders, moisture and temperature change their behavior to a larger degree than is to be expected with the more common metal feedstock.
Manufacturers have addressed this by bringing the temperature of the feedstock chambers to elevated levels, therefore reducing or eliminating the moisture effects. This in turn creates the problem of reliably measuring the feedstock at these elevated temperatures to give an accurate picture of their properties during use. And in addition the problem of moisture influence during storage prevails.
In this work we will show measurements of polymeric feedstock under actual processing parameters and delve into the problems arising during drying and storage.
It will consist of measurements with temperature and moisture controlled ring shear cells as well as data from our novel temperature-controlled stirrer-based cell with fluidized methods.

While most metal feedstock for additive manufacturing (with the noted exception of aluminium alloys) show little influence of temperature and moisture on the behavior, some examples will be given on the degradation of metal feedstock during prolonged moist storage.

608833 Characterization of Powders Spreadability in the Framework of Additive Manufacturing Processes

Abstract:

Metallic and polymer powders are widely used in several Additive Manufacturing (AM) processes, with for example Selective Laser Melting (SLM) and Selective Laser Sintering (SLS). During such operations, successive thin layers of powder are created with a ruler or with a rotating cylinder. Each layer is partially sintered or melted with an energy beam. The layer thickness defines the vertical resolution of the printer, a thin layer leads to a better resolution. In order to obtain a thin layer, the powder is as fine as possible. Unfortunately, when the grain size decreases, the cohesiveness increases and the flowability decreases. The flowability must be good enough to obtain homogenous successive powder layers. The quality of the parts build with AM is thus directly related to the powder flowing properties.

The measurement method based on the rotating drum was used in this paper. It allows indeed to assess the powder spreadability without applying a compressive load during the powder testing what fits with the conditions seen by powder in the AM processes.

Four metallic powders (AlSi7Mg, Scallmaloy, Inconel, Fine inconel) have been studied. Powders spreadability has been determined with the GranuDrum instrument, a rotating drum enabling to quantify the influence of cohesion, shear-thinning/shear-thickening as well as thixotropic behavior. Influence of powders triboelectricity (GranuCharge) and compaction dynamic (GranuPack) have also been investigated. These measurements have been correlated to the powder performance in a SLM 3D printer, where a CCD camera is used to take several snapshots at different recoater speeds (30, 80 and 160mm/s).

594921 Dispersing Nano- and Micro-Sized Portlandite Particulates Via Electrosteric Exclusion at Short Screening Lengths

Abstract:

In spite of their high surface charge (zeta potential $\zeta = + 34$ mV), aqueous suspensions of portlandite (calcium hydroxide: $Ca(OH)_2$) exhibit a strong tendency to aggregate, and thereby present unstable suspensions. While a variety of commercial dispersants seek to modify the suspension stability and rheology (e.g., yield stress, viscosity), it remains unclear how the performance of electrostatically and/or electrosterically based additives is affected in aqueous environments having either a high ionic strength and/or a pH close to the particle's isoelectric point (IEP). We show that the high native ionic strength (pH $\approx$ 12.6, IEP: pH $\approx$ 13) of saturated portlandite suspensions strongly screens electrostatic forces (Debye length: κ^{-1} = 1.2 nm). As a result, Coulombic repulsion alone is insufficient to mitigate particle aggregation and affect rheology. However, a longer-range geometrical particle-particle exclusion that arises from electrosteric hindrance caused by the introduction of comb polyelectrolyte dispersants is very effective at altering the rheological properties and fractal structuring of suspensions. As a result, comb-like dispersants that stretch into the solvent reduce the suspension's yield stress by 5x at similar levels of adsorption as compared to linear dispersants, thus enhancing the critical solid loading (i.e., at which jamming occurs) by 1.4x. Significantly, the behavior of diverse dispersants is found to be inherently related to the thickness of the adsorbed polymer layer on particle surfaces. These outcomes inform the design of dispersants for concentrated suspensions that present strong charge screening behavior.

601791 Modeling Complex Particle Interactions in Dry Powder Inhaler

Abstract:

Dry powder inhaler (DPI) delivers active pharmaceutical ingredient (API) particles to the airways and/or lungs [1]. Larger carrier particles are utilized to agglomerate with small API particles so as to improve fluidization, and then deagglomerate to release API particles for drug delivery [2]. Agglomeration and deagglomeration in DPI are affected by complex particle interactions, including van der Waals' cohesion [3] and electrostatic interaction [4]. Computational fluid dynamics coupled with discrete element method (CFD-DEM) has the advantage of explicitly capturing these particle-particle interactions but is computationally costly [5]. In order to make the simulation affordable, we undertake a two-step approach, particle softening and particle coarsening. First, we soften the particles in the simulation so as to reduce simulation time step. Second, we develop a coarsening approach in which one simulates a subset of representative API particles while still tracking all the carrier particles.

Towards this end, we conduct micro-scale tests using a carrier particle decorated with many API particles. These tests involve collision of a decorated carrier with a wall or another decorated carrier. In these simulations we consider both the van der Waals cohesion and electrostatic interaction between charged carrier and API particles. Through these tests, we examine if, and if so how, the model parameters should be adjusted such that particle softening and coarsening do not lead to falsification of the fraction of API released. For example, it is known that the Hamaker constant must be modified upon softening to ensure that the effective coefficient of restitution is preserved in head-on collision [6]. We find that some modification to rolling friction is also necessary when one coarsens the simulations. In a surprising result, we find that attractive interaction between the API particles translates to an effective increase in the carrier-API attraction, which would be underestimated upon coarsening; a method to correct for this effect is also proposed.

We complement these microscale tests with full DPI simulations to assess the performance of our coarse-grained modeling approach.

[1] Islam N, Gladki E. Dry powder inhalers (DPIs)—a review of device reliability and innovation. *Int. J. Pharm.* 2008; 360(1-2): 1.

[2] Newman S P, Chan, H K. In vitro/in vivo comparisons in pulmonary drug delivery. *J Aerosol Med Pulm Drug Deliv*. 2008; 21(1): 77.

[3] Yang J, Wu C Y, Adams M. Numerical modelling of agglomeration and deagglomeration in dry powder inhalers: A review. *Curr. Pharm. Des.* 2015; 21(40): 5915.

[4] Yu J, Wong J, Ukkonen A, Kannosto J, Chan H.K. Effect of Relative Humidity on Bipolar Electrostatic Charge Profiles of dry Powder Aerosols. *Pharm. Res.* 2017;34(8):1707-15.

[5] Nguyen D, Remmelgas J, Björn IN, van Wachem B, Thalberg K. Towards quantitative prediction of the performance of dry powder inhalers by multi-scale simulations and experiments *Int. J. Pharm.* 2018; 547(1-2):31-43.

[6] Gu Y, Ozel A, Sundaresan S. A modified cohesion model for CFD–DEM simulations of fluidization. *Powder Technol.* 2016;296:17-28.

602251 Powder Fluidization in Dry Powder Inhalers

Abstract:

Lactose powders are utilized in Dry Powder Inhalers (DPI) as carriers for Active Pharmaceutical Ingredient (API) particles to improve their fluidization [1]. Carrier-carrier and carrier-wall collisions result in the release of API particles from the carrier particles, and they determine the fine particle fraction (FPF) of API leaving the DPI which is of crucial importance in treatment outcomes [2]. The dynamics of carrier particles is affected by their physical properties, fluid turbulence and device geometry [3]. Powder fluidization has been studied in the literature probing these effects in commercial DPIs both experimentally and computationally by Coates et al. [4, 5]. Turbulent channel flows of carrier particles with different properties have been studied experimentally by Mahmoudi et al. [6], measuring the powder evacuation time at various Reynolds numbers.

We study, by means of Computational Fluid Dynamics - Discrete Element Method (CFD-DEM) simulations, the fluidization of carrier particles in turbulent channel flows, comparing the particles evacuation times obtained numerically with the experimental results of Mahmoudi et al. [6]. After this validation study, we examine the effect of device geometry on carrier particle evacuation dynamics. Subsequently, we perform DPI simulations with a representative dose containing both carrier and API particles to probe the effect of device geometry on FPF.

[1] Newman, S. P., Chan, H. K. In vitro/in vivo comparisons in pulmonary drug delivery. J Aerosol Med Pulm Drug Deliv. 2008; 21(1): 77.

[2] A. Kourmatzis, S. Cheng, H.K. Chan, Airway geometry, airway flow, and particle measurement methods: implications on pulmonary drug delivery, Expert Opinion Drug Deliv. 15 (2018) 271–282.

[3] J.S. Shrimpton, M. Danby, Effect of poly-dispersity on the stability of agglomerates subjected to simple fluid strain fields, Powder Technol. 228 (2012) 241–249.

[4] M.S. Coates, H.K. Chan, D.F. Fletcher, et al., Influence of air flow on the performance of a dry powder inhaler using computational and experimental analyses, Pharm. Res. 22 (2005) 1445–1453.

[5] M.S. Coates, H.K. Chan, D.F. Fletcher, et al., Effect of design on the performance of a dry powder inhaler using computational fluid dynamics. Part 2: Air inlet size, J. Pharm. Sci. 95 (2006) 1382–1392.

[6] Mahmoudi, S., Elserfy, K., Cheng, S., Chan, H. K., Hebbink, G., & Kourmatzis, A. (2019). Fluidisation characteristics of lactose powders in simple turbulent channel flows. Experimental Thermal and Fluid Science, 103, 201-213

602833 Bulk Properties Assessment and Drum Capsule Filling Evaluation of High Dose Inhaled Formulations

Abstract:

Inhaled therapeutics are routinely used for the treatment of pulmonary diseases such as asthma, COPD, and cystic fibrosis. In recent years, the need to deliver a high therapeutic dose into the lungs is becoming increasingly more commonplace. In the case of recently approved anti-infectives (e.g. colobreathe, tobramycin DPI, and aerovanc) and Parkinson's therapy (inbrija), for example, the total lung doses are thousands of times greater than those of conventional asthma medications (e.g. ICS, LABA, and LAMA).

Inhaled drug products, typically a blend comprising the active pharmaceutical ingredient (API) and excipients, are often packaged in blisters or standard-sized capsules and subsequently dispersed and delivered to the lungs by a specially designed inhaler. To promote patient compliance, administering the required lung dose in as few inhalations as possible is desirable. Minimizing inhalations, combined with the fill volume constraint of standard capsules, often leads to the need for maximizing the drug loading in each dose.

This work presents the challenges associated with the process of filling a model high drug loading formulation into capsules. These challenges arise from the drug product's cohesive nature, which is a consequence of the size of the API payload: in order to reach the lung regions relevant to drug absorption, a common approach is to micronize the API to a specific size range. While certain excipients can be blended with the API to improve filling capability, the fact remains that the amount that can be added is limited due to the reasons explained above.

We studied the effects of drug loading, excipient particle size, the amount of force normalizing agents (e.g. magnesium stearate) in the formulation, the order of excipient addition during blending, blending method (high- vs low-shear), and surface modifying agents (e.g. leucine and tri-leucine) on the flowability and compressibility of the API/excipient blends. High- and low-shear mixing were achieved using a Turbula blender and a ProCepT, respectively. Filling experiments were performed on select blends using a Harro Hofliger Drumlab, a lab-scaled drum-filling machine. The following drum-filling parameters were studied: bore size, vacuum pressure, and drum/stirrer offset. Blend properties were subsequently correlated to the filling performance quantified by the consistency of dosage units.

We found that the blend properties are influenced primarily by the drug loading, excipient particle size, and the order of excipient addition during the blending process. The effects of force normalizing agent are complex—there appears to be an optimum amount of force normalizing agent that maximizes flowability and minimizes compressibility; however, increasing or decreasing the amount past this level will reverse these effects. Unit dosage consistency, on the other hand, was most sensitive to the drum filling vacuum pressure, with higher vacuum pressure generally resulting in more consistent dosage units. Operating the drum filler at its maximum pressure, however, will limit the versatility of the equipment because under such a constraint the fill weight will be a function of only the drum bore size and the blend compressibility. Perhaps as expected, poor dosage unit consistency resulted when filling poorly flowing and highly compressible materials (in this study, these properties were exhibited by the blends with the highest drug loading). This observation was the result of a powder 'bridge' that formed on the rotating stirrer inside the drum filler hopper, which in turn prevented the product from reaching the drum bore. Lastly, we also used DEM simulations to explore how to potentially redesign the drum filling equipment to optimize the filling process for cohesive and compressible materials.

609343 Predictive Approach to Estimate Viability of Dry Coating for CQA Enhancement of Pharmaceutical Powders

Abstract:

Loose powder packing and powder flowability are both important critical quality attributes (CQAs) to consider in relation to pharmaceutical tablet manufacturing. Such CQAs impact the efficiency of various unit operations such as feeding, blending, conveying, and compaction. Fine pharmaceutical powders (<100 μm) often have inadequate packing and flow, due to high cohesion among the individual particles. In this work, ability to predictively enhance such properties, specifically, the bulk density, expressed by powder packing porosity, and flowability, expressed through flow function coefficient (FFC) after nano-silica dry coating, was examined. This was done through computation of the granular Bond number and cohesion force estimated via multi-asperity contact model for twenty different pharmaceutical powders. Despite variations in size (10–225 μm), particle size distribution, aspect ratios (1-3.5), material density, and dispersive surface energy, the bulk density and FFC improvements after dry coating followed predictive trends. However, there were a few outliers for the porosity prediction model, Microcrystalline Cellulose (MCC) excipients were outliers, for which their behavior was assessed through analysis of their shape, surface morphology and specific surface areas (SSA). Consequently, approaches to improve the multi-asperity models for better estimation of the Bond number were examined. The work also developed guidelines for what class of powder materials may provide more accurate prediction. The investigation included experimental results of optimum nano-silica amount, and trends of CQA improvement after dry coating, and comparison of nano-silica glidants. Most impressively, the bulk density and FFC enhancement were as high as 70-90 % for finer, more cohesive powders, indicating that dry coating works well for the powders for which it is needed the most. This ability to estimate the expected enhancements after dry coating *a priori* can guide decision making regarding whether or not dry coating is necessary or would be beneficial.

599124 Investigating the Impact of Texture Modifiers on the Flow Properties of Cosmetic and Personal Care Formulations

Abstract:

Across numerous industries, it is necessary to introduce additives and modifiers to formulations in order to improve their processability and final product quality. Within the cosmetics and personal care sectors, these additions can be made for a variety of reasons. In addition to improving process performance, additives can be introduced to improve consumer experience, such as product texture, spreadability on skin, appearance and scent, increase the stability of expensive fragrance components and influence substantivity on the skin or hair.

Despite several advantages of introducing additives or making adjustments to the formulation, it is important to understand how these changes affect the bulk flow properties throughout preceding unit operations. In this study, the powder flow properties of spherical silica texture modifiers are investigated. The purpose of this silica modifier is to improve the spreadability, 'skin feel' and appearance on skin of a powder or cream product, and can even provide SPF boosting. These performance parameters can be influenced by changes in surface properties and oil absorption ability of the silica bead. The relationship between these Critical Quality Attributes (CQAs) and the flow properties of the dry component is evaluated to understand whether compromise is required to simultaneously optimise process efficiency, yield and product quality.

Powder flow properties were evaluated using an FT4 Powder Rheometer® (Freeman Technology Ltd), a universal powder flow tester that provides automated measurements of dynamic, bulk and shear properties. These parameters were used to quantify the influence of surface area and oil absorption on the flow properties of the texture modifiers. Clear correlations were observed with several measured parameters quantifying performance across a range of different stress and flow regimes. In addition, the reproducibility of test methods across different instruments and locations was demonstrated.

This study highlights the wide range of process considerations that need to be made when adjustments are made to a formulation. Improvements in product quality and consumer experience through the introduction of additives and modifiers must be balanced with process efficiency. This is particularly important when the majority of process stages handle the formulation in a different physical state to the end product. Properties that are beneficial to liquid flow and performance may not be relevant to powder behaviour and vice versa.

601621 Scaling-up of Magnetite Nanoparticles Production for Drug Delivery Applications Via Static Mixers: Role of Turbulent Reacting Flow

Abstract:

Over the two decades, the nanomaterials field has tremendously expanded to enable basic and applied research studies. In this regard, today there are approximately 300,000 nanomaterials, more than 840,000 articles published, and about 24,000,000 patents worldwide[1]. Some of these nanomaterials have reached industrial-scale production or implementation. For example, graphene sheets are industrially manufactured to produce sensors to detect the presence of contaminants in water[2]. Another example is TiO2 nanoparticles, which represent over 70% of total pigment production worldwide[3]. Despite the growing interest and potential of nanotechnology, most of these materials are still only produced at the lab scale. In the particular case of iron oxide paramagnetic nanoparticles, or magnetite, no reports are available about the production at the commercial scale for biomedical or pharmacological applications. This considerably limits the possibility of reaching commercial applications in the medical and diagnosis fields. This includes magnetic resonance contrast for high-resolution medical imaging, magnetic hyperthermia induction, diagnosis by targeted detection in cancer, detoxification of poison gases, and drug delivery[4-5]. To assure large-scale production the synthesis and purification of magnetite require high yields and low manufacturing costs. At the lab scale, magnetite can be synthesized by several different methods, including sol-gel[6], sonochemical[7], thermal decomposition[7], and the widely used chemical co-precipitation[8]. Each of them leads to different particle size distribution, crystalline contents, surface morphology, and porosity[6]. Despite the low polydispersity, relatively high production yield, and uniform morphology of magnetite obtained via thermal decomposition or ultrasound impact, implementation at the large-scale is rather complicated and expensive. Alternatively, the sol-gel or co-precipitation methods show ease for implementation, relatively high-yields, and most importantly low associated costs. This suggests that these processes are well-suited for scaling-up[8]; however, they pose a major challenge regarding the dependency with respect to the mixing conditions, which in turn, control the growth and nucleation processes of magnetite[8]. These conditions need to be addressed in a dynamic manner during the process to assure that the obtained materials comply with the requirements in terms of particle size distribution and morphology. In the case of drug delivery applications, which are of interest to our research group, the average particle size should be around 100-200 nm.

Here, we propose to tackle this major issue by maintaining proper mixing conditions via multiphase turbulent reacting flow during the synthesis of magnetite. The first step for the scaling-up of the co-precipitation method was to run the synthesis at the bench scale. In this case, the synthesis proceeded in a semi-batch mode of operation in a 3L stainless steel reactor equipped with a Rushton impeller, and a peristaltic pump to control the addition of the iron chlorides to the reaction mixture. The reaction started by adding the iron (II) and (III) chlorides to a sodium hydroxide solution (5M) at a rate of 10 mL/min while maintaining vigorous agitation at 600 RPM. The temperature was maintained at 90°C throughout the course of the reaction. The obtained material showed a wide size distribution that peaked at around 1 um, which fails to comply with the requirement for the intended application in drug delivery. According to our observations, this larger distribution appears to be related to poor mixing conditions where shear stress was insufficient. We propose to address this issue by locating a static mixer followed by a venturi system and a second static mixer on a recycling pipe that is constantly taking the reaction mixture and returning it after passing by the in-line mixing device (Figure 1). As a result, we hypothesize that this assembly should suffice to produce a turbulent reacting flow regime where growth and nucleation lead to the material with the required size. The proposed system was first evaluated *in silico* with the aid of COMSOL Multiphysics V.5.3. We considered two different types of mixers, one with a corrugated topology (Figure 2A) and second with a helicoidal one (Figure 2B). In COMSOL we implemented the chemical reactions and the transport of diluted species modules, which allowed us to estimate shear stress, input flows, pressure drop, and the proper location on the pipe. The computational domain of the complete assembly using the helicoidal mixer is shown in Figure 2C, which was discretized via about 600,000 free tetrahedral elements with a refinement in the edges prior to running the simulation to achieve convergence.

As the simulations indicated high-stress levels for the corrugated plates static mixer we, therefore, proceeded to manufacture such assembly. Experimental tests at the bench-scale confirmed a sharper particle size distribution with an average of about 120 nm. As opposed to the initial synthesis, the obtained material complies with the requirements of our application in drug delivery. Further tests at the pilot plant scale will be needed to adjust geometrical parameters of the mixers for evaluating yields and particle size distributions.

References:

1. Nano – A comprehensive nanotechnology database by Nature Research. Available at: https://nano.nature.com/?gclid=CjwKCAjwg6b0BRBMEiwANd1_SKvZ8lVUeaTdZsvdALYyV1NHSI2kWbM-r0GiShpPADbSBWU5o_1ooxoC6AcQAvD_BwE.(Accessed:6th April 2020)
2. Arfin, T. & Tarannum, A. Engineered nanomaterials for industrial application: An overview. in Handbook of Nanomaterials for Industrial Applications 127–134 (Elsevier, 2018). doi:10.1016/B978-0-12-813351-4.00006-7
3. Larue, C. et al. Fate of pristine TiO2 nanoparticles and aged paint-containing TiO2 nanoparticles in lettuce crop after foliar exposure. J. Hazard. Mater. 273, 17–26 (2014).
4. Varadan, V. K. Nanomedicine : design and applications of magnetic nanomaterials, nanosensors, and nanosystems. TA - TT - (Wiley, 2008).
5. Latorre, M. & Rinaldi, ; Carlos. Applications of Magnetic Nanoparticles in Medicine: Magnetic Fluid Hyperthermia. prhsj.rcm.upr.edu (2009).
6. Edebali, S., Oztekin, Y. & Arslan, G. Metallic engineered nanomaterial for industrial use. in Handbook of Nanomaterials for Industrial Applications 67–73 (Elsevier, 2018). doi:10.1016/B978-0-12-813351-4.00004-3
7. Nee Koo, K., Fauzi Ismail, A., Hafiz Dzarfan Othman, M., Rahman, M. A. & Zhong Sheng, T. Preparation and characterization of superparamagnetic magnetite (Fe3O4) nanoparticles: A short review. / Malaysian Journal of Fundamental and Applied Sciences 15, (2019).
8. Paliwal, R., Jayachandra Babu, R. & Palakurthi, S. Nanomedicine Scale-up Technologies: Feasibilities and Challenges. doi:10.1208/s12249-014-0177-9

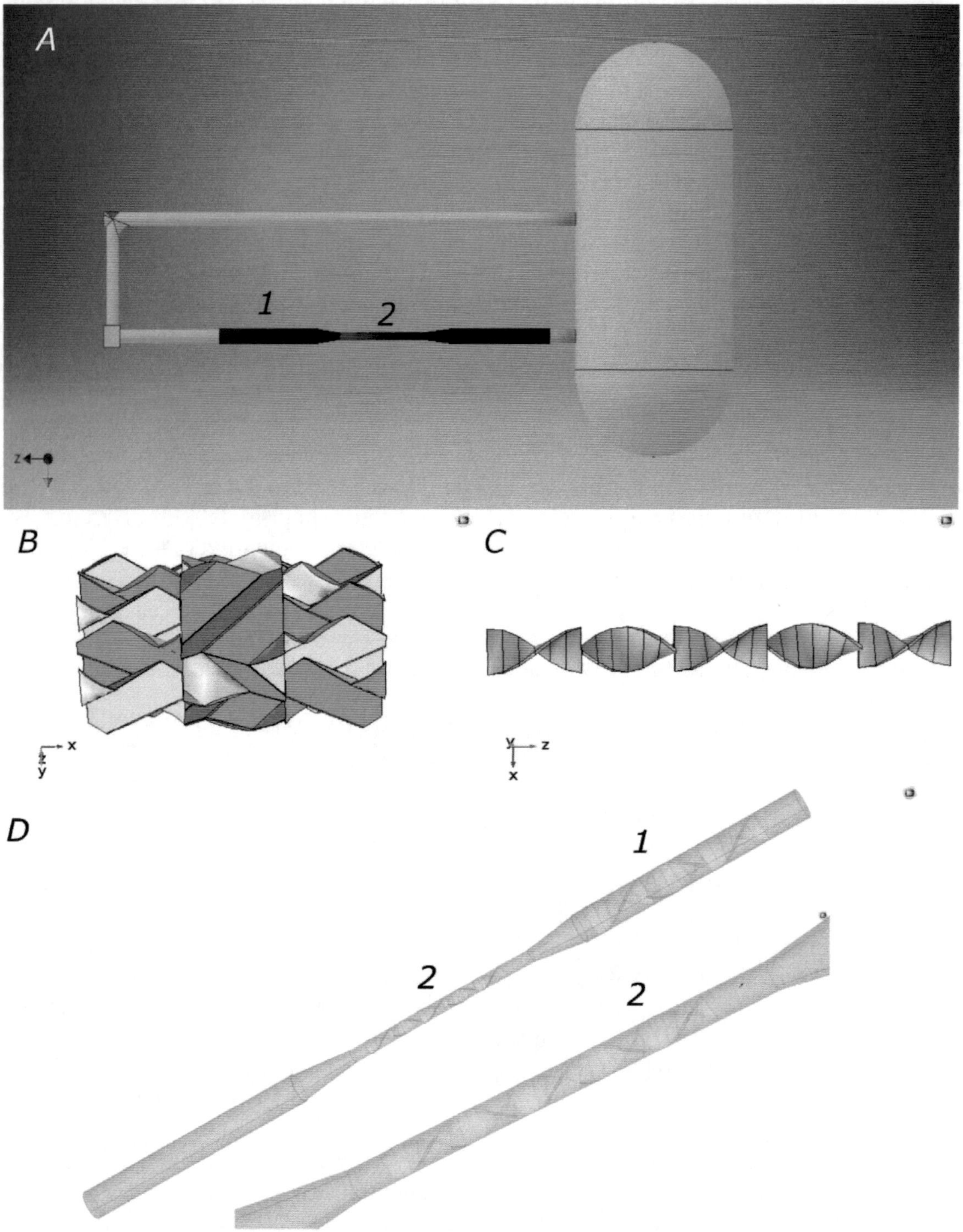

Figure 1. Evaluated static mixers and the proposed assembly for in-line installation (A) proposed assembly for in-line production of magnetite. The figure shows the reaction vessel connected to an external loop where the static mixer are impulsed by a centrifugal pump (squared bottom corner). The loop outlet is placed at the bottom of the vessel in order to facilitate separation processes. Sodium hydroxide is added to the mixer at low flow rates at point 1. (B) Corrugated static mixer. (B) Helicoidal static mixer(C) Complete static mixer/venturi/static mixer assembly to increase the shear stress. 1 represents the first static mixer, 2 represents diameter reduction with the second static mixer.

602015 PAT Controlled Crystallization of Energetic Materials

Abstract:

The high explosive crystal quality, size and morphology can significantly affect their initiation sensitivity. Properties such as crystal shape, size, voids and inclusions can be directed by controlling the crystallization conditions. Improved control of process parameters is required to achieve the desired crystal characteristics since the control of these properties is difficult at the production scale of high explosives and results in large particle size distribution, requiring milling to achieve different class sizes and other potential material rework. When optimizing the size and shape of the crystals during batch crystallization, it is often preferred to operate within the metastable zone to avoid undesired nucleation. In this work RDX seeds are generated using ultrasound at different saturation temperature within the MSZ. Those seeds are controlled to follow a supersaturation profile within the MSZ to promote the growth rate and keep the rate of nucleation low. Process Analytical Technology (PAT) was used to control the crystallization process through real-time measurements using ATR-FTIR, FBRM and PVM tools.

600770 Computer-Aided Molecular Dcsign for Chemical and Energy Applications

Abstract:

The energy efficiency and technical difficulties of an industrial chemical process are often determined by the physicochemical properties of materials.[1] It is desirable to design the right structure of chemicals to meet the targeted material properties, such as the adsorption or emission spectrum of light. However, the research and development (R&D) of functional chemicals has been based largely on researcher's experiences and experimental trial-and-error. With the advances in computing power and algorithm, Computer-Aided Molecular Design (CAMD)[2, 3] technique can now serve as an auxiliary methodology to improve the efficiency of R&D. There are two elements in CAMD: First is the methodology for the prediction of molecular properties, including thermodynamic models, quantum mechanical calculation (QM), and molecular simulation etc.; The second part of CAMD is the optimization algorithm that could automatically search molecules with given requirements of molecular properties, including genetic algorithm (GA), simulated annealing algorithm (SA) etc.[4] We have developed a new molecular data structure (MDS)[5] that allows for flexible creation of new chemical structures with GA. In this presentation, we will demonstrate the application of the MDS in three different applications: *(1) Finding new organic solvents of specified value of octanol-water partition function (K_{ow}).*[2, 3] *(2) Finding new organic solvents of specified value of LUMO-HOMO the LUMO-HOMO gap. (3) Finding new ionic liquids (ILs) for CO2 capture.*[6] We show that CAMD could provide new chemicals that have better or comparable performance with well-known/commercialized specialty chemicals.

References

1. Sandler, S., Chemical, Biochemical, and Engineering Thermodynamics. **2006**.
2. Hsu, H. H.; Huang, C. H.; Lin, S. T., Fully Automated Molecular Design with Atomic Resolution for Desired Thermophysical Properties. *Industrial & Engineering Chemistry Research* **2018,** 57, (29), 9683-9692.
3. Achenie, L.; Venkatasubramanian, V.; Gani, R., *Computer-Aided Molecular Design : Theory and Practice*. 1st ed ed.; Elsevier: Netherlands, 2003; Vol. 12.
4. Austin, N. D. Tools for Computer-Aided Molecular and Mixture Design. Dissertation, Carnegie Mellon University, 2017.
5. Hsu, H.-H.; Huang, C.-H.; Lin, S.-T., New Data Structure for Computational Molecular Design with Atomic or Fragment Resolution. *Journal of Chemical Information and Modeling* **2019,** 59, (9), 3703-3713.
6. Wang, J.; Song, Z.; Cheng, H.; Chen, L.; Deng, L.; Qi, Z., Computer-Aided Design of Ionic Liquids as Absorbent for Gas Separation Exemplified by CO2 Capture Cases. *ACS Sustain. Chem. Eng.* **2018,** 6, (9), 12025-12035.

594347 Lignin Carbon Composites As Renewable Components in Electrochemical Devices: Synthesis, Characterization & Modeling

Abstract:

Lignin is a low-cost, renewable resource that, under controlled processing conditions, can be converted into a nanostructured carbon composite with crystalline and amorphous domains. Features of the nanostructure, including the size of the graphitic nanocrystallites and the volume fraction of the crystalline domain, can be tailored through the choice of processing temperature and lignin feedstock. A collaborative experimental and computational research effort has investigated the processing-structure-property-performance relationship for these materials and their application as anodes in lithium-ion batteries. The atomic structure has been investigated using synchrotron x-ray and neutron scattering, complemented by electron microscopy. Interpretation of the scattering patterns has been accomplished via atomistic Molecular Dynamics (MD) simulation and Hierarchical Decomposition of the Radial Distribution Function (HDRDF). The HDRDF approach decomposes the total RDF into a set of atomistic and mesoscale components, which can be rapidly parameterized in terms of structural descriptors including nanocrystallite size and shape, crystalline volume fraction and total density. MD simulations of lithiated and sodiated systems reveal distinct binding mechanisms and explain variations in charging capacity of composites as a function of processing temperature. Electrochemical performance is measured via evaluation of coin cells under a range of operating conditions. The observation that lignin carbon composites demonstrate storage capacities in excess of the theoretical capacity of graphite anodes is explained in terms of a different binding mechanism in these carbon composites.

600606 Efficient Mesoscale Simulation Model for Self-Assembly in Polymeric Systems

Abstract:

The self-assembly in various polymeric systems is the core principle to many of advanced nanotechnologies including energy applications. Success of most of these applications utilizing the self-assembly relies on how well one can adjust, and switch the shape, size and arrangement direction of self-assembled structures. While theoretical and numerical studies have provided valuable insights in understanding the underlying physical principles in the self-assembly of polymeric, often complicated interactions, non-flat free surface, high sensitivity on various system parameters, wide range of length and time scales related with self-assembled structures limit the usages of existing theoretical/numerical models. Moreover, many systems are in metastable states, thus kinetics, not just thermodynamics, plays a key role for the ability of a polymeric material to self-assemble into a desired state. In this presentation, I would like to present our efforts on overcoming the limitation of existing models and extending the scope of numerical approaches to describe experimentally observed microstructures, to predict new mesophases, to examine the suitability of different polymeric systems, as well as to provide the kinetic routes between various microphases in polymeric self-assembly.

606636 Per-Atom Identification: Structure/Dynamics of Supercritical Fluids

Abstract:

At the liquid-gas critical point, the boiling curve terminates and the two phases merge into a single phase of supercritical fluid (SCF). SCF shows an anomalous blend of liquid-like and gas-like traits, which not only leads to industrial applications, but also raises questions on the fundamental physics that governs its exotic physicochemical behaviors. In this work, we answer this question by identifying liquid-like and gas-like molecules coexisting in the SCF, captured by machine learning analysis on simulation data. The deltoid coexistence region encloses the Widom line, and it may therefore be termed the Widom delta. The number fractions are found to follow simplified two-state thermodynamics model, and are closely related to the magnitude of supercritical anomaly which originates from the fluctuation between microstates. Furthermore, we provide evidence that the microstate number fraction is a more fundamental descriptor of supercritical fluid compared to conventional order parameters, e.g. bulk density. These results strongly support modeling the SCF as a mixture of two interchangeable microstates, of which macroscopic properties arise from the microscopic partitioning of the space by molecules.

595769 Computational Study of Perovskite Materials for Highly Efficient Optoelectronic Applications

Abstract:

Due to their superior photophysical properties, metal-halide perovskite (MHP) materials and their nanostructures are receiving great attention. MHPs have demonstrated not only dramatic improvements in photoelectric conversion efficiency, but also high external quantum efficiencies, promising the use of MHPs for solar-cells and light-emitting diodes (LEDs). For successful deployment of MHPs as next-generation optoelectronic applications, however, photophysical properties of MHP materials still need to be further optimized and improved. Toward this end, a fundamental understanding is becoming essential in order to interrelate compositions, structural phases, material stability, and photo-dynamics of MHPs. We thus perform state-of-the-art first-principles-based simulations to unveil complicated relationships among material properties, which will be outlined in this talk. We anticipate that our computational studies will provide theoretical and mechanistic guidelines to develop highly efficient and stable MHP materials for optoelectronic applications.

606608 Tailored Coupling of Biomineralized CdS Quantum Dots to Reduced Graphene Oxide to Realize Ambient Synthesis of a High-Performance Hydrogen Evolution Photocatalyst

Abstract:

Nanocomposite photocatalysts offer a promising route to efficient and clean hydrogen production. However, the multi-step, high temperature, solvent based syntheses typically utilized to prepare these photocatalysts can limit their scalability and sustainability. Bio-synthetic routes for photocatalyst production offer an attractive room temperature and aqueous phase alternative, but often result in poorly performing catalysts. We have developed a method to produce CdS quantum dots (QDs) supported on reduced graphene oxide (rGO) using enzyme-based syntheses and a ligand exchange mediated self-assembly method for a highly efficient hydrogen evolution photocatalyst. All preparation steps are carried out in an aqueous environment at ambient temperature. Size-controlled CdS QDs are prepared through a biomineralization process involving the enzyme-mediated turnover of L-cysteine to H2S in aqueous solutions of Cd-acetate. Graphene oxide is reduced by enzymatically produced H2S to form rGO. Exchange of cysteamine (CA) for the native L-cysteine ligand capping the CdS QDs, as confirmed by zeta potential measurements and ATR-FTIR spectroscopy, enables exploitation of electrostatic interactions to drive assembly of the positively charged CA-capped CdS (CdS/CA) onto negatively charged rGO. UV-vis spectroscopy and TEM imaging offer insight into the CdS loading and spatial distribution on the rGO. The short CA linker molecule allows for efficient charge transfer from CdS to rGO, increasing exciton lifetime and, subsequently, photocatalytic activity. The visible light hydrogen evolution rate of the resulting CdS/CA/rGO photocatalyst was measured at 3300 μmol hr-1 g-1, which is nearly double the rate of the unsupported CdS/CA. This represents, to our knowledge, one of the highest reported rates for a CdS/rGO nanocomposite photocatalyst. Beyond advancing the state-of-the-art in benign, scalable synthesis of high-performance photocatalysts, the ligand exchange strategy developed here offers a potentially generalizable approach for assembling other high-performing, nanocomposite photocatalysts under ambient conditions.

608106 CO_2 Capture and Transport Behaviors of Encapsulated Liquid-like Nanoparticle Organic Hybrid Materials (NOHMs)

Abstract:

The development of hybrid CO_2 capture sorbents has provided a transformative solution to novel water-less and water-lean CO_2 capture solvents that have been challenged with their high viscosities. A number of novel CO_2 capture solvents including ionic liquids (ILs) and nanoparticle organic hybrid materials (NOHMs) have been developed to capture CO_2 from various industrial sources. While NOHMs show thermodynamic advantages for stable CO_2 capture such as strong CO_2 binding energy resulting from functionalized amine groups, great oxidative thermal stability due to silica nanoparticles as a support, low regeneration energy, and negligible vapor pressure, the overall CO_2 capture process is often limited by mass transfer as a result of high viscosity. Therefore, the application of NOHMs in large-scale CO_2 capture requires an innovative design of novel solvent carriers. In order to overcome these issues, we have designed and developed a hybrid CO_2 capture sorbent system based on a unique encapsulation technique, Solvent Impregnated Polymers (SIPs). By immobilizing NOHMs solvents as microemulsions inside a gas permeable polymer (PDMS) matrix, the CO_2 capture rate was significantly improved compared to the bulk NOHMs solvent due to the increase in the gas-liquid interfacial areas and the gas permeability. The governing mechanism of CO_2 capture using encapsulated NOHMs was investigated based on kinetic modeling. The experimental findings suggest that the encapsulation of NOHMs can allow for a flexible design of CO_2 capture units ranging from a fixed bed to a fluidized bed as the sorbents can be treated as a particulate system. The effect of moisture on the CO_2 capture behavior was also investigated as well as the recyclability of the encapsulated NOHMs.

Prepared by LLNL under Contract DE-AC52-07NA27344

594141 Hybrid Nanocrystal-Amorphous Solid Dispersions (HyNASDs): A New Class of Materials for Bioavailability Enhancement of Poorly Soluble Drugs

Abstract:

Drug nanoparticle-laden microparticles (nanocomposites) and amorphous solid dispersions (ASDs) have been commonly used to deliver poorly soluble drugs and enhance their solubility, dissolution rate, and bioavailability. A major shortcoming of drug nanocomposites as compared with drug ASDs is their limited supersaturation capability in dissolution, whereas ASDs have various issues about physical stability of the metastable form of the drug (amorphous). Here, we prepared drug hybrid nanocrystal–amorphous solid dispersions (HyNASDs) and compared their performance to ASDs. A wet-milled griseofulvin (GF, model poorly soluble drug) nanosuspension and a GF solution, both containing the same dissolved polymer–surfactant (SDS: sodium dodecyl sulfate) with 1:1, 1:3, and 1:5 GF:polymer mass ratios, were spray-dried. Hydroxypropyl cellulose (HPC) and Soluplus (Sol) were used as matrix-forming polymers. XRPD, DSC, and Raman spectroscopy reveal that ASDs were formed upon spray-drying the solution-based feed, whereas nanocomposites and nanocomposites with >10% amorphous content, HyNASDs, were formed with the nanosuspension-based feed. Sol provided higher GF relative supersaturation in the dissolution tests than HPC owing to Sol's stronger intermolecular interactions and miscibility with GF and its recrystallization inhibition. Besides the higher kinetic solubility of GF in Sol, presence of GF nanoparticles vs. micron-sized particles in the nanocomposites enabled fast supersaturation. SDS provided enhanced wettability, allowing for fast supersaturation from Soluplus-based HyNASDs (~300% within 20 min), while higher Soluplus loading led to higher supersaturation. This study demonstrates successful preparation of fast supersaturating HyNASDs, which renders nanoparticle formulations competitive to ASDs in bioavailability enhancement of poorly soluble drugs.

608921 Leveraging Hollow Fibre Membranes and High Boiling Point Organic Solvents in the Production of Enteric Amorphous Solid Dispersions

Abstract:

Leveraging hollow fibre membranes and high boiling point organic solvents in the production of enteric amorphous solid dispersions

Rute Mota, Rui C. Silva, João Vicente

In recent years up to 90 % of the drugs in the discovery pipeline are poorly soluble molecules which poses challenges for bioavailability and dosage form development. Furthermore, the physical and chemical properties of these candidate compounds result in poor solubility in aqueous and volatile organic solvents which hinders the development of amorphous solid dispersion (ASD) using spray drying (SD), while at the same time possessing melting points too high to use hot-melt extrusion (HME): such compounds are typically designated as “brick dust” [1].

In the literature several technologies can be found to deal with poorly soluble APIs: nanomilling or complexation with cyclodextrins fall within the non-ASD alternatives; co-precipitation; supercritical CO2 and KinetiSol [2], as an alternative ASD technology. One such strategy has been developed where nano or micro particles of API-polymer composite precipitate are produced via co-precipitation process using microfluidization. This technology offers the advantage of precise control of particle morphology and generates particles with surface areas that can be 10 times greater than those obtained using HME or SD [3]. Although this technology circumvents the low solubility aspect of the “brick dust” molecules it has some the following drawbacks: a) a suspension with very low solids load and, thus, longer drying process times; b) with “brick dusts” finding a solvent for both API and polymer in amounts for economically viable throughput usually means using organic polar solvents with high boiling points, such as dimethylacetamide (DMA) or dimethylformamide (DMF) or dimethyl sulfoxide (DMSO) which poses challenges to spray dry [3]. These drawbacks amount to longer process times with low throughput which may not be financially appealing in early stage development.

The work presented herein consists on a co-precipitation process with the inclusion of a cross-filtration hollow fibre membranes unit as a straightforward and scalable technology focusing on producing amorphous solid dispersions with “brick dust” compounds. The cross-filtration operation consists in a two-step approach: 1) the first step, concentration of the suspension, the removal of excess antisolvent to increase the solids load; 2) the second step, a solvent swap, where the presence of high boiling point organic solvents in the suspension is minimized by replacing it with an aqueous based solvent. This approach widens the co-amorphous technology to include high boiling points organic solvents best suited for “brick dust” while circumventing the need for secondary drying and reducing cycle time.

References

[1] Temtem M, The Coming of Age of Amorphous Solid Dispersions, Pharma's Almanac, 26

October, 2018.

[2] Jermain SV, Brough C, Williams RO 3rd, Amorphous Solid Dispersions and Nanocrystal Technologies for Poorly Water-Soluble Drug Delivery – An Update, Int J Pharm. 2018 Jan 15; 535 (1-2): 379-392.

[3] Duarte Í, Corvo ML, Serôdio P, Vicente J, Pinto JF, Temtem M, Production of nano-solid dispersions using a novel solvent-controlled precipitation process – benchmarking their in vivo performance with an amorphous micro-sized solid dispersion produced by spray drying, Eur J Pharm Sci. 2016 Oct 10;93:203-14.

Figure 1 – Co-Precipitation Process Schematics with cross-filtration membranes setup

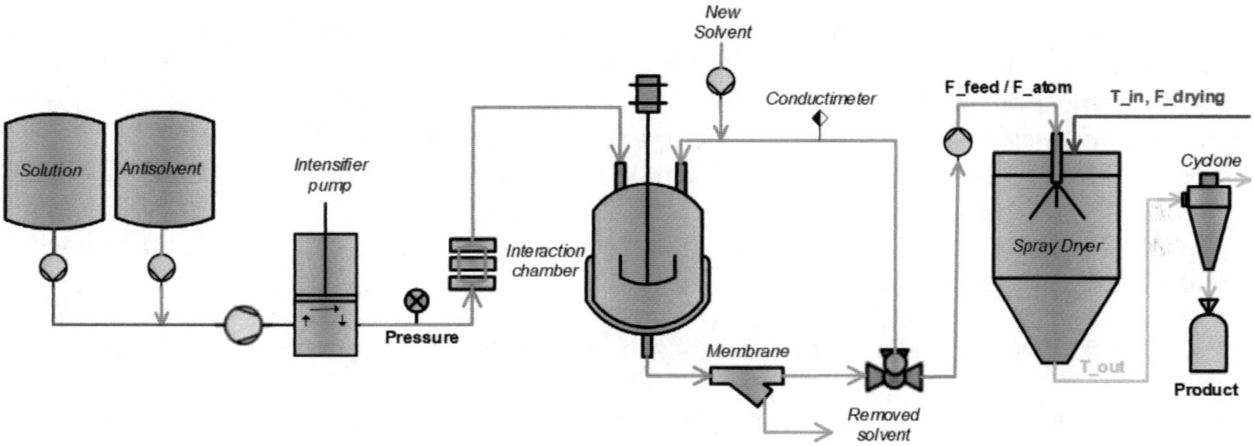

610310 Insight into Amorphous Solid Dispersion Dissolution Behavior: Laser-Diffraction and Raman Spectroscopy Holding Hands

Abstract:

Introduction

Formulation of Biopharmaceutics Classification System (BCS) class II and IV drugs into amorphous solid dispersions (ASDs) is a promising strategy to increase both the apparent aqueous solubility and bioavailability of these drugs. Although drug release from ASDs is still poorly understood, it has been suggested that drug release from ASDs is polymer-controlled at low drug loadings, whereby both drug and polymer release at the same rate (congruently) [1]. However, in contrast, when a higher drug loading is reached, which differs for different drug-polymer combinations, an abrupt decrease in drug release rate may be observed which may be explained by a switch to drug-controlled release, whereby release of drug and polymer is incongruent [2].

In the present work, it is intended to further understand the dissolution performance of different ASD formulations by following the dissolution rate of ASD particles using a laser-diffraction methodology. Furthermore, it will be also studied the possible drug enrichment on the partially dissolved ASD particles by Raman. Intrinsic dissolution was also applied to measure the dissolution rate of the API from the ASD matrix. For this study several ASDs were prepared with two grades of HPMC-AS having different ratios of succinoyl:acetyl groups (L and M) by spray-drying. A BCS class II drug will be used as a model drug.

Materials and methods

Production of ASDs

Amorphous solid dispersions were produced by spray-drying using a mixture of methanol and dichloromethane at 1:4 ratio (w/w), respectively, as solvents. The ASDs were produced at different Drug:HPMC-AS grades (L and M) and at different drug loads (15% and 35 % w/w). For these experiments, a Buchi Mini Spray Dryer B-290 was used with a two-fluid nozzle in closed loop. The spray-dried samples were further dried to remove any residual solvents in a vacuum drying oven at 40 °C.

Physical characterization of ASD

Amorphous solid dispersions were characterized by X-Ray Powder Diffraction (XRPD) to confirm their amorphous nature. Morphology was studied by scanning electron microscopy (SEM) and particle size distribution was evaluated by Laser Diffraction (Sympatec).

ASDs and API dissolution performance

To get an insight into ASDs disintegration/dissolution kinetics, these ASDs particles were added to the dispersion unit of Mastersizer 2000 containing 150 mL of FaSSIF (biorelevant dissolution media). 86 mg and 200 mg of each ASD with 35 wt.% and 15 wt.% drug loads, respectively, were added to this dispersion unit to obtain 0.2 mg/mL as a drug target concentration. Particle size of ASDs was monitored for 4 hours. The reduction of particle size of ASDs during dissolution and colloidal formation was compared with the dissolution rate of the API released from these ASD particles. The drug concentration over time was measured by HPLC.

Results

ASD prepared by spray-drying showed an amorphous nature with no diffraction peaks from crystalline molecule. The particle morphology of the produced ASD was also analyzed by SEM. All the ASDs presented a shriveled morphology (raisin-like particles) which is characteristic of slower spray-drying kinetics. The particle size distribution was similar between different amorphous solid dispersions.

In order to monitor the dissolution performance of the four ASD, the different formulations were suspended in FaSSIF (pH 6.5) in the dispersion unit of the Mastersizer 2000.

In all ASD formulations it was possible to see the appearance of a colloidal population with a submicron size (< 600 nm) after the 240 minutes (Figure 1, A). ASD dissolution showed to be slower in formulations containing higher API loading, since after 60 minutes it is still possible to observe ASD particles in suspension. It's worth mentioning that for formulation - API:HPMC-AS M (35:65) - a full disintegration of ASD did not occur even after 240 minutes.

Raman imaging was performed to the API:HPMC-AS M (35:65) ASD before and after being submitted to the experiment in the Mastersizer equipment. Since part of this sample was not dissolved after 240 minutes, an aliquot was taken from the dispersion unit and centrifuged prior Raman analysis. The intensity of one specific Raman band of the API (1605 cm^{-1}) and the Polymer (2943 cm^{-1}) were monitored in ASD particles before and after the experiment. Results showed that the ratio of 2943 cm^{-1}/1605 cm^{-1} bands is above 1 (red spectra, Figure 1, B) in this ASD particles before being suspended into FaSSIF. However, this ratio showed to invert after the experiment, i.e, the polymer specific band at 2943 cm^{-1} seems to reduce faster than API specific band suggesting an absence of a concomitant dissolution of both elements when this ASD is suspended in FaSSIF. This is consistent with a surface drug enrichment phenomenon [2], suggesting faster dissolution of the polymer in relation to the API.

Intrinsic dissolution confirms that the dissolution rate of ASDs is aligned with the dissolution of ASDs particles observed in laser-diffraction experiment. It is possible to note that formulations containing 35% of API present slower dissolution rates than the ones containing 15% of API. Again, API:HPMC-AS M (35:65) formulation showed to have the lowest performance (Figure 1, C), which seem to be related with a drug enrichment of ASD particles in this particular formulation.

Conclusions

The results presented in this work indicate that dissolution of ASD particles (by laser-diffraction) give an important insight on the API dissolution rate. In detail, it was interesting to note that formulations which showed to disintegrate/dissolve faster leading to a stable colloidal suspension in laser-diffraction test were also the formulations with the highest and fastest API dissolution rate. This proves that their improved performance is related with the ASD dissolution mechanism. The faster dissolution of ASD particles in 15% drug load formulations suggests possible polymer-controlled dissolution which is associated with low drug loadings formulations [2]. The Raman analysis of API:HPMC-AS M (35:65) particles after being suspended/stirred in FaSSIF for 240 min suggested a faster dissolution of the polymer in comparison to the API, leading to particles enriched in API. This explains the slowest dissolution rate observed in this formulation. Therefore, it can be concluded that API dissolution performance is impacted not only by the drug load but also by the HPMC-AS grade. As a future work, the different hydrogen bonding between the API and the 2 different grades of HPMC-AS with varying ratios of succinoyl:acetyl groups will be explored. This study also shows that laser diffraction and Raman spectroscopy pose as orthogonal techniques to understand the dissolution mechanisms of ASDs.

REFERENCES

Tres F, Posada MM, Hall SD, Mohutsky MA, Taylor LS. Mechanistic understanding of the phase behavior of supersaturated solutions of poorly water-soluble drugs. *Int J Pharm.* 543(1-2): 2018

Saboo S, Kestur US, Flaherty DP, Taylor LS. Congruent release of drug and polymer from amorphous solid dispersions: insights into the role of drug-polymer hydrogen bonding, surface crystallization and glass transition. *Mol Pharm.* 17(4): 2020.

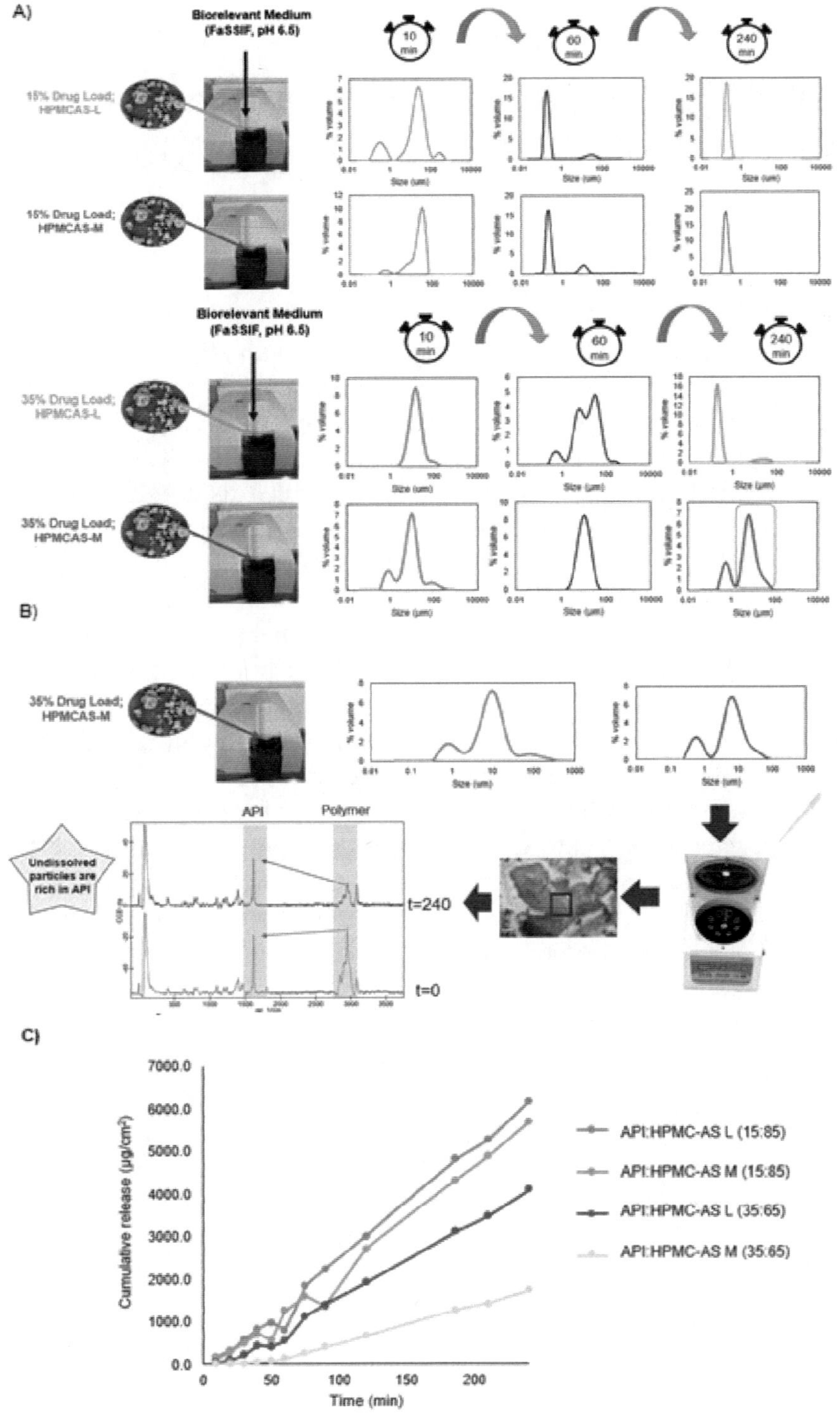

Figure 1:

A) Particle size distribution as function of time when ASDs were suspended in FaSSIF (pH=6.5). From top to bottom 1) API:HPMC-AS L (15:85); 2) API:HPMC-AS M (15:85); 3) API:HPMC-AS L (35:65); 4) API:HPMC-AS M (35:65).

B) Raman spectra of API:HPMC-AS M (35:65) ASD particles before and after being suspended into FaSSIF media in Mastersizer 2000 equipment.

C) Surface normalized dissolution rates of API from 4 different ASD formulations in FaSSIF (pH=6.5).

602798 Multi-Factor Modeling to Predict Significant Factors in Tablet Disintegration and Dissolution Using the FBRM

Abstract:

We have previously demonstrated the application of Focused Beam Reflectance Measurement (FBRM) for analysis during immediate-release tablet disintegration to evaluate changes in particle size and count. This approach could distinguish changes in tablet particle size during disintegration due to varying properties, such as tablet hardness. Here we present a proof-of-concept study demonstrating the generation of a predictive statistical model of a drug product design of experiments (DoE) using parameterization of FBRM measurements taken during tablet disintegration. The FBRM was used to monitor the disintegration of tablets generated in an 18-run DoE evaluating five variables at three levels. The DoE evaluated API properties as well as dry granulation and tableting conditions. Statistical models were generated using the DoE input variables fit to predict FBRM measurement parameters. Robust, multi-factor models identified that several FBRM measures such as total particle count, time of maximum particle count, mean chord (particle) length, or bounded particle size ranges were dependent on several DoE inputs. Furthermore, these factors identified via FBRM disintegration corresponded with the significant factors of dissolution prediction models. In contrast, a USP disintegration method failed to provide a predictive model with multiple significant factors. The FBRM approach to monitoring disintegration allows for a more comprehensive understanding of tablet performance and provides greater insight into the impact of upstream material properties or processing parameters. Material properties and process parameters ultimately impacting dissolution could be predicted earlier and in a material-sparing way. It further offers an orthogonal measurement to gain a better understanding of tablet dissolution explained in a framework of underlying particle behavior.

608277 Melt-Coating of APIs with Surfactant for Drug Dissolution Enhancement

Abstract:

For a solid dosage form, medicine is dispensed based on the dissolving of the active ingredient in the gastrointestinal tract. The significance of the dissolution rate on drug clinical performance has been recognized for a long time. One of the most critical quality attributes was assessed depends on its dissolution behavior. Unfortunately, significant numbers of FDA approved drugs and developmental drugs are hydrophobic and poorly soluble. It has been documented that approximately 40% of approved drugs and 90% of drugs under development are categorized as molecules of low water-solubility and low bioavailability. Due to discovering new drugs is increasing difficulty and become a non-effective way to approach therapeutic requirements, enhancing the solubility of these available drugs is considerably investigated. The prevalent approach for dissolution improvement can be ionizing the drug molecules, producing salt from, micronizing API particles, and using solid dispersion techniques. However, these techniques may require complex unit operations, create unstable solid forms, or generate APIs of poor manufacturability. In this work, we will propose a novel approach in which a small amount of surfactant (< 10% of API weight) is coated on the surface of API particles by applying shearing and heat simultaneously. In this way, we are able to enhance the dissolution of hydrophobic and insoluble API without changing its original solid state.

In our work, three poorly soluble APIs (Ibuprofen, Carbamazepine, and Fenofibrate) and two low-melting-point surfactants (Poloxamer407 and Cetylpyridinium Chloride) are selected. Three specific case studies are performed. In each case study, a twin-screw extruder is used to perform melt-coating because it can apply shearing and heat simultaneously. At a certain temperature, the surfactant can partially melt, then smears and coats on the surface of API particles. Then, the hydrophobic APIs presents a new hydrophilic surface. The particle morphology is photographed using SEM, which shows that the surfactant successfully coats on API particles.

Then, the dissolution test is performed for both treated powder and finished product (tablet and capsule) and compared with the physical mixture and no-surfactant case. Results show that the melt-coating of the surfactant considerably enhances the dissolution rate. The treated product can release drugs 2-5 times faster than the physically mixed product. In contrast, the physical blending surfactant has a marginal effect on dissolution in comparison to the no-surfactant case.

608802 Cocrystallization Process Development of Benzoic Acid – Apremilast Cocrystal

Abstract:

In this work, we have designed several new cocrystals and solvates of a pharmaceutical molecule apremilast. Out of these new multicomponent forms, the cocrystal with benzoic acid is particularly promising due to its improved intrinsic dissolution rate (approximately 400%) and comparable thermal properties to marketed apremilast B. The crystal structure of this cocrystal was solved using single-crystal X-ray diffraction to provide further insight into the cocrystal formation. Cocrystallization process was explored to simplify further scale up. Utilizing the ternary phase diagram, constructed by the combination of specific experiments and a solid-liquid equilibria model, we were able to significantly decrease the experimental efforts compared to traditional approaches such as discontinuous isoperibolic thermal analysis (DITA). Methylehtyl ketone was chosen as a solvent for the cocrystallization experiment. This was performed in a 100ml reactor, were we investigated the impact of process parameters included cooling rate, stirring speed, impeller, concentration and seeding on the size and morphology of the apremilast-benzoic acid cocrystal. Obtained results confirmed strong impact of operating conditions on the crystals sizes and shapes. These crystals were consequently tested during dissolution and it was observed that significant improvements in the dissolution rate can be achieved by varying the crystal sizes and well as shapes.

601650 Plasma Processing for the Enhancement of Energetic Potential of Boron Nanoparticles

Abstract:

The oxidation of metals and metalloids (B, Al, Mg) is a highly exothermic reaction, a property that can be used to enhance the energy content of liquid fuels through the addition of a small volume fraction of nanoparticles. The strong tendency of these materials to oxidize leads to the formation of few nanometers thick surface oxide layer, which for particles in the nanometer size range may represent a quarter of the total particle volume or more. A requirement for the efficient use of energetic nanoparticles is to remove this oxide layer and passivate the surface against re-oxidation. Here we demonstrate the enhancement in the gravimetric energy density of boron nanoparticles (70 nm) using non-thermal plasma processing. Hydrogen plasma was used to reduce the native oxide layer from the surface of boron nanoparticles as a result of surface reactions between excited species of hydrogen and boron oxide. The surface reduction of oxide minimizes the diffusion barrier for the oxidizer during the oxidation and combustion process of the boron nanoparticles. This was followed by in-situ passivation of the boron particles to prevent re-oxidation of nanoparticles. Passivation was done by plasma-enhanced chemical vapor deposition (PECVD) of perfluorodecalin. After performing the series of experiments for different time durations, we found that 120 minutes of hydrogen plasma treatment with 15 minutes of PECVD treatment resulted in a 19% increase in the gravimetric energy density of boron nanoparticles. Analytical techniques like HRTEM, STEM-EDS in high angle annular dark field, and XPS were employed to characterize the reduction in the surface oxide. The reduction in the oxide leads to increase in gravimetric energy density which is shown by characterizing the untreated and plasma- treated particles using thermogravimetric analysis and differential scanning calorimetry. PECVD improved the storage life of the nanoparticles in ambient conditions by providing excellent passivation against air and humidity for 60 days. Thus, the application of in-situ hydrogen plasma and PECVD leads to the amelioration of energetic performance and storage life of boron nanoparticles, making them conducive for nanoenergetic applications.

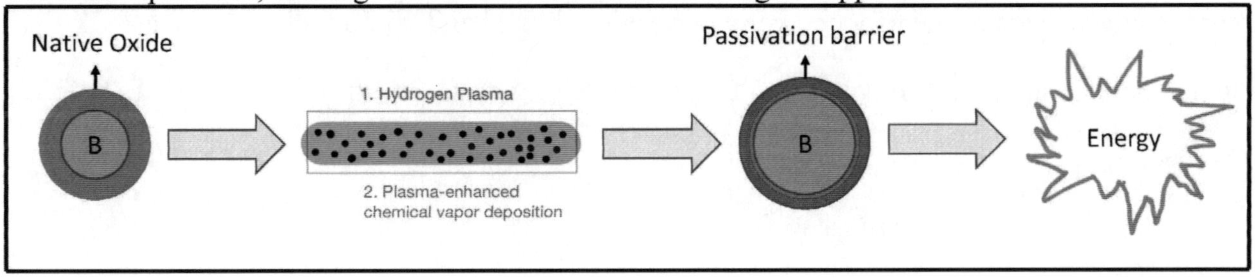

601679 Synthesis of Nanocomposites of Boron and Magnesium for Energetic Applications

Abstract:

Boron (B) has been considered as a promising reactive material for energetic applications such as solid rocket fuels and liquid propellants because of its potential ability to release enormous energy (58 KJ/g) during oxidation reactions compared to those of other metals like Aluminum (31 KJ/g) and Magnesium (25 KJ/g). Reducing the B particles into nanoparticulate form can lead to the improved oxidation kinetics owing to the resulting high specific surface area. However, also the high reactivity of B in nanoparticulate form reduces their energetic performance due to the formation of native oxide layer, which covers the significant mass fraction of the nanoparticle and acts as a diffusion barrier for the oxidizer. We need to channelize the transport of oxidizer by some means along with the reduction of native oxide layer. Magnesium (Mg) is an active metal with excellent combustion characteristics and comparatively lower melting and boiling points. So, if we use Mg along with B in energy systems, the preignition of Mg can induce the ignition of B by elevating the environment temperature and reducing the native boron oxide layer to form metallic B. This can enhance the oxidation kinetics of the B nanoparticles. Here, we are reporting the synthesis of nanocomposites of B and Mg having B in the core and Mg and Mg boride on the surface. We hypothesized that Mg could enhance the energy release from B nanoparticles along with their thermal stability. To prove our hypothesis, we performed various characterization experiments on Mg/B nanocomposites. Characterization was done with XPS and XRD methods to analyze the chemical bonds and the phases present in the nanocomposites. STEM-EDS in high angle annular dark-field (HAADF) was performed to demonstrate the elemental distribution of Mg and B in the nanocomposite. Thermal analysis (TGA, DSC) results provide information about the improved energy release, enhanced thermal stability, and higher reaction kinetics during the oxidation of Mg/B nanocomposites when compared to B oxidation under similar conditions. Also, particle size analysis using dynamic light scattering was performed to know about the particle size and its distribution. So, the Mg/B nanocomposite concept can be applied to the cutting edge design of the nanoenergetic materials for solid and liquid fuels.

601609 Computational Modeling of Gas-Particle Flows: From Kinetic Theory to Turbulence Models and Beyond

Abstract:

Professor Hrenya has done groundbreaking research on many facets of gas-particle flows starting from her early work on turbulence modeling in risers to her more recent contributions in the kinetic theory of granular flows (KTGF), and her associated computational studies using Eulerian-Eulerian (EE) and Eulerian-Lagrangian (EL) methods. Thanks to her efforts, and that of many others, we now have a good understanding of the range of applicability and the limitations of EE and EL methods, as well as the (high) computational requirements needed to obtain accurate predictions from such approaches. Indeed, even more than when she did her work on turbulence modeling, the need for multi-scale models to address real-world industrial problems is obvious due to the impossibility of solving KTGF models without a closure for the unresolvable flow structures.

In this presentation, our recent work on comparing fully resolved EE and EL simulations for vertical risers will be presented. The flow conditions were chosen such that in the absence of the solid particles, the gas flow is turbulent albeit with a modest Reynolds number. In prior work, it had been demonstrated, using an EL code that follows every particle in the flow, that as the mass loading of the solids increases a transition from turbulent channel flow to a quasi-laminar flow occurs. Then, upon further increase in the solids mass loading, the riser flow exhibits so-called cluster-induced turbulence (CIT) whose statistics are driven by the presence of large clusters of particles, and not by the usual mean shear production mechanisms seen in turbulent channel flows. The principal objective of this study was to assess the ability of an EE code based on the KTGF to predict the flow regimes seen in the EL simulations, and it was shown that except for very dilute flows with large Stokes number particles the EE and EL approaches yield nearly the same flow statistics.

As time allows, our work on developing a second-order turbulence model for vertical risers will be described. Due to the strong anisotropy of the Reynolds stresses in CIT, it is necessary to formulate Reynolds-stress models for both phases, which are strongly coupled through the drag terms for the mean momentum and fluctuating kinetic energies. In general, this turbulence model is able to capture the regime transition from turbulent channel flow to CIT. However, at the higher mass loadings needed for CIT, the fully developed CIT riser flow is non-stationary in time due to the continuing presence of large clusters. In some sense, this result was to be expected due to the relatively slow time scales associated with the cluster dynamics, which cannot be separated from the turbulence integral time scales. But, from a computational standpoint, time-dependent simulations (as with the KTGF model) must still be performed, which are still cost prohibitive for most industrial applications.

608824 A Numerical Investigation into the Correlation Function of Inelastic Hard Spheres Using Event Driven Particle Dynamics

Abstract:

The modelling of a collection of interacting particles has attracted the attention of a number of researchers worldwide, and various methodologies have been proposed [1, 2, 3, 4]. In the 80s, the development of the kinetic-collisional theory for granular flow (KTGF) [5] represented a major breakthrough in the modelling of the hydrodynamics of particulate flow. Despite the great success of KTGF, in practical situations, the flow of particles presents complexities that cannot be described by the KTGF alone. Many complicated phenomena happen at the meso-scale, the scale between "small" and "large". As an example, clusters or agglomerates are formed at a length scale between particle size and equipment size; particle-particle interactions, other than inelastic and instantaneous collisions (e.g. adhesion) might be necessary to capture the observed behaviour. Additionally, a number of functions and parameters appearing in continuum models require accurate approximations and physical understanding.

In this work we present the study of the radial correlation function which describes how the particle density varies as a function of the distance from a fixed particle. Such function is of great importance: in granular media, collisional interactions at the microscopic level can change the value of the correlation function at contact. We construct the radial correlation function in one, two, and three dimensions for various packing fractions and coefficients of restitution, and consider the effect of external friction on the result, using novel applications of event driven particle dynamics. This systematic quantification allows us to implement a novel continuum model (dynamical density functional theory), for which the value at contact is a required input; the results show that the properties of the correlation function depend strongly on the parameters considered. We have also demonstrated that including these accurate, microscopically-determined values in continuum models is crucial in correctly describing granular media at the meso-macroscale.

[1] P. A. Cundall and O. D. L. Strack, *Géotechnique*, vol. 29, no. 1, pp. 47–65, 1979.

[2] A. Hospital, J. R. Goñi, M. Orozco, and J. L. Gelpí, "*Advances and applications in bioinformatics and chemistry: AABC*, vol. 8, p. 37, 2015.

[3] R. Carmona, P. Del Moral, P. Hu, and N. Oudjane, vol. 12, pp. 1–46. 03 2012.

[4] J. Hansen and I. Donald, *Theory of simple liquids*. Academic Press, 1991.

[5] Lun, C.K.K, S.B. Savage, D.J. Jeffrey, N. Chepurniy, J. Fluid Mech., 140, 223-256, 1984.

609671 Using Discrete Element Simulations to Predict the Large By Understanding the Small

Abstract:

Particle flows are ubiquitous in nature (e.g., avalanches, volcanic eruptions, and planetary rings) and are common in processes used by a wide range of industries, such as energy, agriculture, and chemical processing. Despite the prevalence of processes involving particle transport, the behavior of flowing particles is often poorly understood, and improved predictive capabilities are needed. Particle flows tend to be chaotic, and subtleties that occur at small length scales (e.g., a single particle) can significantly impact large-scale flow behavior. As such, empirical tools can be unreliable when extrapolated to new systems, and fundamental modeling approaches will play a key role in developing better design tools that do not rely solely on costly experimentation. This presentation will discuss two applications of discrete element simulations. In the first application, discrete element simulations are used to elucidate heat transfer mechanisms to flowing particles and develop continuum closures suitable for modeling large scale systems. In the second application, discrete element simulations are used to help develop a kinetic theory for complex granular flows with non-spherical particles.

596440 Using Machine Learning Methods to Understand and Predict Riser Flow Characteristics

Abstract:

Machine learning methods were applied to circulating fluidized bed (CFB) riser data. The goals were to (i) provide insights on various fluidization phenomena through determining the relative dominance of the process variables, and (ii) develop a model to provide predictive capability in the absence of first-principles understanding that remains elusive. The Random Forest results indicate radial position had the most dominant influence on local mass flux and species segregation, overall mass flux was the most dominant for local particle concentration, while no variable was particularly dominant or negligible for the local clustering characteristics. Furthermore, the Neural Network can be trained to provide good predictive capability, without any mechanistic understanding needed, if a sufficiently large dataset is used for training and if the input variables fully account for all the effects at play. This study underscores the value of machine learning methods in fluidization to advance understanding and provide adequate predictions.

613837 Computational Fluid Dynamics on Cerebras CS-1

Abstract:

There is an emerging processor technology that eliminates many of the performance limitations that are inherent in existing supercomputers. Existing supercomputers only deliver a very low fraction (0.5-3.1%) of their peak floating-point performance on Computational Fluid Dynamics (CFD) codes. This performance limitation stems from low bandwidth and high latency of the off-die data access. Emerging wafer-scale processors have the potential to eliminate these performance limiters and could become cost-effective "single chip supercomputers". The first wafer-scale processor to arrive on the market is the Cerebras Wafer Scale Engine (WSE), and the first system powered by such a processor is Cerebras CS-1. While the WSE was developed to be an artificial intelligence engine, its hardware arrangement is also ideal for structured CFD codes. Thus, NETL, in partnership with Cerebras, has been exploring the possibility of performing MFIX (open-source multiphase CFD code) simulations on CS-1. This presentation will provide preliminary findings from the exploratory study. Cerebras and NETL have developed the first (bare-bones) CFD solver to run on CS-1 and have found that the CS-1 could conduct fast, cost-effective CFD computations. We will discuss the CFD-code implementation, current performance, and future development plans.

613921 Critical Analysis of the Foundation for Various Physical Models Used in Fluid-Particle CFD

Abstract:

Computational fluid dynamics (CFD) has the potential to become a promising predictive toolbox for industrial operations. Most industrial operations involve particle technology in at least one part of the process, so the benefits of accurately predicting the flow of solid particles with CFD is significant. Computational fluid (and particle) dynamic frameworks are essentially composed of software that solve the equations that govern motion, e.g., continuum-based mass, momentum, and energy balances are solved in the two-fluid method (TFM) approach to simulating fluid-particle flow. To reliably predict the particle flow, the particle-particle and fluid-particle forces must be accurately modeled. In some cases, first-principle description of these interactions is well-known. For instance, collisional stresses in TFM resulting from binary, spherical particle collisions forces are well described. Additionally, drag force on a single particle (fluid-particle interaction) can be predicted accurately with various empirical correlations. However, as will demonstrated with this work, not all particle-particle and fluid-particle interactions are well known. For example, fundamental and physical modeling approaches of particle-particle cohesion and drag on multiple particles has evaded the particle technology for decades. As a result, reliable and accurate model predictions of large-scale equipment remains elusive. Here, a critical analysis of the foundation for various physical models used in CFD is discussed. For instance, much of our understanding of cohesive particles flow is based on empirical classifications (and sometimes anecdotal experiences) of material behavior in one type of unit operation. Additionally, the physical limitations of validating the homogenous drag models that underpin the fluid-particle model predictions. In this work, a critical examination of the foundation for modeling various physical interaction is discussed as a necessary first step to developing physical models for predicting particle behavior.

613919 Virtual Mass of an Oscillating Sphere

Abstract:

Virtual mass or added mass describes the inertia of the fluid around a submerged body that acts to resist a relative acceleration between the body and the fluid. When the body is a sphere, this mass was determined to be ½ of the mass of the displaced fluid from the inviscid flow theory, under the condition that the relative acceleration is slow and the radius of the sphere is small such that the sphere develops a steady response to the acceleration. This result since then became a broadly accepted term in the equation of motion for spherical particles submerged in fluids.

Not all flows satisfy the condition of this classical result. In this study, we report that for a sphere performing small-amplitude oscillations in gases, its virtual mass is not surprisingly at odds with the classical expression. Sphere was suspended by a spring in a closed cell, the pressure of which was controlled and varied. We triggered oscillation of the sphere magnetically, and then measured the frequency of free oscillation, from which the virtual mass of the sphere was extracted. We found that for three gases CO_2, N_2 and CH_4 at pressures up to 96 bar virtual masses were consistently greater than that from the classical correlation by 30-40%. Simulations of oscillatory flow around sphere show that the flow still possesses an inviscid core, which explains that the observed virtual masses were nearly independent of the viscosities. The shape of the inviscid core, however, is changed by oscillation. Modeling of multiphase flows with strong virtual mass effects could benefit from improving our knowledge on virtual mass. We will also discuss a few practical applications of our oscillatory flow.

613918 The Role of Friction in Pattern Formation in Pulsed Fluidized Beds

Abstract:

Christine Hrenya's work beautifully demonstrates the importance of introducing the correct physics of gas-solid systems to represent phenomena at multiple scales, all the way down to particle surface roughness at the nanoscale. Her work also involves measurements to validate and parameterize physical models that advance our fundamental understanding of fluidization yet, in addition, it allows us to practically apply these findings in industrial settings. In this presentation, I embrace a similar philosophy in our study of dynamically structured pattern formation in pulsed fluidized beds. The different conditions under which regular patterns appear in shallow granular layers and deep bubbling beds are used to compare the assumptions underpinning different modelling approaches with experimental observations. In particular, the effect of particle friction on pattern nucleation and propagation in dynamically structured fluidized beds is investigated, showing no appreciable effect in pattern nucleation, but a key role in a bubble pattern's propagation. These dynamic viscoplastic flows involve spatiotemporally, periodically structured zones that are dense and quasi-stationary, as well as other, more dilute ones that remain active and mobile, and are responsible for micromixing. A detailed investigation offers new insights into the fundamentals of fluidization at mesoscopic scales, incorporating microscopic information, and relevant to simulate macroscopic gas-solid systems.

613917 Hybrid Continuum-Discrete Approach to Model Industrial Scale Gas-Solids Reacting Flows

Abstract:

Developing new and efficient commercial scale multiphase reactors is extremely difficult as one needs to consider the complex interactions over a wide range of both temporal and spatial scales encountered in these systems (from molecules to macroscale). Therefore, it is important to codify and automate the knowledge collectively acquired so far, reserving human resources for the creative solutions that build upon codifying past learnings. Computational science (algorithms, theory and modeling, computer science, etc.) combined with exponential growth in computing hardware has great potential to provide us valuable tools for improving the effectiveness of existing plants but also design new reactors. Industry is in desperate need of fast simulation tools that are comparable to the accuracy of the single-phase reacting flows to advance the adoption of multiphase flow reactors and primarily to reduce the risk of scale-up. On the other hand, academia and national laboratories are focused on improving numerical methods, submodels or certain physics to push the state-of-the-art and predictability of the models. However, there is tremendous need for rapid integration of various capabilities that exploit latest computer hardware improvements, robust model reduction techniques, etc. that would improve the computational tools available to the industry to deliver the next generation of efficient and cost-effective reactors. One way to address this divide is to develop mechanistic upscaling framework to coarse-grain information from more detailed modeling approaches to fast models accessible to the industry with relatively minimal information loss.

I will present such an approach using a gas-phase fluidized bed polymerization reactor as an example. First, I will provide an overview of the various models currently used at the different scales in gas-solids reacting flows. The coupling across the scales will be introduced through few different approaches:

1. Discrete-Continuum Coupling using Discrete Element Method for particles
2. Upscaling data from Discrete Element Modeling results to continuum based Computational Fluid Dynamics (CFD)
3. Upscaling information from CFD to simplified reactor network models

In conclusion, the presentation will elucidate the approaches and opportunities in bridging the gap between academic advances and industry needs to model and advance our understanding of gas-solids reacting flow reactors.

613916 An Index to Characterize Particle Mixing at Different Scales

Abstract:

Characterizing the mixing and segregation of particles which may differ in type, size or density is important in many chemical engineering applications, and has been extensively studied for more than half a century. This has resulted in a large number of mixing indices (~40) which are used, with varying levels of success, to characterize mixing in granular and gas-solid flows. In spite of the extensive literature on this topic, none of the existing mixing indices explicitly account for the fact that the level of mixing or segregation can vary depending on the scale at which the system is observed, which is conveniently characterized by a sampling volume. It is useful to classify these scales into three ranges: (i) the *microscale* representing sampling volumes with a characteristic length scale in the range O(1-10 d_p), where d_p is the particle diameter, (ii) the *mesoscale* representing sampling volumes with a characteristic length scale in the range O(10-100 d_p), and (iii) the *macroscale* representing sampling volumes on the scale of the device. Here we describe a new mixing index that explicitly accounts for the phenomenon of *scale-dependent* mixing, and is based on the average solid volume fraction fields corresponding to the different particle types obtained from multifluid multiphase computational fluid dynamics (mCFD) simulations or experiment. The performance of this scale-dependent mixing index (SDMI) is characterized using synthetic fields as well as data obtained from multifluid mCFD simulations of biomass mixing in sand in a fluidized bed reactor. Care is taken to distinguish grid-independence of the mCFD results from scale-dependence in the mixing index, resulting in a robust approach which can be applied to well-resolved as well as potentially under-resolved simulations or experimental data. Extension of the SDMI to discrete particle counts encountered in other multiphase flow simulation approaches such as Euler-Lagrange and particle-resolved direct numerical simulations is also discussed.

613915 Fluidization of Non-Standard Particles

Abstract:

When using fluidized beds as catalytic reactors, the particle size and shape can be optimized to get good fluidization behaviour. When using fluidized beds to produce or process particles, there is typically much less freedom in adapting the particle properties such that they show good fluidization behaviour. In recent years, more attention has been given to fluidization of "problematic" bed materials, such as wet particles, particles in the micron-range, nanoparticles, flakes, and rods. In this presentation, I will share some recent insights in the fluidization of such bed materials, and discuss approaches to improve their fluidization behaviour.

AUTHOR INDEX

AIChE
120 Wall Street, FL 23
New York, NY 10005-4020

ISBN 978-1-7138-2298-1

CURRAN ASSOCIATES INC.
proceedings.com